Introduction to Cake Filtration

Analyses, Experiments and Applications

Introduction to Cake Filtration

Analyses, Experiments and Applications

Chi Tien

Department of Biomedical and Chemical Engineering
Syracuse University
New York, USA

ELSEVIER

AMSTERDAM • BOSTON • HEIDELBERG • LONDON • NEW YORK • OXFORD
PARIS • SAN DIEGO • SAN FRANCISCO • SINGAPORE • SYDNEY • TOKYO

Elsevier B.V.
Radarweg 29
P.O. Box 211, 1000 AE Amsterdam
The Netherlands

Elsevier Inc.
525 B Street, Suite 1900
San Diego, CA 92101-4495
USA

Elsevier Ltd
The Boulevard, Langford Lane
Kidlington, Oxford OX5 1GB
UK

Elsevier Ltd
84 Theobalds Road
London WC1X 8RR
UK

First edition 2006

Library of Congress Cataloging in Publication Data
A catalog record is available from the Library of Congress.

British Library Cataloguing in Publication Data
A catalogue record is available from the British Library.

ISBN-13: 978-0-444-52156-9
ISBN-10: 0-444-52156-9

∞ The paper used in this publication meets the requirements of ANSI/NISO Z39.48-1992 (Permanence of Paper). Printed in The Netherlands.

To

B.F. Ruth, H.P. Grace, F.M. Tiller, M. Shirato

and

All other earlier workers of Cake Filtration Studies

Preface

The idea of writing this volume came to me almost two decades ago shortly after I became seriously involved with cake filtration studies. By all account, cake filtration is an important solid/fluid separation process and has been widely applied in the process, chemical and mineral industries. It was (still is) one of the topics discussed in almost all undergraduate, unit operations texts since the publication of the first edition of *Principles of Chemical Engineering* in 1927. However, there are only a few books and monographs devoted exclusively to the subject and most of them are aimed at applications. The purpose of the present book is to give an introductory and yet fairly comprehensive account of cake filtration as a physical process in a more fundamental way. Hopefully, it will provide people who contemplate to do research and development work in cake filtration with a source of information and get them quickly on track.

This book is divided into three parts. Part I deals with cake filtration analyses using different approaches including the conventional theory of cake filtration, analysis based on the solution of the volume-averaged continuity equations and treatment of cake filtration as a diffusion problem. Dynamic simulation of cake filtration which examines both filtration performance and cake structure and its evolution is also included. Descriptions and discussions of cake filtration experiments, the procedures used and the various methods used for the determination of cake properties constitute Part II. In Part III, three fluid/particle separation processes which feature cake formation and growth together with other phenomena are discussed.

As stated earlier, I have prepared this book for the purpose of initiating those who are interested in cake filtration research and development work including students who plan to do their theses in this area. In order to gain a wider audience, the background information necessary to comprehend the materials presented is kept to a minimum. The level is consistent with what is taught at an accredited B.Sc. degree program in chemical, civil and mechanical engineering. It should therefore be possible to adopt the book as a text or part of text for graduate courses dealing with separation or solid/fluid separation, even though, strictly speaking, it is not written as a text.

There is another reason for writing this book. During the past two decades, we have seen considerable discussions and debates about the future of chemical engineering as a profession and as a discipline. Numerous suggestions and plans on chemical engineering education and research have been advanced for the purpose of restoring the profession to its past glory. Somewhat overlooked in these efforts is the fact that the viability of any profession as a field of study depends, to a large degree, upon its appeal to talented young people on account of the intellectual challenges and practical relevancy it poses. In this regard, while the topic of unit operations is recognized as a core subject of the chemical engineering discipline, a search of library and publication catalogues reveals that most of texts and monographs dealing with this topic, but not on an elementary level, were published more than three or four decades ago, thus giving the impression and creating the perception that the discipline has reached its maturity a long time ago. It is therefore not surprising that as a subject of study, chemical engineering nowadays is not able to attract a sufficiently large number of talented students as it once did. It is hoped that writing a book such as this one may, in a very small measure, contribute to rectify the prevailing erroneous impression.

A major part of this book is based on the studies of fluid/particle separation I conducted during the past 20 years at Syracuse University and the National University of Singapore. I would like to acknowledge the significant roles played by my former students and colleagues in these studies: Professor R. Bai, Professor M.S. Chiu, Professor Y.-W. Jung, B.V. Ramarao, Dr K. Stamatakis, Professor R.B.H. Tan, Dr S.-K. Teoh, and Professor C.-H. Wang. I am particularly indebted to R. Bai for his tireless efforts in obtaining some of the numerical results of cake formation and growth included in this book. I should also add that the countless hours of stimulating discussions on cake filtration and related problems I had with B.V. Ramarao during the past decade were certainly one of the major rewards of writing this book.

Finally, I would like to thank my former and present publishing editors, Anouschka Zwart and Louise Morris of Elsevier, for their efforts and assistance which made prompt publication of this book possible, Kathy Datthyn-Madigan for her keyboard skill in typing and assembling the manuscript and last but not the least, my wife, Julia, for all the help and support she has given me for the past four and a half decades.

Chi Tien

Contents

–1–

INTRODUCTION

Notation

k_1	empirical constant of Equation (1.1) (t^{1-k} m^{k-2})
k_2	empirical constant of Equation (1.1) (-)
t	time (s)
V	cumulative filtrate volume (m)

Cake filtration as a process is used for separating the two phases (solid and liquid) of a suspension from each other. The specific purpose of the separation varies from case to case, including the recovery of the solid (discarding the suspending liquid), clarifying the liquid (discarding the solid) or recovering both. It is a long-standing engineering practice and has been widely used in the chemical, process and mineral industries. The principle of filter press operation can be traced to the ancient practice of squeezing juice through cloth in sugar manufacturing (Wakeman, 1972). Similarly, many of the filtration devices used nowadays may claim their origins of more than a century ago. At the same time, new development and inventions of cake filtration systems (both hardware and software) continue and abound because of the importance of solid/liquid separation technology to our manufacturing economy.

The critical role of solid/liquid separation in industrial applications can be seen from the examples shown in Figs 1.1 and 1.2, which give the flow sheets of producing raw sugar and sugar refining (King, 1980). As shown in these figures, there are altogether 11 different classes of separation steps: sedimentation (clarifier), filtration (scum filter and pressure filter), centrifugation (centrifuges for both raw and refined sugar), screening (classification by crystal size), expression (milling rolls for dewatering), washing and leaching, precipitation (lime tanks), evaporation (evaporator), crystallization (vacuum pans), adsorption (char filters) and drying (granulators); among which four classes – sedimentation, filtration, centrifugation and expression belong to the category of solid/liquid separation. This example, by no means, is an exception. The importance of and the reliance on solid/liquid separation technology can be found in a number of industries including mineral processing, paper making, and water and waste water treatment.

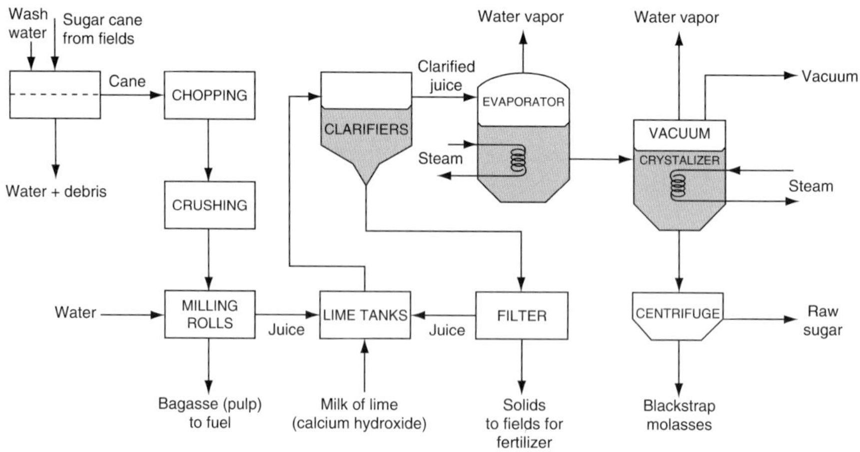

Figure 1.1 Flow sheet of raw sugar production. (King, 1980. Reprinted by permission of McGraw-Hill Inc.)

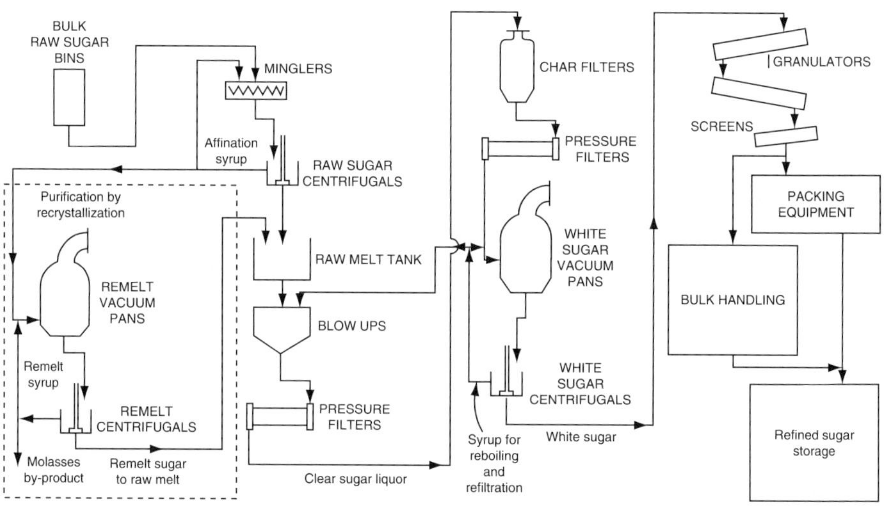

Figure 1.2 Flow sheet of sugar refining. (King, 1980. Reprinted by permission of McGraw-Hill Inc.)

1.1 CAKE FILTRATION AS A SEPARATION PROCESS

A simple schematic representation illustrating the working of a separation process is shown in Fig. 1.3. Through the application of a separating agent which may be either energy or matter or both, a feed stream is split into several streams of different

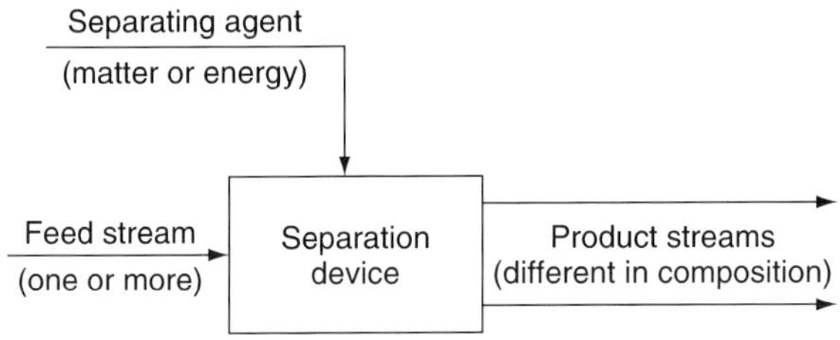

Figure 1.3 A general representation of the separation process. (King, 1980. Reprinted by permission of McGraw-Hill Inc.)

compositions. With a different concept, Giddings advanced the premise that separation of a mixture is caused by the relative displacement of the various components involved (1991). Accordingly, cake filtration may be viewed as a process employing an agent consisting of energy (which causes the flow of suspension) and matter (filter media). The relative displacement between suspending liquid and suspended particles of a cake filtration process results from the particle-exclusion effect of the filter medium used. Similar selective displacement may also be caused by gravity (in sedimentation), rotation (in centrifugation) and mechanical force (in consolidation).

Over the years, a large number of solid/liquid separation processes have been developed and they are too numerous to be mentioned individually. Generally speaking, the most commonly used ones include cake filtration, depth (deep bed) filtration, cycloning, thickening, flocculation, and consolidation. The relationship among these processes can be seen from the classification scheme proposed by Tiller (1974). This scheme shown in Fig. 1.4 is based on Tiller's idea that solid/liquid separation can be viewed as a system consisting of one or more subsystems: (1) pretreatment to facilitate subsequent processing, (2) solid concentration to increase particle content, (3) solid separation, and (4) post-treatment to further enhance the degree of separation and improve product quality. Based on this classification scheme, cake filtration is applied mainly for the separation or recovery of suspended particles from suspensions of relatively high solid content. This is consistent with what is shown in the flow sheet of sugar manufacturing (see Figs 1.1 and 1.2).

1.2 CAKE FILTRATION VS. DEEP BED FILTRATION

Cake filtration and deep bed filtration share a common feature of using the same kind of separating agent. However, the roles played by filter medium in these two processes are different. In cake filtration, the filter medium acts as a screen so that particles of the suspension to be treated are retained by the medium, resulting in the formation of filter cakes. In contrast, in deep bed filtration, separation is effected through particle deposition throughout the entire depth of the medium. In other words, the individual

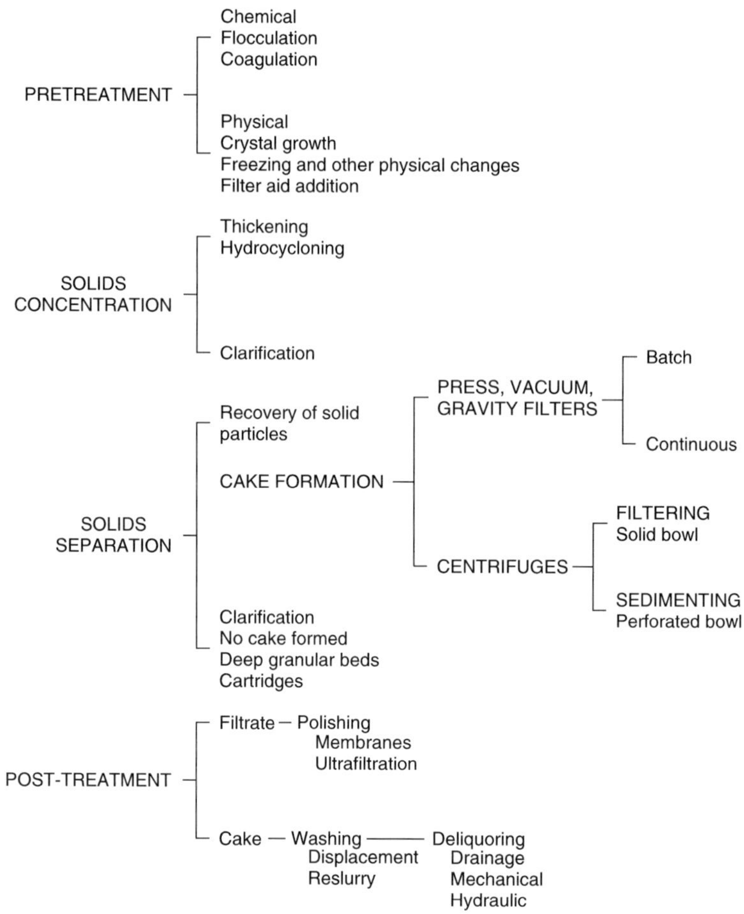

Figure 1.4 Stages of solid/liquid separation according to Tiller.

entities constituting the medium act as particle collectors. Consequently, cake filtration is also known as surface filtration while deep bed filtration is often referred to as depth filtration. A schematic illustration of this difference is shown in Fig. 1.5.

Qualitatively speaking, the most important factor in determining cake formation is the relative medium pore size to particle size. The empirical 1/3 law suggests that cake formation commences if the particle size is 1/3 of the size of the medium pore. While this law may not be exact, sufficient empirical evidence exists indicating the occurrence of cake formation if the particle size and the medium pore size are of the same order of magnitude.

In terms of applications, cake filtration is used to treat suspensions of relatively high solid content while deep bed filtration is applied to clarify suspensions of low

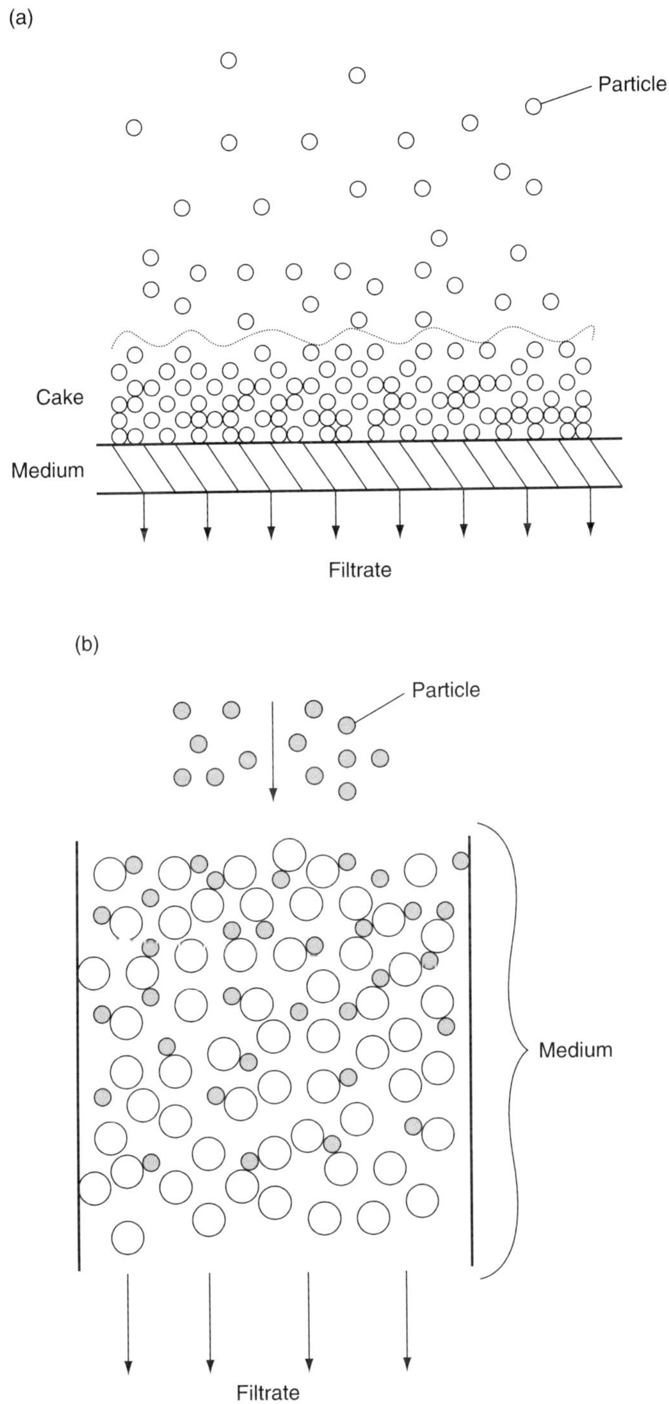

Figure 1.5 Cake filtration (a) vs. deep bed filtration (b).

particle concentration.[1] However, this difference has become blurred with the advent of cross-flow membrane filtration. Taking advantage of the relatively thin cakes formed and the capability of fabricating membranes of small thickness, clarification by membrane filtration has become increasingly popular especially in food and beverage industries.

Over the years, investigators have speculated the mechanism and the manner of cake formation. Beginning with Hermans and Bredee (1935), the so-called "laws of filtration" were advanced. Based on the manner in which particle deposition takes place, cake filtration was classified into four different types: complete blocking, intermediate blocking, bridging and standard blocking. The dynamics of filtration is given as

$$\frac{d^2t}{dV^2} = k_1 \left(\frac{dt}{dV} \right)^{k_2} \tag{1.1}$$

where V denotes the cumulative filtrate volume, t the time, and k_1 and k_2 are empirical constants.

It was suggested that the value of k_2 characterizes the types of cake formation with $k = 0, 1, 1.5$ and 2 corresponding to bridging (proper cake filtration), intermediate blocking, standard blocking and complete blocking, respectively.

Hermans of Bredee's formulation was based entirely on intuitive argument with some arbitrary assumptions. Both complete blocking and bridging lead to cake formation. Furthermore, since in practical situations, the medium pores and particles are likely not uniform in sizes, different types of deposition may take place simultaneously. On a more fundamental level, cake formation, in principle, can be examined in detail through simulation studies (see Chapter 5). Besides some historical interests, the significance of the so-called laws of filtration is rather limited.

1.3 CAKE FILTRATION VS. CROSS-FLOW FILTRATION

Traditionally, cake filtration is carried out with the direction of the feed (suspension) flow coinciding with that of the filtrate flow and cake growth taking place along the opposite direction. However, one may carry out cake filtration by passing the suspension to be treated along the filter medium such that the direction of the filtrate flow is normal to that of the suspension flow. This type of operation may therefore be referred to as cross-flow filtration. In contrast, the term "dead end filtration" is often used to describe the traditional operation in which both the suspension and filtrate flow along the same direction (see Fig. 1.6).

Significant and successful developments of membrane technology during the past three decades have made available classes of materials (polymeric, ceramic and metal) suitable

[1] For using deep bed filtration in water treatment, the particle concentration of the feed stream is often limited to 100 parts per million.

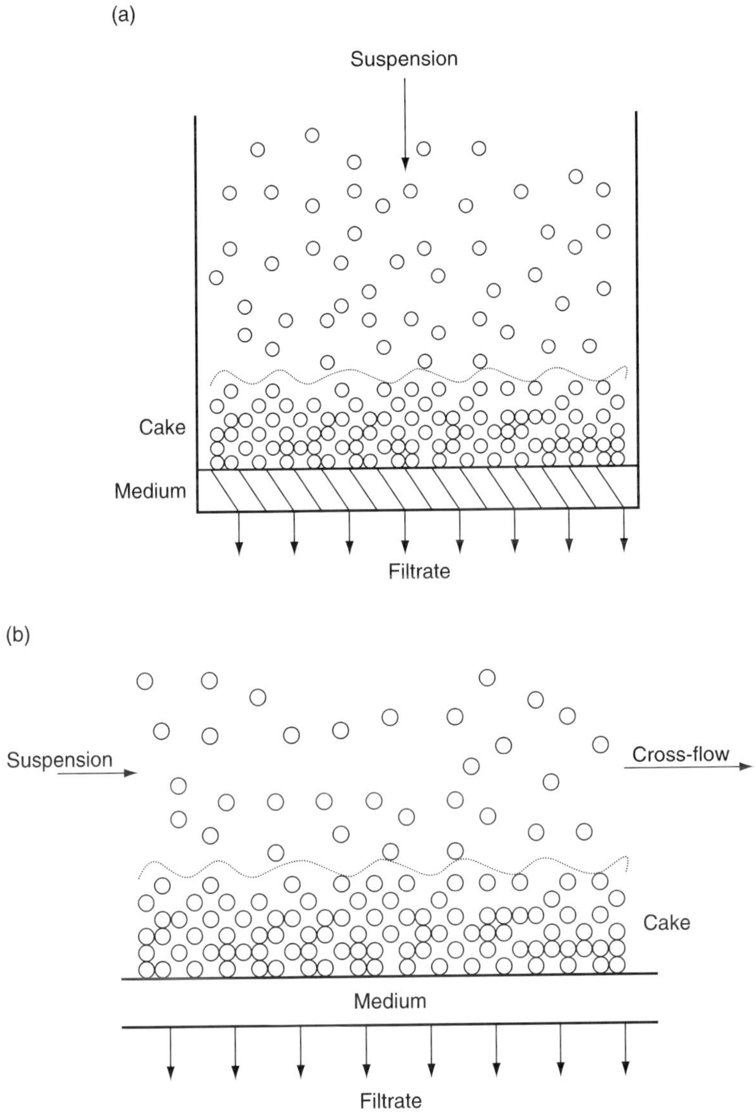

Figure 1.6 Dead-end filtration (a) vs. cross-flow filtration (b).

as filtration media and with them, various types of devices for filtration applications. The very nature of membrane modules developed so far has made it practical to carry out filtration in the cross-flow mode. These operations have found applications in removing particles from liquid streams. The size of the particles removed ranges from submicrons (as low as 10 nm) to microns depending upon the types of membranes used (ultrafiltration vs. microfiltration membranes).

In both dead-end and cross-flow filtration, particle separation leads to the formation of filter cakes which contribute resistance to filtrate flow. There are, however, significant differences between the two types of operations. Regular pressure cake filtration (dead-end) may operate under relatively high pressure (10^3 kPa) with filtration velocity of the order of 10^{-4}–10^{-3} m s^{-1}. The thickness of the cake formed is of the magnitude of 10^{-2} m. In contrast, the operating pressure (the so-called transmembrane pressure) in cross-flow membrane filtration, in most cases, is not more than 100 kPa (often much lower). The thickness of the cake formed is thin (less than 10^{-3} m) and the filtration velocity is below 10^{-4} m s^{-1}. Equally significant is the difference of the hydraulic resistance of the filter medium used. For medium used in traditional cake filtration equipment (belt filters and diaphrange filters), R_m is of the order of 10^{11} m^{-1} while for micromembrane filtration, R_m is of the order of 10^{12} m^{-1}. Because of these differences, the assumption commonly used in cake filtration of neglecting medium resistance is no longer valid in cross-flow membrane filtration.[2] Accordingly, the need of properly accounting for the effect of medium surface clogging is imperative. These problems will be discussed later (see Chapter 8, Section 8.2.).

1.4 FILTRATION CYCLE

Actual operation of cake filtration equipment may be divided into several phases including filtration, consolidation, washing, deliquoring and cake discharge. The exact number of the phases involved depends upon the type of the equipment used, the kind of suspension to be treated and the specific purpose of the operation. Three examples of filter cycles given by Wakeman and Tarleton (1999) are shown in Fig. 1.7.

The physics of these different phases of operation may be similar or different. The phenomena of filtration and consolidation can be described on a common basis embracing problems such as liquid flow through saturated porous medium undergoing growth (in filtration) or reduction (consolidation). They are different from cake washing, which is largely a problem of mass transfer (for example, to reduce the entrained liquid or other impurities). Similarly, deliquoring by air or compressed gas flow reduces filter cake from the saturated state to unsaturated state and is governed by laws different from those of filtration and consolidation. It is, however, safe to say that the filtration phase represents the essential part of the operation of all types of filtration cycles. It is this part of the operation which has occupied the major interest of investigators during the past several decades.

The present monograph is intended to provide an introduction to the analysis and study of the formation and growth of filter cakes in cake filtration. Simple (conventional) and more exact (complex) analyses of cake growth are described and outlined in the first part

---

[2] Fradin and Field (1999) found from their microfiltration experiments that the medium resistance was always greater than the cake resistance by a factor of 2.

of this book. Experiments and measurements which validate the analyses and provide the necessary information for analyses are given in Part 2. Discussions of experimental results also reveal the deficiencies of the present analysis and suggest areas of studies for investigators.

As cake formation and growth are often present in various engineering processes, analyses of such processes require the combination and incorporation of theories governing

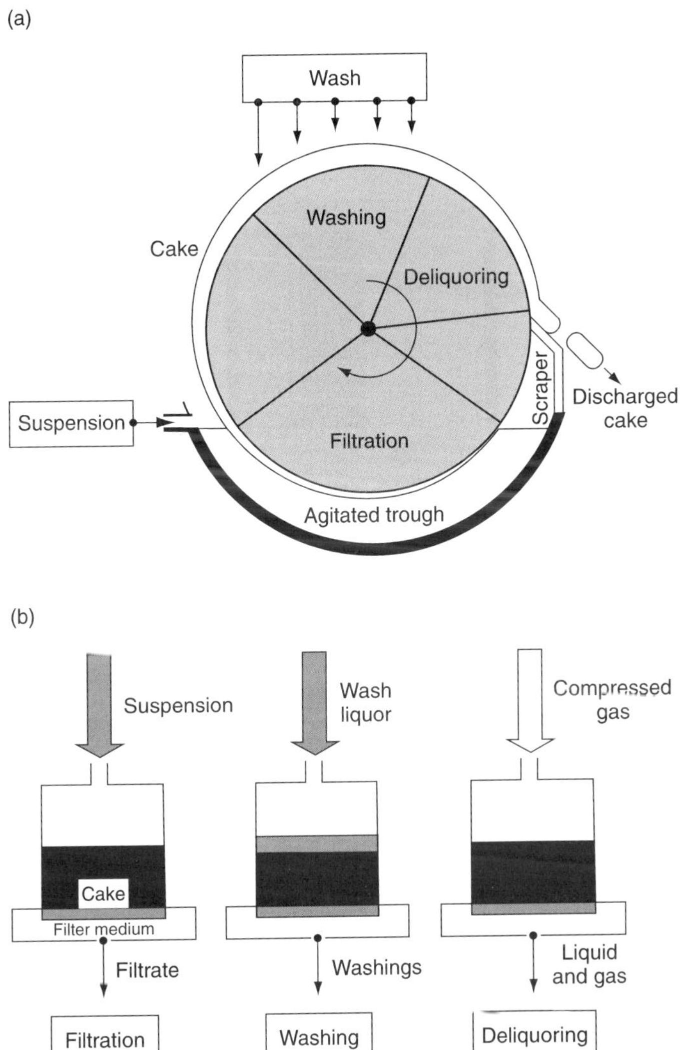

Figure 1.7 Three kinds of filtration cycle: (a) rotary drum filter cycle; (b) Nutsche filter cycle; (c) diaphragm and frame and plate filter press cycle. (Wakeman and Tarleton, 1999. Reprinted by permission of Elsevier.)

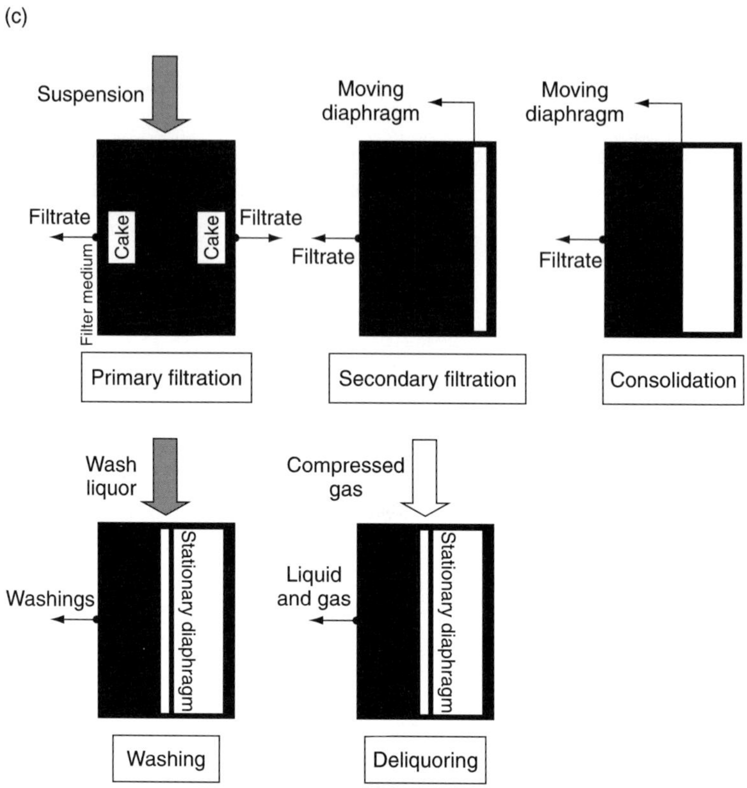

Figure 1.7 *Continued.*

cake growth together with other relevant information. The last chapter of this book presents three such examples for the purpose of illustration.

REFERENCES

Fradin, B. and Field, R.W., *Sep. Pur. Tech.*, **16**, 25 (1999).

Gidding, R.W., *Unified Separation Science*, John Wiley & Sons, New York (1991).

Hermans, P.H. and Bredee, H.L., *Rec. Trav. Chim. des Pays-Bas*, **54**, 680 (1935).

King, C.J., *Separation Processes*, second edn, McGraw-Hill Inc., New York (1980).

Tiller, F.M., *Chem. Eng.*, April 29, 117 (1974).

Wakeman, J.W., "Pressure Filtration", in *Solid-Liquid Separation*, ed. L. Svarovsky, Butterworths, Amsterdam (1972).

Wakeman, J.W. and Tarleton, E.S., *Filtration: Equipment Selection Modeling and Process Simulation*, Elsevier Advanced Technologies, Oxford (1999).

PART I
Analyses

Synopsis

Application of cake filtration for solid/liquid separation is often carried out in several stages, including cake formation and growth, cake dewatering, and cake washing. The generic problems encountered in cake formation and growth and those of cake dewatering are rather similar while cake washing is concerned primarily with mass transfer in porous media. Accordingly, cake washing will not be discussed here and the main emphasis is placed on cake formation and growth.

Over the past eight decades, a substantial number of studies on cake filtration have appeared in the literature. These studies were made using various assumptions with more restricted ones in the earlier investigations (resulting in simpler results) and relaxations of some of these assumptions in later studies (therefore yielding more exact results and more complete information). The presentations to be given below may appear to be in a chronological order although providing a historical account of cake filtration studies is not the main purpose of this book.

The materials presented may be grouped into the following categories: the conventional cake filtration theory, which was developed during the first half of the last century and remains to be the mainstay in design and scale-up of cake filtration systems, the solution of the averaged continuity equation based on the multiphase flow theory, and the treatment of cake filtration as a diffusion problem. Also included is a chapter on simulation of cake filtration and structure. This aspect of study has attracted attention only in recent years. The results available are rather limited and more speculative than those of the first three topics.

–2–

THE CONVENTIONAL THEORY OF CAKE FILTRATION

Notation

A	defined by Equation (2.2.16a) ($m^2\,s^{-1}$)
C_1	defined by Equation (2.3.14) (-)
c	particle concentration of gas stream (kg/m^3)
$f'(\varepsilon_s)$	relationship between p_ℓ and p_s [see Equation (2.1.11)]
g	gravitational acceleration ($m\,s^{-2}$)
i	numbers of drainage surfaces
k	cake permeability (m^2)
k^o	cake permeability at the zero stress state (m^2)
L	cake thickness or height of solid/liquid mixture undergoing expression (m)
L_1	height of solid/liquid mixture undergoing expression at the end of the filtration period (m)
L_m	fictitious solid/liquid mixture height to account for membrane resistance (m)
L_o	initial height of solid/liquid mixture undergoing expression (m)
L_∞	ultimate solid/liquid mixture height undergoing expression (m)
$\overline{m}$	wet to dry cake mass ratio
p_a	quantity appearing in Equations (2.3.5) and (2.3.8) (Pa)
p_ℓ	liquid (filtrate) pressure (Pa)
p_o	operating pressure (Pa)
p_s	compressive stress (Pa)
$(p_s)_{eff}$	defined by Equation (2.5.1.18) (Pa)
p_{s_m}	compressive stress at cake/medium interface (Pa)
Q	constant filtration rate ($m\,s^{-1}$)
q_ℓ	superficial liquid velocity ($m\,s^{-1}$)
q_{ℓ_m}	value of q_ℓ and the cake/medium interface ($m\,s^{-1}$)
q_{ℓ_o}	values of q_ℓ at $r = r_o$ ($m\,s^{-1}$)
q_s	superficial liquid velocity ($m\,s^{-1}$)

R_m	medium resistance (m^{-2})
r_c	radial position of cake/suspension interface (m)
r_ℓ	radial position of liquid front (m)
r_{ℓ_o}	initial value of r_ℓ (m)
r_o	radius of centrifuge bowl (m)
r_s	radial position of suspension/clean liquid interface (m)
$\tilde{r}$	value between r_c and r_o
s	particle mass fraction of suspension (-)
t	time (s)
t_m	fictitious time to account for V_m or L_m (s)
$(u_s)_g$	particle velocity in gravitational sedimentation ($m\,s^{-1}$)
V	cumulative filtrate volume per unit medium surface area (m)
V_m	fictitious filtration value to account for medium resistance (m)
W	cake mass per unit bowl length ($kg\,m^{-1}$)
w	cake mass per unit medium (or bowl) surface area ($kg\,m^{-2}$)
w_o	total solid mass of a solid liquid mixture subject to expression (kg/m^2)
x	distance away from medium (m)

Greek letters

α	specific cake resistance ($m\,kg^{-1}$)
$[\alpha_{av}]_{P_{s_m}}$	average specific cake resistance defined by Equation (2.1.13)
α_{av}	average specific cake resistance defined by Equation (2.5.1.9) ($m\,kg^{-1}$)
$\overline{\alpha}$	average specific cake resistance defined by Equation (2.5.1.14) ($m\,kg^{-1}$)
β	exponent of Equation (2.3.8)
Δp_c	pressure drop across cake (Pa)
Δp_m	pressure drop across medium (Pa)
δ	exponent of Equation (2.3.5)
ε	cake porosity (-)
ε_s	cake solidosity (-)
$\overline{\varepsilon}_s$	average cake solidosity defined by Equation (2.2.10) (-)
$\overline{\overline{\varepsilon}}_s$	average cake solidosity defined by Equation (2.1.8) (-)
ε_s^o	cake solidosity at the zero stress state (-)
ε_{s_o}	particle volume fraction of feed suspension (-)
μ	fluid viscosity (Pa s)
ρ	filtrate density ($kg\,m^{-3}$)
ρ_s	particle density ($kg\,m^{-3}$)
ω	angular acceleration (s^{-2})

Analysis of cake filtration began with the work of Ruth (1933a,b, 1935a,b). The pioneer work of Ruth's together with a number of subsequent studies (Ruth, 1946; Grace, 1953a,b,c; Tiller, 1953, 1955, 1958; Tiller and Huang, 1961; Tiller and Cooper, 1962; Tiller and Shirato, 1964; Tiller *et al.*, 1972; Tiller and Lu, 1972, Tiller and Green, 1973;

Tiller and Yeh, 1985, 1987; Okamura and Shirato, 1955a,b; Shirato and Okamura, 1956; Shirato and Aragaki, 1969; Shirato *et al.*, 1969a,b, 1971a, 1987) have yielded a body of knowledge, which is commonly referred to as the conventional theory of cake filtration. The conventional theory enables simple analysis of cake filtration and easy prediction of filtration performance. It is widely applied in design calculations, scale-up and data interpretation.

A general presentation of the conventional theory is given in the following sections. It is based on the studies mentioned above, with minor changes for better consistency and comparisons with the more exact analyses to be discussed in later chapters.

2.1 BASIC EQUATIONS

A schematic diagram depicting cake filtration is shown in Fig. 2.1. A fluid/particle suspension with a known particle size and concentration (e.g. volume fraction of particles, ε_{s_0}) under pressure flows toward a medium (septum). It is assumed that the suspended particles cannot penetrate into the medium and are retained on the upstream side of the medium to form a cake. The suspending fluid passes through the medium as filtrate. The thickness of the cake increases with time as filtration proceeds.

The one-dimensional continuity equations of particles and fluid are:

$$\frac{\partial q_\ell}{\partial x} = -\frac{\partial \varepsilon_s}{\partial t} \tag{2.1.1a}$$

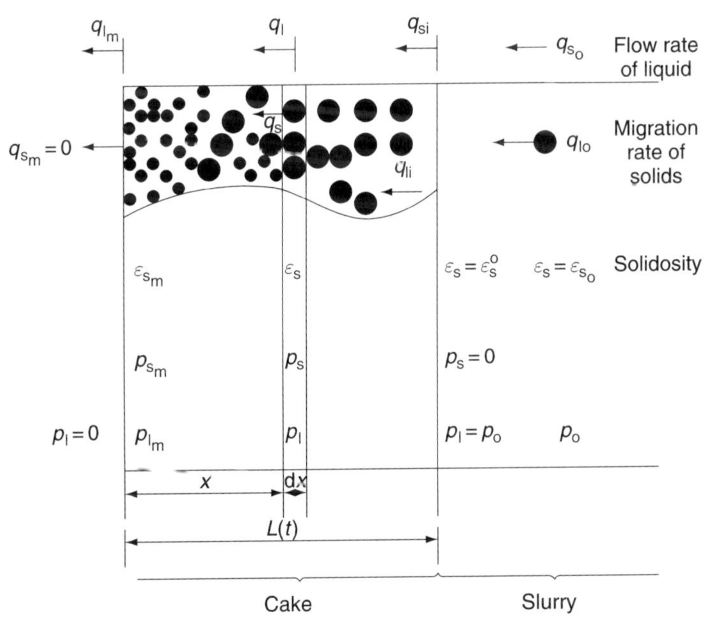

Figure 2.1 Schematic diagram depicting cake formation and growth.

For $0 < x < L(t)$

$$\frac{\partial q_s}{\partial x} = \frac{\partial \varepsilon_s}{\partial t} \tag{2.1.1b}$$

where x is the distance away from the medium, ε_s is the cake solidosity (or particle volume fraction of cake), L is the cake thickness, and q_ℓ and q_s are the superficial liquid and particle velocities, respectively. q_ℓ and q_s are in the opposite direction of x. Adding Equation (2.1.1a) to (2.1.1b), one has

$$\frac{\partial}{\partial x}(q_\ell + q_s) = 0 \tag{2.1.2}$$

or the sum of q_ℓ and q_s, at any instant, is constant across the entire cake thickness.

The major assumptions of the conventional theory are:

(1) The particle velocity, q_s, is negligible. With this assumption and on account of Equation (2.1.2), q_ℓ is constant across the entire cake thickness at any instant.
(2) The liquid flow obeys Darcy's law, or

$$q_\ell = \frac{k}{\mu}\frac{\partial p_\ell}{\partial x} \tag{2.1.3}$$

where k is the cake permeability, μ the liquid viscosity and p_ℓ the pore liquid pressure.

From Equation (2.1.3), one may write

$$\mu \rho_s q_\ell \varepsilon_s \, dx = k \rho_s \varepsilon_s \, dp_\ell \tag{2.1.4}$$

where ρ_s is the particle density.

The cake specific resistance, α, is defined as

$$\alpha = (k\rho_s\varepsilon_s)^{-1} \tag{2.1.5}$$

With the assumption that q_s is negligible, $\partial q_\ell/\partial x = 0$, or the liquid velocity across the cake is constant. Referring back to Equation (2.1.1a), this implies that $\partial\varepsilon_s/\partial t = 0$. In other words, the results of the conventional theory are those of the pseudo steady state solution of the continuity equation of Equation (2.1.1a), which is the theoretical basis of the conventional theory.

Note that at the cake/suspension interface $x = 0$, p_ℓ is equal to p_o, the applied (or operating) pressure. At the downstream side of the medium, p_ℓ may be assumed to be zero. If the pressure drop across the cake and that across the medium are Δp_c and Δp_m, one has

$$p_o = (\Delta p_c) + (\Delta p_m) \tag{2.1.6}$$

and

$$p_\ell = \Delta p_m = p_o - \Delta p_c \quad \text{at} \quad x = L(t) \tag{2.1.7}$$

Now, integrating Equation (2.1.4) across the cake (i.e. $x = 0$ to $x = L(t)$), noticing that q_ℓ is constant as already stated

$$\mu q_\ell \int_o^L \rho_s \varepsilon_s \, dx = \int_{p_o - \Delta p_c}^{P_o} \left(\frac{1}{\alpha}\right) dp_\ell$$

The integral $\int_o^L \rho_s \varepsilon_s dx$ gives the mass of the dry cake per unit medium area, w. If $\bar{\bar{\varepsilon}}_s$ denotes the average cake solidosity defined as

$$\bar{\bar{\varepsilon}}_s = \frac{1}{L} \int_o^L \varepsilon_s \, dx \tag{2.1.8}$$

The first integral of Equation (2.1.7) may be written as

$$\int_o^L \varepsilon_s \rho_s \, dx = \bar{\bar{\varepsilon}}_s \rho_s L = w \tag{2.1.9}$$

For the second integral of Equation (2.1.7), unless α is constant (or the cake is incompressible), its evaluation requires the knowledge of the relationship between α and p_ℓ. It may appear as a simple matter to write

$$\int_{p_o - \Delta p_o}^{p_o} \left(\frac{1}{\alpha}\right) dp_\ell = \left(\overline{\frac{1}{\alpha}}\right) \Delta p_c \tag{2.1.10}$$

where $(\overline{1/\alpha})$ is the average value $(1/\alpha)$ over p_ℓ ranging from $p_o - \Delta p_c$ to p_o. Although Equation (2.1.10) is mathematically correct, it does not address the issue of the nature of the dependence of α on p_ℓ. Equation (2.1.10) may be applied to determine $(\overline{1/\alpha})$ from filtration experimental results, but $(\overline{1/\alpha})$ so determined is nothing more than a fitting parameter. Its significance in characterizing cake structure remains unanswered.

From the definition of α [see Equation (2.1.5)], it is a quantity which depends upon the extent of the compactness of a cake and can be expected to be a function of the compressive stress of the cake phase. But what is the compressive stress? One may hypothesize that during filtration, liquid flowing through a cake imparts drag force on the cake particles, giving rise to a compressive stress p_s. This hypothesis was first made by Walker *et al.* (1937) almost seventy years ago. The same conclusion can be reached by applying the multiphase flow theory.[1] Tien *et al.* (2001) showed that in cake filtration,

[1] This point will be discussed in detail later in Chapter 3.

the liquid pressure p_ℓ and cake compressive stress p_s may be related by the following expression

$$\frac{\partial p_\ell}{\partial p_s} = f'(\varepsilon_s) \tag{2.1.11}$$

Using Equation (2.1.11), the second integral of Equation (2.1.7) becomes

$$\int_{p_o-\Delta p_c}^{p_o} \left(\frac{1}{\alpha}\right) \mathrm{d}p_\ell = \int_{o}^{p_{s_m}} (-f') \left(\frac{1}{\alpha}\right) \mathrm{d}p_s \tag{2.1.12}$$

The cake compressive stress vanishes at the cake/suspension interface and p_{s_m} is the cake compressive stress at the cake/medium interface (a value of p_s corresponding p_ℓ being $p_o - \Delta p_c$). Furthermore, Equation (2.1.12) may be rewritten as

$$\int_{p_o-\Delta p_m}^{p_o} \left(\frac{1}{\alpha}\right) \mathrm{d}p_\ell = \int_{o}^{p_{s_m}} (-f') \left(\frac{1}{\alpha}\right) \mathrm{d}p_s = \frac{\Delta p_c}{[\alpha_{av}]_{p_{s_m}}}$$

and

$$[\alpha_{av}] = \frac{\Delta p_c / p_{s_m}}{\left[\int_{o}^{p_{s_m}} \left(\frac{1}{\alpha}\right)(-f')\,\mathrm{d}p_s\right] \Big/ p_{s_m}} \tag{2.1.13}$$

The definition of Equations (2.1.10) and (2.1.13) may appear very similar. In fact, for the case with $f' = -1$, they become identical. But Equation (2.1.13) gives a physical significance of α_{av} which cannot be seen from Equation (2.1.10). The structure of a cake, which depends upon the degree of its compactness is determined by the compressive stress. One may expect α to be a function of p_s and this information can be obtained by forming cakes under different degree of compression and determining their porosity and permeability (The so-called compression-permeability measurement which will be discussed later in Chapter 7). With the knowledge of α vs. p_s and the relationship between p_ℓ and p_s [i.e. Equation (2.1.11)], $[\alpha_{av}]$ can be determined readily from Equation (2.1.13). This rather fundamental difference between Equation (2.1.13) and (2.1.10), unfortunately, is often overlooked even by workers of cake filtration.

Combining Equations (2.1.7) and (2.1.13), one has

$$q_\ell = \frac{1}{\mu} \frac{\Delta p_c}{w[\alpha_{av}]_{p_{s_m}}} \tag{2.1.14}$$

For the permeation across the medium, q_ℓ may be expressed as

$$q_\ell = \frac{\Delta p_m}{\mu R_m} \tag{2.1.15}$$

where R_m is the flow resistance of the medium.

Combining Equations (2.1.14) and (2.1.15) yields

$$q_\ell = \frac{\Delta p_c + \Delta p_m}{\mu\{w[\alpha_{av}]_{p_{s_m}} + R_m\}} = \frac{p_o}{\mu\{w[\alpha_{av}]_{p_{s_m}} + R_m\}} \tag{2.1.16}$$

The above expression is the basic equation of the conventional cake filtration theory. It simply states that the instantaneous filtration rate (q_ℓ) is directly proportional to the pressure applied p_o, and inversely proportional to the flow resistance composed of the resistances due to the cake and medium. One may begin with Equation (2.1.16) as the starting point of the conventional theory. The more elaborate and seemingly redundant derivations presented above are intended to demonstrate a theoretical basis of Equation (2.1.16) and provide a comparison between the conventional theory with more exact analyses presented in latter chapters.

The performance of cake filtration may be seen from the volume of filtrate collected, and the solid particles recovered per unit medium surface area, or V and w as functions of time. The instantaneous filtrate rate q_ℓ is simply

$$q_\ell = \frac{dV}{dt} \tag{2.1.17}$$

Equation (2.1.16) has two dependent variables, V and w. It is possible to express w in terms of V. If one denotes s as the particle mass fraction of the suspension to be filtered, and $\overline{m}$, the overall mass ratio of wet to dry cake ($\overline{m}$, therefore, is an indication of the average cake porosity), one has

$$s = \frac{w}{V\rho + \overline{m}w}$$

or

$$w = \frac{V\rho s}{1 - \overline{m}s} \tag{2.1.18}$$

and $\overline{m}$, by definition, is given as

$$\overline{m} = 1 + \frac{\int_0^L \rho(1-\varepsilon_s)\,dx}{\int_0^L \rho_s\,\varepsilon_s\,dx} = 1 + \frac{\rho(1-\bar{\bar{\varepsilon}}_s)}{\rho_s\,\bar{\bar{\varepsilon}}} \tag{2.1.19}$$

and ρ is the filtrate density.

Substituting the above expressions into Equation (2.1.16), one has

$$q_\ell = \frac{dV}{dt} = \frac{p_o}{\mu\left\{[\alpha_{av}]_{\Delta p_c}\frac{V\rho s}{1-\overline{m}s} + R_m\right\}} \tag{2.1.20}$$

Equation (2.1.20) is the equivalent of Equation (2.1.16). The difference between the two expressions is that Equation (2.1.20) has only one dependent variable, namely, V if $\overline{m}$ can be treated as a constant. This point will be further discussed later.

2.2 EXPRESSIONS OF CAKE FILTRATION PERFORMANCE

The primary operating variable of filtration is the pressure applied. The operating pressure may be kept constant (constant pressure filtration); kept according to a particular manner i.e. p_o is a specified function of time (variable pressure filtration); or so kept that the rate of filtration is constant (constant rate filtration). For the last case, the volume of the filtrate collected is specified, the operating pressure required to sustain the specified rate is the information to be sought.

(a) *Constant pressure filtration*
Equation (2.1.20) may be rewritten as

$$\mu s\rho(1-\overline{m}s)^{-1}[\alpha_{av}]_{\Delta p_c} V\frac{dV}{dt} + \mu R_m \frac{dV}{dt} = p_o \tag{2.2.1}$$

Integrating the above expression with the initial condition $V=0,\ t=0$ and the assumption that R_m remains constant, one has

$$\mu s\rho\,\overline{(1-\overline{m}s)^{-1}[\alpha_{av}]_{p_{s_m}}}\frac{V^2}{2} + \mu R_m V = p_o t \tag{2.2.2}$$

where

$$\overline{(1-\overline{m}s)^{-1}[\alpha_{av}]_{p_{s_m}}} = \frac{2}{V^2}\int_o^V (1-\overline{m}s)^{-1}[\alpha_{av}]_{p_{s_w}} V\,dV \tag{2.2.3}$$

(b) *Variable pressure filtration*
If p_o is a function of time, the relationship between V and t obtained from the integration of Equation (2.1.20) becomes

$$\mu s\rho\,\overline{(1-\overline{m}s)^{-1}[\alpha_{av}]_{p_{s_m}}}\frac{V^2}{2} + \mu R_m\, V = \int_o^t p_o(t)dt \tag{2.2.4}$$

(c) *Constant rate filtration*
If the filtration rate is kept constant, or $q_\ell = Q = \text{constant}$, one has

$$V = Qt \tag{2.2.5}$$

and the required operating pressure p_o is

$$p_o = \mu Q\{s\rho(1-ms)^{-1}[\alpha_{av}]_{p_{s_m}} Qt + R_m\} \tag{2.2.6}$$

The extent of cake growth is given by the history of cake thickness or L vs. t. By definition, one has

$$\varepsilon_{s_o}(L+V) = \bar{\bar{\varepsilon}}_s L \tag{2.2.7}$$

where $\varepsilon_{s_o} = \dfrac{s/\rho_s}{(s/\rho_s)+((1-s)/\rho)}$ is the particle volume fraction of the feed suspension. $\bar{\bar{\varepsilon}}_s$ is the average cake solidosity [see Equation (2.1.8)].

For constant pressure filtration, substituting Equation (2.2.7) into (2.2.2), one has

$$\mu s \rho\, \overline{(1-\overline{m}s)^{-1}[\alpha_{av}]_{p_{s_m}}}\, \frac{1}{2}\left(\frac{\bar{\bar{\varepsilon}}_s}{\varepsilon_{s_o}}-1\right)^2 L^2 + \mu R_m \left(\frac{\bar{\bar{\varepsilon}}_s}{\varepsilon_{s_o}}-1\right) L = p_o t \tag{2.2.8}$$

Another piece of information of interest is the pressure drop across the cake Δp_c. Using Equation (2.1.8), one has

$$\frac{\Delta p_c}{p_o} = \frac{\mu s \rho\, \overline{(1-\overline{m}s)^{-1}[\alpha_{av}]_{p_{s_m}}}\, V}{\mu s \rho\, \overline{(1-\overline{m}s)^{-1}[\alpha_{av}]_{p_{s_m}}}\, V + \mu R_m} \tag{2.2.9}$$

Equations (2.2.2), (2.2.4) and (2.2.6) are the main features of the conventional cake filtration theory. Equation (2.2.2), in particular, is widely used in design calculations and data interpretation, and is known as the "parabolic law" of constant pressure filtration if the quantity $\overline{(1-\overline{m}s)^{-1}[\alpha_{av}]_{p_{s_m}}}$ can be treated as a constant. The assumptions leading to this conclusion are:[2]

(i) During the course of filtration, both the cake solidosity profile and the pressure drop across the cake and, therefore, the cake compressive stress at the cake/medium interface, p_{s_m}, undergo continuous changes. The overall wet to dry cake mass ratio, $\overline{m}$, as defined by Equation (2.1.19) is a function of the solidosity profile [see $\bar{\bar{\varepsilon}}_s$ defined in Equation (2.1.8). Since the conventional theory does not give a simple expression of $\bar{\bar{\varepsilon}}_s$, it is often assumed to be the same as the stress-averaged solidosity, $\bar{\varepsilon}_s$, defined as

$$\bar{\varepsilon}_s = \frac{1}{p_{s_m}} \int_0^{p_{s_m}} \varepsilon_s \mathrm{d}p_s \tag{2.2.10}$$

(ii) Evaluating $\overline{(1-\overline{m}s)^{-1}[\alpha_{av}]p_{s_m}}$ requires the information of the variations of $\overline{m}$ and $[\alpha_{av}]_{p_{s_m}}$ with V (or time) [see Equation (2.2.3)]. The expression $[\alpha_{av}]_{p_{s_m}}$ is a function of Δp_c or p_{s_m} according to Equation (2.1.13). With (i), $\overline{m}$ can also be considered as a function of p_{s_m}. If the medium resistance R_m is relatively modest,

[2] These assumptions are often overlooked by investigators in applying the conventional theory.

one may assume that both $\overline{m}$ and $[\alpha_{av}]_{p_{s_m}}$ approach their respective ultimate values (corresponding to the condition, $\Delta p_c \simeq P_o$) rapidly. For the most part of a filtration run, one has

$$\overline{(1-\overline{m}s)^{-1}[\alpha_{av}]_{p_{s_m}}} = (1-\overline{m}s)^{-1}\alpha_{av} \tag{2.2.11}$$

and α_{av} is given as

$$\alpha_{av} = \frac{p_o}{\int\limits_o^{p_{sm}(p_o)} \left(\frac{1}{\alpha}\right)(-f')\,\mathrm{d}p_s} \tag{2.2.12}$$

with both α_{av} and $\overline{m}$ evaluated at $\Delta p_c \simeq P_o$. Equation (2.2.2) now becomes

$$\mu s\rho\;[(1-\overline{m}s)_{\Delta p_c=p_o}][\alpha_{av}]_{\Delta p_c=p_o}\frac{V^2}{2} + \mu R_m V = p_o t$$

or

$$t/V = \left(\frac{1}{2p_o}\right)\mu s\rho[(1-\overline{m}s)^{-1}_{P_{s_m}(P_o)}][\alpha_{av}]_{P_{s_m}(P_o)}V + \mu R_m/p_o \tag{2.2.13}$$

It should be emphasized that the above expression is obtained by simplifying the result of the conventional theory, i.e., Equation (2.2.2). The simplification was made by assuming the wet to dry cake mass ratio, $\overline{m}$, to be constant and the average specific cake resistance can be approximated by assuming the pressure drop across the cake being the same as the operating pressure.

Alternatively, if one expresses the medium resistance to be equivalent to that of a fictitious cake layer corresponding to a cumulative filtrate volume of V_m, Equation (2.1.20) may be rewritten as

$$q_\ell = \frac{\mathrm{d}V}{\mathrm{d}t} = \frac{p_o}{\mu[\alpha_{av}]_{p_{s_m}(p_o)}\dfrac{\rho s}{1-\overline{m}s}(V+V_m)} \tag{2.2.14a}$$

and

$$V_m = \frac{1-\overline{m}s}{\rho s(\alpha_{av})_{p_{s_m}(p_o)}}R_m \tag{2.2.14b}$$

Integrating Equation (2.2.14a) subject to the same assumptions used in obtaining Equation (2.2.12), one has

$$(V+V_m)^2 = A(t+t_m) \tag{2.2.15}$$

where

$$A = \frac{2p_o(1-\overline{ms})}{\mu[\alpha_{av}]_{\Delta p_c=p_o}\rho s} \tag{2.2.16a}$$

and

$$t_m = \frac{V_m^2}{A} = \frac{\mu[\alpha_{av}]_{\Delta p_c=p_o}\rho s}{2p_o(1-\overline{ms})}V_m^2 \tag{2.2.16b}$$

One may rewrite Equation (2.2.15) as

$$\frac{t}{V} = \frac{V}{A} + \frac{2V_m}{A} \tag{2.2.17}$$

According to Equation (2.2.13) [or its equivalent Equation (2.2.17)], plotting constant pressure filtration data in the form of t/V vs. V yields a straight line with a slope of $(1/2p_o)\mu s\rho[\alpha_{av}]_{\Delta p_c=p_o}[(1-\overline{ms})_{\Delta p_c=p_o}]^{-1}$ and an intercept of $\mu R_m/p_o$; thus providing a simple and straightforward procedure for determining $[\alpha_{av}]_{p_{s_m}(p_o)}$. This procedure has become the commonly used method of determining the specific cake resistance.

A somewhat different approach of utilizing the conventional cake filtration theory in obtaining specific cake resistance was suggested by Shirato *et al.* (1987). These investigators argued that instead of Equation (2.2.13) [or (2.2.17)], one may employ Equation (2.1.20) written in the form

$$\frac{dt}{dV} = \frac{\mu\rho s}{p_o(1-\overline{ms})}[\alpha_{av}]_{\Delta p_c}V + \frac{\mu R_m}{p_o} \tag{2.2.18}$$

on the ground that in using Equation (2.2.18), the uncertainty of identifying the exact moment when $t=0$ is removed. This point will be discussed later in Chapter 7.

The validity of neglecting the medium resistance in evaluating $\overline{(1-\overline{ms})^{-1}[\alpha_{av}]_{p_{s_m}}}$ [i.e. Equation (2.2.11)] can be seen from the experimental data shown in Fig. 2.2a,b. These results were obtained from filtration experiments of 2% $CaCO_3$ aqueous suspensions under constant pressure ($p_o = 1$ bar) with a medium composed of a number of Whatman No. 1 filter papers (Teoh *et al.*, 2005). The results of V vs. t are shown in Fig. 2.2a, and the same results are shown in the format of t/V vs. V in Fig. 2.2b. It is clear that for the majority part of the data point, the linearity of t/V vs. V is observed. It is also clear that the assumption of replacing Δp_c with p_o is incorrect during the initial period of filtration regardless of the magnitude of the medium resistance.[3] However, if filtration experimental data were taken after sufficient elapse of time, one would indeed observe the linearity of t/V vs. V for all the data collected, thus giving credence to the validity of the assumptions stated here.

[3] This is obvious since the resistance to liquid flow initially is offered only by the medium. Surprisingly, this rather obvious fact is sometimes overlooked by some investigators.

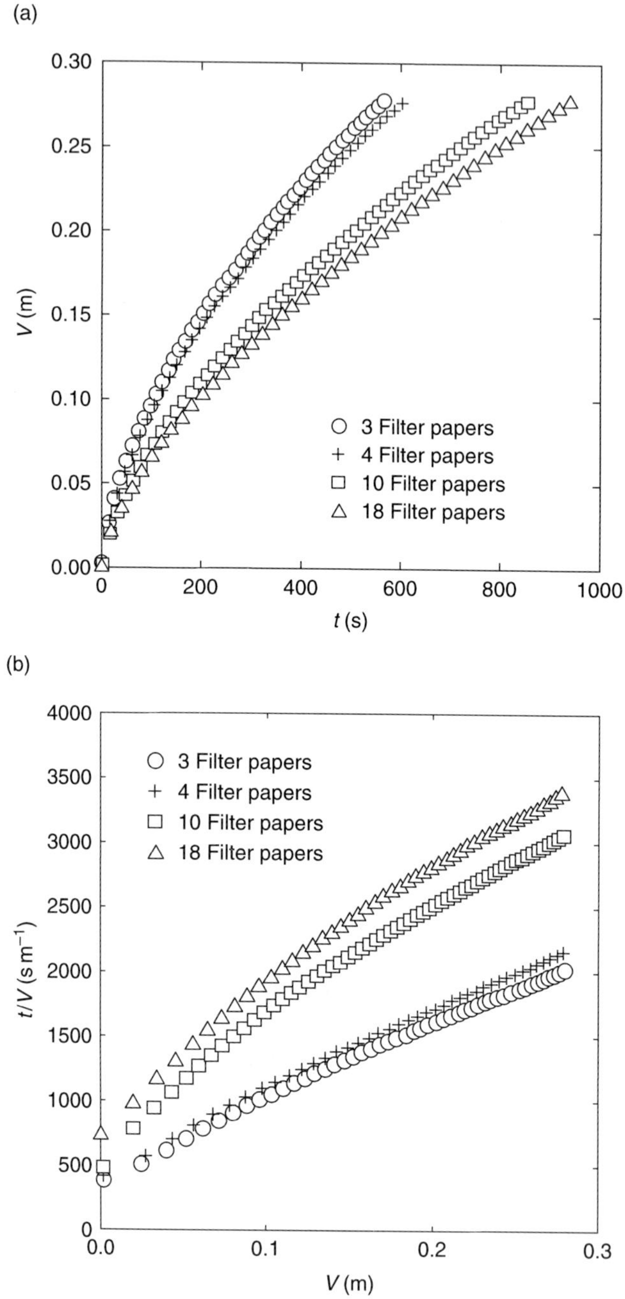

Figure 2.2 Constant pressure filtration of 2% (wt) $CaCO_3$ suspension at 1 bar; filter medium composed of a number of Whatman No. 1 Filter paper (Teoh, 2003) (a) V vs. t; (b) t/V vs. t. (Teoh, 2003. Reprinted by permission of Dr S.-K. Teoh.)

2.3 PRESSURE, COMPRESSIVE STRESS, SOLIDOSITY PROFILES, AND OTHER EXPRESSIONS

The pressure (or compressive stress) distribution across a cake can be found approximately in the following manner. From Equations (2.1.3) and (2.1.11), one has

$$\mu q_\ell \, \mathrm{d}x = k \, \mathrm{d}p_\ell = (k)(f')\mathrm{d}p_s \tag{2.3.1}$$

The integrated form of the above expression is

$$\int_o^x \mu q_\ell \, \mathrm{d}x = \int_{p_{s_m}}^{p_s} (k)(f')\mathrm{d}p_s \tag{2.3.2}$$

and

$$\int_o^L \mu q_\ell \, \mathrm{d}x = \int_{p_{s_m}}^{o} (k)(f')\mathrm{d}p_s \tag{2.3.3}$$

Dividing Equation (2.3.2) by Equation (2.3.3), one has

$$\frac{\int_o^x \mu q_\ell \, \mathrm{d}x}{\int_o^L \mu q_\ell \, \mathrm{d}x} = \frac{\int_{p_s}^{p_{s_m}} (k)(f')\mathrm{d}p_s}{\int_o^{p_{s_m}} (k)(f')\mathrm{d}p_s} \tag{2.3.4}$$

The left hand side quantity of Equation (2.3.4) can be approximated as x/L based on the assumption of the conventional theory [i.e. Equation (2.1.2)] with $q_s \simeq 0$ and q_ℓ being independent of x. For the right hand quantity, the permeability k is a function of the compressive stress, f' is also a function of the compressive stress (since f' is a function of ε_s), and the quantity $\left(\int_{p_s}^{p_{s_m}} kf'\mathrm{d}p_s\right)/\left(\int_o^{p_{s_m}} kf'\mathrm{d}p_s\right)$ is a function of p_s or ε_s. In other words, Equation (2.3.4) is an implicit function of the solidosity profile. This expression can be greatly simplified if the relationship between the cake permeability k and the compressive stress can be expressed as

$$k = k^o \left(1 + \frac{p_s}{p_a}\right)^{-\delta} \tag{2.3.5}$$

where k^o is the cake permeability at the zero-stress state and p_a and δ are empirical constants; and if

$$f' = -1 \tag{2.3.6}$$

Equation (2.3.4) then becomes

$$\frac{x}{L} = \frac{\left(1 + \frac{\Delta p_c}{p_a}\right)^{1-\delta} - \left(1 + \frac{p_s}{p_a}\right)^{1-\delta}}{\left(1 + \frac{\Delta p_c}{p_a}\right)^{1-\delta} - 1} = 1 - \frac{\left(1 + \frac{p_s}{p_a}\right)^{1-\delta} - 1}{\left(1 + \frac{\Delta p_c}{p_a}\right)^{1-\delta} - 1}$$

which, upon rearrangement, gives the following expression for the compressive stress profile

$$\frac{p_s}{p_a} = \left\{1 + \left(1 - \frac{x}{L}\right)\left[\left(1 + \frac{\Delta p_c}{p_a}\right)^{1-\delta} - 1\right]\right\}^{\frac{1}{1-\delta}} - 1 \tag{2.3.7}$$

If the solidosity–compressive stress relationship also follows the power-law expression, i.e.

$$\varepsilon_s = \varepsilon_s^o \left(1 + \frac{p_s}{p_a}\right)^{\beta} \tag{2.3.8}$$

Combining Equations (2.3.8) and (2.3.7) yields the solidosity profile expression

$$\frac{\varepsilon_s}{\varepsilon_s^o} = \left\{\left[\left(1 + \frac{\Delta p_c}{p_a}\right)^{1-\delta} - 1\right]\left(1 - \frac{x}{L}\right) + 1\right\}^{\frac{\beta}{1-\delta}} \tag{2.3.9}$$

which is the same expression given by Tiller *et al.* previously (1999).[4]

With $f'(\varepsilon) = -1$, $p_\ell + p_s = \Delta p_c$,[5] the liquid pore pressure profile can be readily obtained from Equation (2.3.7) to be

$$\frac{p_\ell}{p_a} = \frac{\Delta p_c}{p_a} - \left[1 + \left\{\left(1 + \frac{\Delta p_c}{p_a}\right)^{1-\delta} - 1\right\}\left(1 - \frac{x}{L}\right)\right]^{\frac{1}{1-\delta}} + 1 \tag{2.3.10}$$

Once p_ℓ vs. x is known, an estimate of the liquid velocity across the cake q_ℓ vs. x can be easily made. From Equation (2.1.2), noting $q_s = 0$ at $x = 0$ (i.e. particles do not penetrate into the medium), one has

$$q_\ell + q_s = q_\ell|_m \tag{2.3.11}$$

where the subscript "m" denotes the cake/medium interface or $x = 0$.

If both the particle and fluid are under motion, a generalization of Darcy's law [i.e. Equation (2.1.3)] gives (Shirato *et al.*, 1969a)

$$\frac{q_\ell}{1 - \varepsilon_s} - \frac{q_s}{\varepsilon_s} = \frac{k}{\mu(1 - \varepsilon_s)}\left(\frac{\partial p_\ell}{\partial x}\right) \tag{2.3.12}$$

---

[4] Equation (21) of Tiller *et al.* (1999) after correcting its typographical error is the same as Equation (2.3.9).

[5] If medium resistance is negligible, $\Delta p_c = p_o$.

The pore liquid pressure gradient can be obtained for Equation (2.3.10):

$$\frac{\partial p_\ell}{\partial x} = \frac{p_a}{1-\delta}\frac{C_1}{L}\left[1 + C_1\left(1 - \frac{x}{L}\right)\right]^{\frac{\delta}{1-\delta}} \tag{2.3.13}$$

and

$$C_1 = \left(1 + \frac{\Delta p_c}{p_a}\right)^{1-\delta} - 1 \tag{2.3.14}$$

Substituting Equation (2.3.13) into Equations (2.3.11) and (2.3.12), one has

$$q_\ell + q_s = \frac{k^o\left(1 + \frac{\Delta p_c}{p_a}\right)^\delta}{\mu}\frac{p_a}{1-\delta}\frac{C_1}{L}(1 + C_1)^{\frac{\delta}{1-\delta}} \tag{2.3.15}$$

$$q_\ell - q_s = \frac{k^o\left(1 + \frac{\Delta p_c}{p_a}\right)^\delta}{\mu}\frac{p_a}{1-\delta}\frac{C_1}{L}\left[1 + C_1\left(1 - \frac{x}{L}\right)\right]^{\frac{\delta}{1-\delta}} \tag{2.3.16}$$

Equations (2.3.15) and (2.3.16) can be solved simultaneously to give the results of q_ℓ vs. x and q_s vs. x. Alternate procedures for estimating internal velocity profiles have also been developed by a number of investigators (Shirato *et al.*, 1969a; Tiller *et al.*, 1999; Lee *et al.*, 2000). A summary of the results given above is presented in Table 2.1.

2.4 AN IMPROVED PROCEDURE FOR CALCULATING FILTRATION PERFORMANCE

The commonly used procedures for calculating cake filtration performance at constant pressure according to the conventional theory is stated in Section 2.2: namely, with the operating conditions and the constitutive relationships (i.e., k, ε_s and α vs. p_s) known, approximate values of $\bar{\bar{\varepsilon}}_s$ can be obtained from Equation (2.2.10) from which $\overline{m}$ can be estimated according to Equation (2.1.19) assuming that $\Delta p_c = p_o$. Similarly, $[\alpha_{av}]_{p_{s_m}}$ can be calculated from Equation (2.1.13) with $\Delta p_c \simeq p_o$. The quantity $\overline{(1 - \overline{m}s)^{-1}[\alpha_{av}]_{p_{s_m}}}$ becomes constant and cake filtration performance, namely, V vs. t and L vs. t, can be found according to Equations (2.2.4) and (2.2.8). Although this procedure is widely used in practice, its accuracy is questionable especially during the initial period of filtration (or more precisely, during the period when the flow resistance due to filter medium R_m remains to be a substantial part of the total resistance). Recently, Tien and Bai (2001, 2003) proposed an improved procedure of calculating filter performance taking into account the variation of Δp_c with time and without the approximation of $\bar{\bar{\varepsilon}}_s$ by

Table 2.1

Main results of the conventional cake filtration theory

I. Basic equations

$$q_\ell = \frac{k}{\mu}\frac{\partial p_\ell}{\partial x} \qquad (2.1.3)$$

$$q_{\ell_m} = \frac{\Delta p_m}{\mu R_m} = \frac{p_o}{\mu s\rho(1-\overline{m}s)^{-1}[\alpha_{av}]_{\Delta p_c}V+\mu R_m} \qquad (2.1.15), (2.1.20)$$

where

$$[\alpha_{av}] = \frac{\Delta p_c/p_{s_m}}{\left[\int_o^{p_{s_m}}\left(\frac{1}{\alpha}\right)(-f')\mathrm{d}p_s\right]/p_{s_m}} \qquad (2.1.13)$$

$$\overline{m} = 1+\frac{\rho(1-\bar{\bar{\varepsilon}}_s)}{\rho_s\bar{\bar{\varepsilon}}_s} \qquad (2.1.19)$$

where

$$\bar{\bar{\varepsilon}}_s = \frac{1}{L}\int_o^L \varepsilon_s \mathrm{d}x \qquad (2.1.8)$$

II. Expressions of filter performance

Constant pressure filtration

$$\mu s\rho\overline{(1-\overline{m}s)^{-1}[\alpha_{av}]_{p_{s_m}}}\frac{V^2}{2}+\mu R_m V = p_o t \qquad (2.2.2)$$

where

$$\overline{(1-\overline{m}s)^{-1}[\alpha_{av}]_{p_{s_m}}} = \frac{2}{V^2}\int_o^V (1-\overline{m}s)^{-1}[\alpha_{av}]_{p_{s_w}} V\mathrm{d}V \qquad (2.2.3)$$

$$\mu s\rho\overline{(1-\overline{m}s)^{-1}[\alpha_{av}]_{p_{s_m}}}(1/2)\left[\frac{\bar{\bar{\varepsilon}}_s}{\varepsilon_{s_o}}-1\right]^2 L^2 + \mu R_m\left[\frac{\bar{\bar{\varepsilon}}_s}{\varepsilon_{s_o}}-1\right]L = p_o t \qquad (2.2.8)$$

Constant rate filtration

$$q_{\ell_m} = Q, \quad V = QT \qquad (2.2.5)$$

$$p_o = \mu Q\{s\rho(1-ms)^{-1}[\alpha_{av}]_{p_{s_m}}Qt+R_m\} \qquad (2.2.6)$$

III. Other expressions of constant pressure filtration

Pressure drop across filter cake

$$\frac{\Delta p_c}{P_o} = \frac{\mu s\rho\overline{(1-\overline{m}s)^{-1}[\alpha_{av}]_{p_{s_m}}}V}{\mu s\rho\overline{(1-\overline{m}s)^{-1}[\alpha_{av}]_{p_{s_m}}}V+\mu R_m} \qquad (2.2.9)$$

Compressive stress profile[a]

$$\frac{p_s}{P_a} = \left\{1+\left(1-\frac{x}{L}\right)\left[\left(1+\frac{\Delta p_c}{p_a}\right)^{1-\delta}-1\right]\right\}^{\frac{1}{1-\delta}}-1 \qquad (2.3.7)$$

Solidosity profile[a]

$$\frac{\varepsilon_s}{\varepsilon_s^o} = \left\{\left[\left(1+\frac{\Delta p_c}{p_a}\right)^{1-\delta}-1\right]\left(1-\frac{x}{L}\right)+1\right\}^{\frac{\beta}{1-\delta}}$$

[a] Obtained under the conditions, $f'(\varepsilon_s) = -1$, k *vs.* p_s and ε_s vs. p_s given by Equations (2.3.5) and (2.3.8), respectively.

Equation (2.2.10). A step-by-step description of the improved procedure is presented below.

(i) Applying the commonly used procedure described before, the relationships of V vs. t and L vs. t are determined from Equations (2.2.4) and (2.2.8) assuming that $\Delta p_c = p_o$ and $\bar{\bar{\varepsilon}}_s$ is the same as $\bar{\varepsilon}_s$.

(ii) Based on the results of (i), the pressure drop across the filter cake Δp_c as a function of time can be estimated from Equation (2.2.9).

(iii) Using the results of (i) and (ii), the solidosity profile (ε_s vs. x) can be found from Equation (2.3.9). With the solidosity profiles known, the values of $\bar{\bar{\varepsilon}}_s$ and $\overline{m}$ at various times (or V) can be found from Equations (2.1.8) and (2.1.19).

(iv) Based on the results of Δp_c vs. t obtained in (iii), the values of $[\alpha_{av}]_{p_{s_m}}$ at various time (or V) can be calculated from Equation (2.1.13).

(v) The quantity $\overline{(1-\overline{m}s)^{-1}[\alpha_{av}]_{p_{s_m}}}$ vs. t (or V) can be calculated according to Equation (2.2.3) from the results of (iii) and (iv).

(vi) With the results of (v), new relationships of V vs. t and L vs. t can be found according to Equations (2.2.2) and (2.2.8).

(vii) The procedure stated above may be repeated until the relationships of V vs. t and L vs. t obtained from successive iterations converge within a prescribed limits. Generally speaking, only one iteration is found sufficient.

Sample calculations were made for the purposes of comparing the results obtained from the improved procedure and those from the commonly used procedure and their comparisons with more exact analysis based on the numerical solutions of the averaged continuity equations (for details, see Chapter 3). The problem considered was the filtration of Hong Kong Kaolin suspension and the filtration performance results V vs. t and L vs. t are summarized in Table 2.2. The conditions used for the calculations are also given in the table.

The difference between the results obtained from the two different procedures (the commonly used procedure vs. the improved procedure) is determined by two factors, the magnitude of the medium resistance and the time. The interplay between these two factors are not simple. Generally speaking, for large medium resistance, especially initially, the rate of filtration is controlled by the medium. Consequently, the two procedures gave essentially the same results which agree well with those of the more exact analysis. On the other hand, because of the relatively large error introduced in approximating $\bar{\bar{\varepsilon}}_s$ (defined by Equation (2.1.8)) with $\bar{\varepsilon}_s$ (defined by Equation (2.2.10)), the results of L vs. t predicted by the commonly used procedure differ significantly from those obtained using the improved procedure. It is also worth noting that the improved procedure gives results (both V vs. t and L vs. t) which are very similar to those from the more exact analysis. The same situation was found in a large number of sample calculations (Tien and Bai, 2003). Thus, one may conclude that the improved procedure is indeed viable for estimating cake filtration performance.

Table 2.2

Comparisons of filtration performance calculations based on the conventional theory with those from the procedure of Section 2.4 and the numerical analysis. Filtration of Hong Kong Kaolin suspensions*

t(s)	V $(m^3/m)^2$			L (m)		
	a[†]	b[‡]	c[§]	a[†]	b[‡]	c[§]
$R_m = 3.1056 \times 10^{13}\ m^{-1}$						
10	3.22×10^{-5}	3.22×10^{-5}	3.22×10^{-5}	3.81×10^{-5}	3.23×10^{-5}	4.00×10^{-5}
20	6.44×10^{-5}	6.43×10^{-5}	6.47×10^{-5}	7.79×10^{-5}	4.46×10^{-5}	7.49×10^{-5}
40	1.29×10^{-4}	1.29×10^{-4}	1.29×10^{-4}	1.57×10^{-4}	8.91×10^{-5}	1.57×10^{-4}
100	3.21×10^{-4}	3.20×10^{-4}	3.20×10^{-4}	3.89×10^{-4}	2.22×10^{-4}	3.80×10^{-4}
400	1.28×10^{-3}	1.26×10^{-3}	1.28×10^{-3}	1.46×10^{-3}	8.75×10^{-4}	1.39×10^{-3}
1,000	3.15×10^{-3}	3.07×10^{-3}	3.13×10^{-3}	3.36×10^{-3}	2.13×10^{-3}	3.14×10^{-3}
1,600	4.97×10^{-3}	4.79×10^{-3}	4.92×10^{-3}	5.05×10^{-3}	3.82×10^{-3}	4.71×10^{-3}
$R_m = 3.1056 \times 10^{12}\ m^{-1}$						
10	3.15×10^{-4}	3.07×10^{-4}	3.13×10^{-4}	3.36×10^{-4}	2.13×10^{-4}	3.14×10^{-4}
20	6.14×10^{-4}	5.90×10^{-4}	6.09×10^{-4}	6.10×10^{-4}	4.09×10^{-4}	5.85×10^{-4}
40	1.17×10^{-3}	1.10×10^{-3}	1.14×10^{-3}	1.08×10^{-3}	7.62×10^{-4}	1.02×10^{-3}
100	2.54×10^{-3}	2.35×10^{-3}	2.46×10^{-3}	2.16×10^{-3}	1.63×10^{-3}	2.04×10^{-3}
400	6.89×10^{-3}	6.43×10^{-3}	6.62×10^{-3}	5.44×10^{-3}	4.45×10^{-3}	5.19×10^{-3}
1000	1.22×10^{-2}	1.15×10^{-2}	1.17×10^{-2}	9.39×10^{-3}	7.47×10^{-3}	8.95×10^{-3}
1600	1.60×10^{-2}	1.52×10^{-2}	1.54×10^{-2}	1.22×10^{-2}	1.06×10^{-2}	1.17×10^{-2}
$R_m = 3.1056 \times 10^{11}\ m^{-1}$						
10	1.22×10^{-3}	1.15×10^{-3}	1.17×10^{-3}	9.34×10^{-4}	7.97×10^{-4}	8.95×10^{-4}
20	1.81×10^{-3}	1.74×10^{-3}	1.75×10^{-3}	1.39×10^{-3}	1.20×10^{-3}	1.33×10^{-3}
40	2.66×10^{-3}	2.57×10^{-3}	2.57×10^{-3}	2.02×10^{-3}	1.78×10^{-3}	1.94×10^{-3}
100	4.34×10^{-3}	4.23×10^{-3}	4.21×10^{-3}	3.27×10^{-3}	2.93×10^{-3}	3.15×10^{-3}
400	8.90×10^{-3}	8.77×10^{-3}	8.67×10^{-3}	6.69×10^{-3}	6.07×10^{-3}	6.46×10^{-3}
1000	1.42×10^{-2}	1.40×10^{-2}	1.39×10^{-2}	1.07×10^{-2}	9.73×10^{-3}	1.03×10^{-2}
1600	1.80×10^{-2}	1.78×10^{-2}	1.76×10^{-2}	1.35×10^{-2}	1.24×10^{-2}	1.31×10^{-2}

*Conditions used for calculation:

$$p_o = 100{,}000\,\text{Pa}$$
$$\varepsilon_s = 0.27\left(1 + \frac{p_s}{1370}\right)^{0.09}$$
$$\alpha\ (\text{m kg}^{-1}) = 7.668 \times 10^{11}\left(1 + \frac{P_a}{1370}\right)^{0555}$$

p_s in Pa

a[†] From numerical solution of the continuity equations of both phases.
b[‡] From the conventional theory.
c[§] From the improved procedure.

2.5 EXTENSIONS OF THE CONVENTIONAL THEORY

Cake filtration often manifests itself in a number of solid/fluid separation operations constituting either the entire case or as part of it. Application of the conventional theory to the analysis of these operations may either be straightforward or require considerable adaptations and modifications. In the following, we present three cases to illustrate the use of the conventional theory for analyzing these solid/fluid separation operations.

2.5.1 Centrifugation

Centrifugation is a solid/liquid separation operation in which separation of a suspension is effected by directing the suspension onto the inner surface of a rotating bowl (perforated). The process may be carried out either in batch or continuously, and consists of two steps: filtration and de-liquoring. The discussion given below will be restricted mainly to the application of the conventional theory for the description of the filtration step.

A schematic diagram depicting centrifugation is shown in Fig. 2.3. Figure 2.3 gives both a top and a side view of the operation. In the most general case, the material present within the rotating bowl is composed of three zones; a cake layer formed on the surface of the bowl, $r_c < r < r_o$ where r_c is the location of the cake/suspension interface and r_o is the radius of the bowl; a suspension region, $r_s < r < r_c$ where r_s is the location of the clear liquid/suspension interface; and a clear liquid (suspending liquid) layer, $r_\ell < r < r_s$ where r_ℓ is the air/liquid interface. The locations of these interfaces change with time and depend upon both the filtration and sedimentation rate as well as the feed rate (for the case of continuous operation). The filtration phase terminates when both r_ℓ and r_s become the same as r_c.

Among several earlier workers, Grace (1953c) derived an expression for the instantaneous rate of centrifugal filtration. The derivation of the expression was based partially on physical argument and average cake properties and this expression has widely been used up to the present time (for example, see the recent work of Chan *et al.* (2003)).

The rate expression for centrifugal filtration can be obtained in a manner similar to the derivation given in Section 2.1. For the flow of permeating liquid, if the particle phase velocity is negligible, Equation (2.1.3) may be modified to give

$$q_\ell = -\frac{k}{\mu}\left(\frac{\partial p_\ell^*}{\partial r}\right) \tag{2.5.1.1}$$

where p_ℓ^* is defined as

$$p_\ell^* = p_\ell - \frac{r^2}{2}\omega^2\rho \tag{2.5.1.2}$$

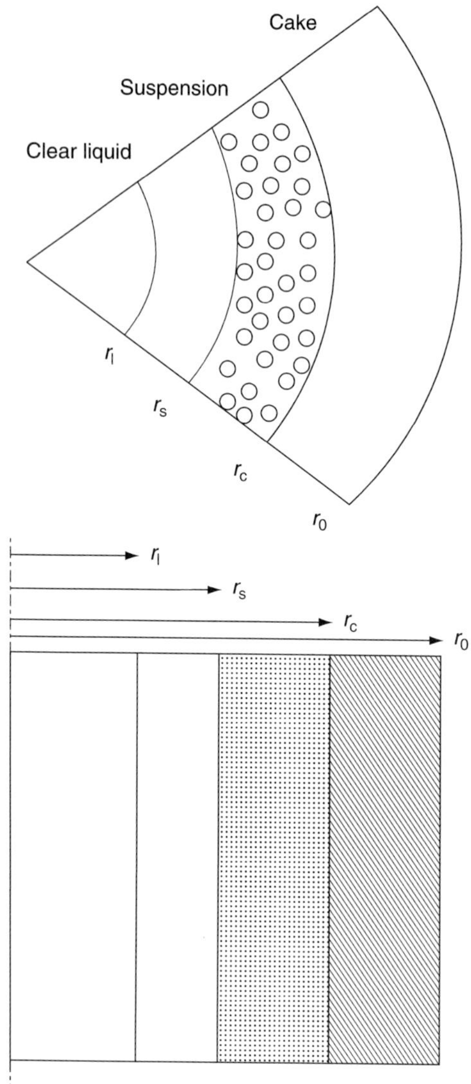

Figure 2.3 Schematic diagram of centrifugation.

The use of p_ℓ^* in the above expression is to account for the centrifugal effect. The presence of the negative sign is due to the fact that in the present case, q_ℓ is assumed to be along the r-direction while in Equation (2.1.3) q_ℓ is along the opposite direction of x.

The counterpart of Equation (2.1.2) for the cylindrical coordinate is

$$\frac{\partial}{\partial r}[r(q_\ell + q_s)] = 0 \tag{2.5.1.3}$$

If the particle velocity is negligible, one has

$$(q_\ell r) = (q_{\ell_o} r_o) = \text{constant} \tag{2.5.1.4}$$

where r is the radial distance, q_ℓ and q_s the radial fluid and particle velocities, and q_{ℓ_o} is the fluid radial velocity across the perforated base surface, or the instantaneous filtration velocity.

Combining Equations (2.5.1.1) and (2.5.1.4) yields

$$\mu(q_{\ell_o} r_o)\frac{dr}{r} = -k\mathrm{d}p_\ell^*$$

or

$$\int_{r_c}^{r_o} \mu(q_{\ell_o} r_o)2\pi r\varepsilon_s\rho_s\mathrm{d}r = \int_{p_\ell^*|_{r_c}}^{p_\ell^*|_{r_o}} -2k\varepsilon_s\rho_s(\pi r^2)\mathrm{d}p_\ell^* \tag{2.5.1.5}$$

The left-hand side of Equation (2.6.1.5) is

$$\mu q_{\ell_o} r_o \int_{r_c}^{r_o} (2\pi r)\varepsilon_s\rho_s\mathrm{d}r = \mu q_{\ell_o} r_o W = \mu q_{\ell_o} r_o(w)(2\pi r_o) \tag{2.5.1.6}$$

Equation (2.5.1.5) becomes

$$\mu q_{\ell_o} r_o^2 w = \int_{(p_\ell^*)_{r_o}}^{(p_\ell^*)_{r_c}} \frac{r^2}{\alpha}\mathrm{d}r \tag{2.5.1.7}$$

where

$$w = \frac{W}{2\pi r_o} \tag{2.5.1.8a}$$

and

$$W = \int_{r_c}^{r_o} 2\pi r\varepsilon_s\rho_s\mathrm{d}r \tag{2.5.1.8b}$$

with W being the dry cake mass per unit bowl length and w the mass of dry cake per unit medium surface area.

If one defines the average specific cake resistance α_{av} to be

$$\alpha_{av} = \frac{r_o^2[(p_\ell^*)_{r_c} - (p_\ell^*)_{r_o}]}{\int_{(p_\ell^*)_{r_o}}^{(p_\ell^*)_{r_c}} \frac{r^2}{\alpha}\cdot\mathrm{d}p_\ell^*} \tag{2.5.1.9}$$

By Equations (2.5.1.2) and (2A.2.14),

$$(p_\ell^*)_{r_c} = (p_\ell)_{r_c} - \frac{\omega^2}{2} r_c^2 \rho \tag{2.5.1.10a}$$

$$(p_\ell^*)_{r_o} = (p_\ell)_{r_o} - \frac{\omega^2}{2} r_o^2 \rho \tag{2.5.1.10b}$$

$$(p_\ell)_{r_c} = p_o + \frac{\rho\omega^2}{2}(r_s^2 - r_\ell^2) + \frac{\rho_{SL}\omega^2}{2}(r_c^2 - r_s^2) \tag{2.5.1.10c}$$

Combining Equations (2.5.1.7), (2.5.1.9) and (2.5.1.10a)–(2.5.1.10c) yields

$$q_{\ell_o} = \frac{p_o + \frac{\omega^2}{2}[\rho(r_s^2 - r_\ell^2) + \rho_{SL}(r_c^2 - r_s^2) + \rho(r_o^2 - r_c^2)] - (p_\ell)_{r_o}}{\mu(\alpha_v w + R_m)} \tag{2.5.1.11}$$

For the filtrate permeation across the medium, one has

$$q_{\ell_o} = \frac{p_\ell|_{r_o}}{(\mu R_m)} \tag{2.5.1.12}$$

Combining Equations (2.5.1.11) and (2.5.1.12) yields

$$q_{\ell_o} = \frac{\frac{\omega^2}{2}[\rho(r_s^2 - r_\ell^2) + \rho_{SL}(r_c^2 - r_s^2) + \rho(r_o^2 - r_c^2)] + p_o}{\mu[w\alpha_{av} + R_m]} \tag{2.5.1.13}$$

Instead of Equation (2.5.1.9), a different definition of the average specific cake resistance $\overline{\alpha}$ may be given as

$$\int_{(p_\ell^*)_{r_o}}^{(p_\ell^*)_{r_c}} \frac{r^2}{\alpha} \mathrm{d}p_\ell = (\tilde{r})^2(1/\overline{\alpha})[(p_\ell^*)_{r_c} - (p_\ell^*)_{r_o}]$$

and

$$\overline{\alpha} = (\tilde{r})^2 \frac{(p_\ell^*)_{r_c} - (p_\ell^*)_{r_o}}{\int_{(p_\ell^*)_{r_o}}^{(p_\ell^*)_{r_c}} \frac{r^2}{\alpha} \mathrm{d}p_\ell^*} \tag{2.5.1.14}$$

where $\tilde{r}$ is a value between r_c and r_o or $r_c < \tilde{r} < r_o$. The two average specific cake resistances are related by the expression

$$\frac{\alpha_{av}}{\overline{\alpha}} = \frac{r_o^2}{(\tilde{r})^2} \tag{2.5.1.15}$$

If $\overline{\alpha}$ is used instead of α_{av}, Equation (2.5.1.13) becomes

$$q_{\ell_o} = \frac{\frac{\omega^2}{2}[\rho(r_s^2 - r_\ell^2) + \rho_{SL}(r_c^2 - r_s^2) + \rho(r_o^2 - r_c^2)] + p_o}{\mu[w\overline{\alpha}(r_c/\tilde{r})^2 + R_m]} \tag{2.5.1.16}$$

If $p_o = 0$ and the suspension is dilute or there is no suspension phase (namely, sedimentation is fast), the above expression becomes

$$q_{\ell_o} = \frac{\frac{\omega^2}{2}\rho(r_o^2 - r_\ell^2)}{\mu[w\bar{\alpha}(r_o/\tilde{r})^2 + R_m]} \tag{2.5.1.17}$$

which is the expression given by Wakeman and Tarleton (1999).

If the numerator of Equation (2.5.1.13) or those of Equations (2.5.1.16) and (2.5.1.17) are considered as the total driving force for liquid flow across the cake, these expressions may be viewed as being equivalent to Equation (2.1.20). The apparent similarity between Equations (2.5.1.13) [or (2.5.1.16), (2.5.1.17)] and (2.1.16) may obfuscate a significant difference between them. The quantity α_{av} appearing in Equation (2.1.16) is a function of the cake material and the liquid pressure drop across the cake. In other words, if the relationships of α vs. p_s and ε_s vs. p_s and the functional form of $f'(\varepsilon)$ are known, and the operating pressure p_o is specified, α_{av} can be readily determined. Equation (2.1.16), therefore, can be used to predict filtration performance if the constitutive relationships are available. Similarly, one may obtain α_{av} from filtration experiments data based on Equation (2.1.16) and use the value for predicting filtration performance of the same system conducted under different conditions.

Using Equation (2.5.1.13) [or (2.5.1.16)] for scale-up is questionable. Its similarity to Equation (2.1.16) has led some investigators to conclude that the linear plot of t/V vs. V (or t is a second order polynomial of V) may be used to determine α_{av} (Wakeman and Tarleton, 1999). However, Sambuichi *et al.* (1987) and, more recently, Chan *et al.* (2003) showed that V, for the most part, was nearly a linear function of time, and the filtration rate (obtained by Chan *et al.*) was approximately constant.

If Equation (2.5.1.13) is to be used for predicting filtration performance, a proper procedure for estimating α_{av} must first be devised. According to Equation (2.5.1.9), for the estimation of α_{av}, the local value of α throughout the cake ($r_c < r < r_o$) is required. Since in centrifugal filtration, the extent of cake compression is effected by the cake compressive stress and the centrifugal force (Grace, 1953c), one may therefore define an effective compressive stress, $(p_s)_{eff}$, to be

$$(p_s)_{eff} = p_s + \frac{\rho_s w r^2}{2} \tag{2.5.1.18}$$

The local specific cake resistance therefore can then be determined from the constitutive relationship and the knowledge of the compressive stress profile.

The compressive stress profile can be obtained from the solution of the relevant equations relating p_ℓ, p_s and the centrifugal forces given in Appendix 2A.2 together with the various interface expressions (r_ℓ, r_s and r_c) given in Appendix 2A.1. These equations may be solved iteratively in a manner described in Appendix 2A.3.

The procedure outlined in Appendix 2A.3 is not straightforward and computation-demanding. To circumvent this difficulty, Chan *et al.* (2003) suggested that α_{av} may be considered as a function of $\Delta[(p_s)_{eff}]$, the range of the effective compressive stress is essentially the same as the centrifugal force and the functional relationship of α_{av} vs. $\Delta[(p_s)_{eff}]$ is similar to that of α_{av} vs. Δp_s obtained from the conventional t/V vs. V plot. The general validity of this approach, however, remains to be tested.

2.5.2 Expression of solid/liquid mixtures

Shirato and Coworkers (1967, 1970, 1971b) defined "expression" as the removal of water from solid/liquid mixtures through the actions of mechanical force. Expression may treat mixtures with solid concentration greater than those encountered in cake filtration (slurry vs. suspension). The operation is often conducted in batch.

A schematic diagram demonstrating the operation of expression is shown in Fig. 2.4. The slurry to be treated is first placed into a cell which is fitted with a permeable septum at the bottom and a movable piston at the top. Mechanical compression is applied through the action of the piston which moves downward as water is progressively squeezed out. The water removed passes through the bottom septum or through both the septum and the piston if the piston is also permeable.

De-watering in expression proceeds in two steps. With the application of mechanical compression, the entire mixture may be assumed to reach a pressure equal to the applied compression. Water is first removed from the mixture in a manner like that of constant pressure filtration with the formation of filter cakes at the draining surfaces. Filtration proceeds until all the particles present initially in the mixture become parts of the cakes formed.

The second step may be referred to as the consolidation step. The cake formed in the first stage undergoes further compression. The liquid velocity (draining rate) is less than that of the filtration period and the effect of the solid velocity becomes more important. This process may continue until the cake reaches equilibrium with the applied compression.

As the first stage of expression operation is similar to that of constant pressure filtration, the results obtained previously (namely, those of Sections 2.2 and 2.3) can be readily applied. The performance of expressions is often described by the volume reduction achieved. Consider a cell of unit cross-sectional area, the initial solid/mixture height being L_o. The height becomes L at an arbitrary instant during the filtration period. $L_o - L$ represents the water removed or the cumulative filtrate per unit septum area which is the same as V. Applying Equation (2.2.15), one has

$$(L_o - L) + L_m = i[A(t + t_m)]^{1/2} \tag{2.5.2.1}$$

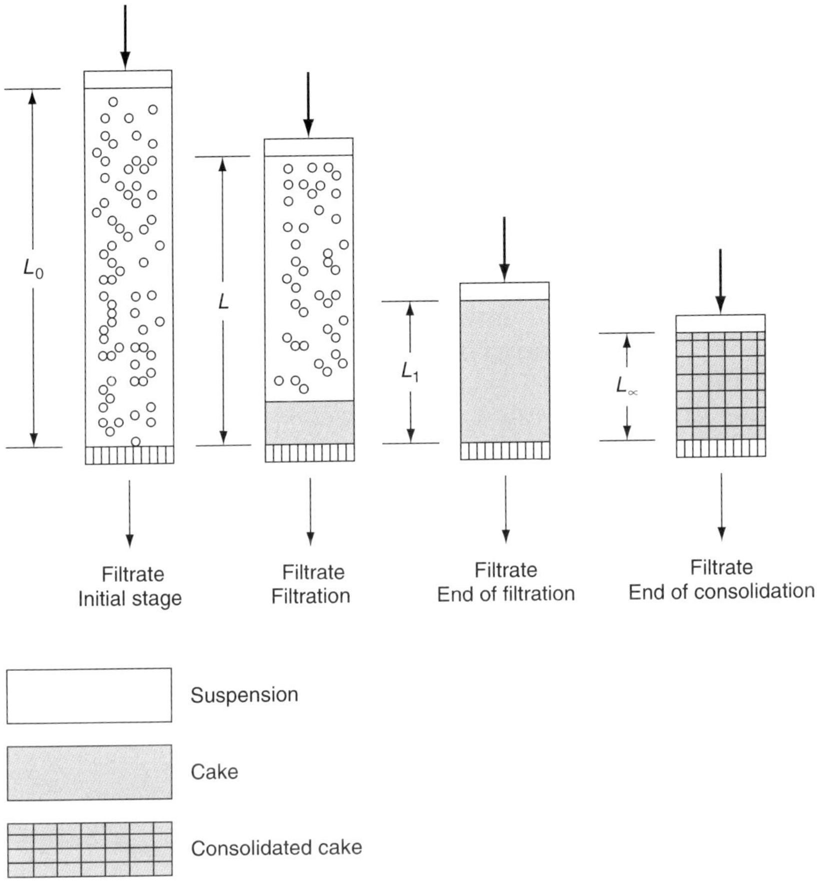

Figure 2.4 Expression of a solid/liquid mixture.

where we have replaced V_m with L_m, the fictitious mixture height equivalent to the septum resistance. The quantity i denotes the number of draining surfaces and is introduced in order to account for the fact that water may flow out of the cell both downward and upward.

If L_1 denotes the mixture height at the end of the filtration step, L_1 based on mass balance may be expressed as

$$L_1 = \frac{w_o}{\rho_s} + (\overline{m} - 1)\frac{w_o}{\rho} = w_o[(1/\rho_s) + (\overline{m} - 1)/\rho] \tag{2.5.2.2}$$

where w_o is the total solid mass (kg/m^2) present in the mixture initially. Strictly speaking, $\overline{m}$, the average wet to dry cake mass ratio, of the above expression should be the value at the time, when L reaches L_1. But as stated before, $\overline{m}$ remains essentially constant and therefore can be considered as a constant. The initial mixture height L_o expressed in w_o is

$$L_o = (w_o/\rho_s)\frac{(s/\rho_s)+(1-s)/\rho}{(r/\rho_s)} = w_o\frac{\rho s+(1-s)\rho_s}{s\rho\rho_s} \tag{2.5.2.3}$$

From Equations 2.5.2.2 and 2.5.2.3, one has

$$\begin{aligned} L_o - L_1 &= w_o\left[\frac{s\rho+(1-s)/\rho_s}{s\rho\rho_s} - \frac{1}{\rho_s} - \frac{\overline{m}-1}{\rho}\right] \\ &= w_o\frac{1-\overline{m}s}{s\rho} \end{aligned} \tag{2.5.2.4}$$

The extent of dewatering during the filtration period can be expressed as $(L_o - L)/(L_o - L_1)$, which can be obtained by dividing Equations (2.5.2.1) by (2.5.2.4). Upon rearrangement, one has

$$\frac{L_o - L}{L_o - L_1} = \frac{i}{w_o}\left[\frac{2p_o s\rho}{\mu(\alpha_{av})_{\Delta p_c}(1-\overline{m}s)}\right]^{1/2}\left[\sqrt{t+t_m} - \sqrt{t_m}\right] \tag{2.5.2.5}$$

In other words, data obtained during the initial period of expression when plotted as $(L_o - L)/(L_o - L_1)$ vs. $i(\sqrt{t+t_m} - \sqrt{t_m})/w_o$ should yield a straight line. This is shown in Fig. 2.5.

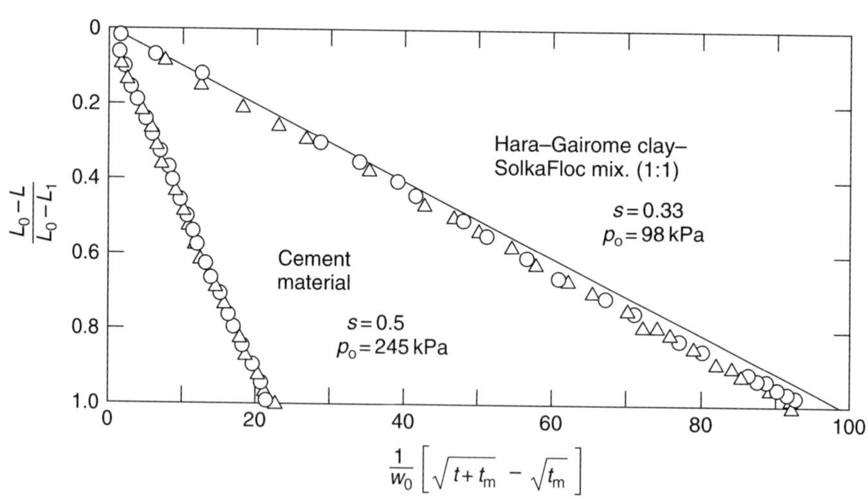

Figure 2.5 Fitting of experimental data according to Equation (2.5.2.5) (Shirato *et al.*, 1987). (Shirato, Murase, Iritani, Tiller and Aliciatore, 1987. Reprinted by permission of Taylor & Francis.)

For the performance of expression during the consolidation step, the conventional theory is not applicable since particle movement is now a dominant feature and the conventional theory is based on the assumption of negligible particle velocity. The problem can be solved through the solution of the appropriate equation of continuity. This will be discussed in Chapter 4.

2.5.3 Optimum operating pressure and the skin layer effect

As seen from Equation (2.1.16), the operating pressure, p_o, can be considered as the driving force of cake filtration. Intuitively, one may expect improvement in cake filtration performance (i.e. higher filtration rate) with an increase of p_o. However, increasing the operating pressure also leads to greater cake compaction and therefore greater cake resistance. For optimum operation, one may expect a trade-off between these two opposing effects.

For certain suspension systems, there exists a threshold operating pressure beyond which higher operating pressure may not be optimal. Tiller and Green (1973) presented a simple analysis identifying the presence of this threshold pressure. Their analysis is based on the conventional theory, as shown below. By rearranging Equation (2.1.4), one has

$$\mu q_\ell \, \mathrm{d}x = \frac{\mathrm{d}p_\ell}{\alpha \rho_s \varepsilon_s} \tag{2.5.3.1}$$

Integrating the above expression from $x = 0$ to $x = L$,

$$\mu q_\ell L = \frac{1}{\rho_s} \int_{\Delta p_m}^{p_o} \frac{\mathrm{d}p_\ell}{\varepsilon_s \alpha} = \left(\frac{1}{\rho_s}\right) \int_0^{p_{s_m}} (-f') \frac{\mathrm{d}p_s}{\varepsilon_s \alpha} \tag{2.5.3.2}$$

where f' is defined by Equation (2.1.11).

Tiller and Green (1973) applied the commonly used assumptions, that $f' = -1$ and the pressure drop across the medium, Δp_m, negligible, the compressible stress at the cake/membrane interface becomes p_o. Equation (2.5.3.2) is now given as

$$\mu \rho_s q_\ell L = \int_0^{p_o} \frac{\mathrm{d}p_s}{\varepsilon_s \alpha} \tag{2.5.3.3}$$

For a given suspension, product of the filtration rate and cake thickness is a monotonic increasing function of p_o. The dependence of the product on p_o is determined by the constitutive relationships (i.e. ε_s vs. p_s and α vs. p_s). The results of three specific systems considered by Tiller and Green are shown in Fig. 2.6, which gives the values of $\mu \rho_s q_\ell L$ vs. p_o for the filtration of suspensions of Solkafloc, Gairomi clay and latex particles. For both Solkafloc and Gairomi clay cakes with p_o up to 10^3 kPa, the product increases

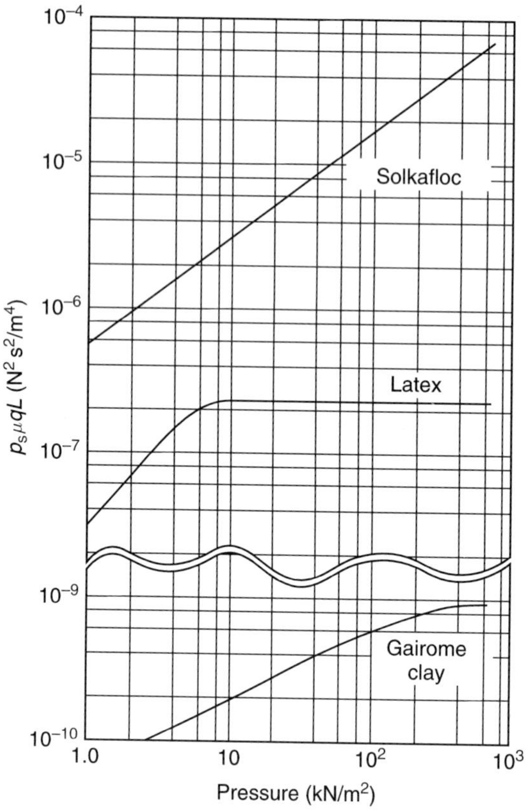

Figure 2.6 Filtration rate – cake thickness vs. time (Tiller and Green, 1973). (Tiller and Green, 1973. Reprinted by permission of The American Institute of Chemical Engineers.)

with the increase of p_o in a nearly linear fashion (on logrithmic coordinate system). On the other hand, for latex cakes, the product first increases with p_o and becomes nearly constant for $p_o > 10$ kPa.

The implication from this nearly constant behavior can be stated as follows. Consider a given system operating at two different operating pressures $(p_o)_1$ and $(p_o)_2$. The cake thickness at a given time, as an approximation, can be considered to be proportional to the cumulative filtrate volume V. If $(L)_{(p_o)_1} > (L)_{(p_o)_2}$, it implies that $(V)_{(p_o)_1} > (V)_{(p_o)_2}$. On the other hand, that $(\mathrm{L})_{(p_o)_1} < (\mathrm{L})_{(p_o)_2}$ means that $(q_\ell)_{(p_o)_1} > (q_\ell)_{(p_o)_2}$ since the product of filtration rate and cake thickness is constant. However, $(q_\ell)_{(p_o)_1} < (q_\ell)_{(p_o)_2}$ can only lead to $(\mathrm{V})_{(p_o)_1} < (\mathrm{V})_{(p_o)_2}$, which is contradictory to the assumption used. The only condition based on the results shown in Fig. 2.6 is that q_ℓ to be independent of the operating pressure for $p_o > 10$ kPa. The reason for this independence arises from the compressive behavior of latex cakes. The constitutive relationships of those systems are given in Figs 2.7a (ε_s vs. p_o) and 2.7b (α vs. p_o). As shown in these two figures, latex cakes are extremely compressible as compared with the other two systems. As a result, the value

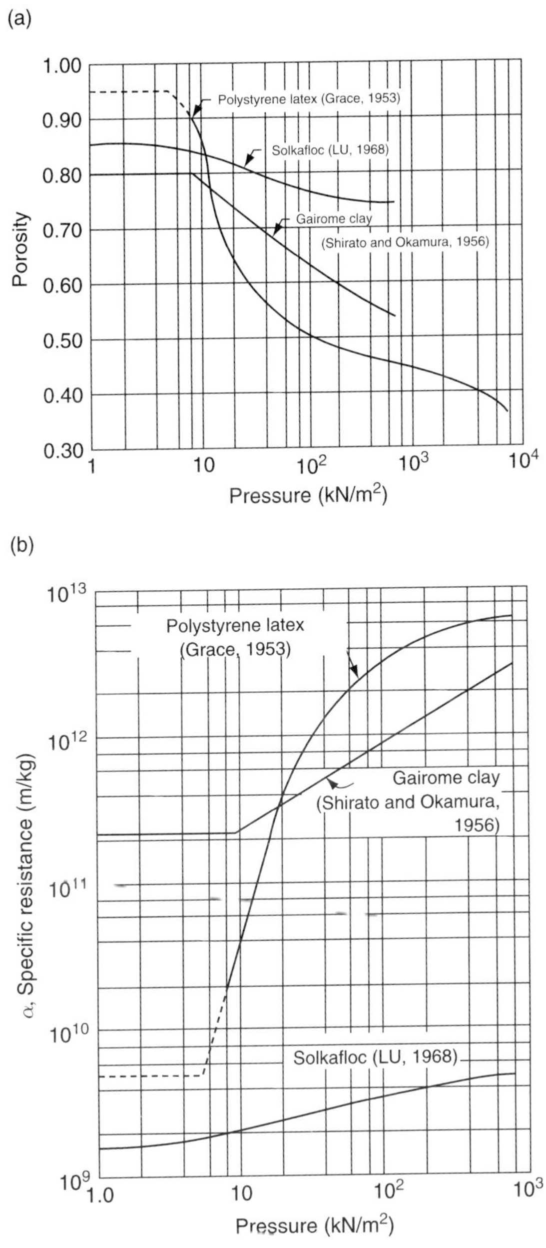

Figure 2.7 Constitutive relationships of latex cakes, Cairome cakes and Salkafloc Cakes (a) $1-\varepsilon_s$ vs. p_s (b) α vs. p_s. (Tiller and Green, 1973. Reprinted by permission of The American Institute of Chemical Engineers.)

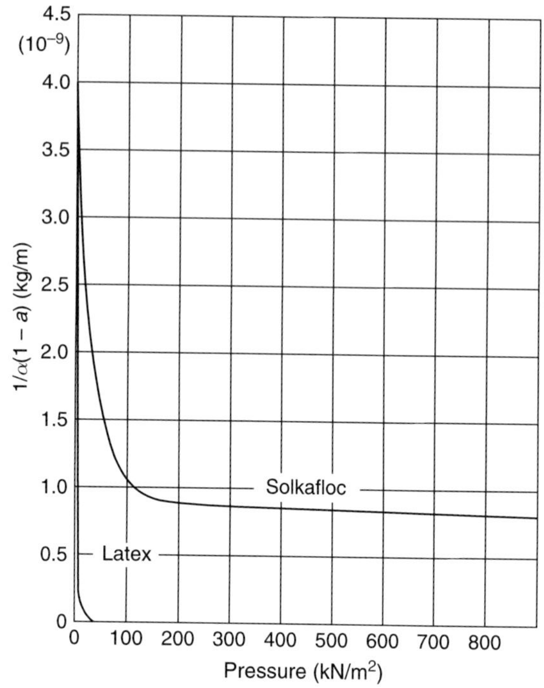

Figure 2.8 Plot of $(\alpha\varepsilon_s)^{-1}$ vs. p_s (Tiller and Green, 1973). (Tiller and Green, 1973. Reprinted by permission of The American Institute of Chemical Engineers.)

of the integral of Equation (2.5.13) reaches a nearly constant value with $p_o = 10\,\text{kPa}$ (Fig. 2.8).

The physical significance of the latex cake behavior can be discerned from the porosity and pressure profiles of Fig. 2.9a,b. These profiles were determined according to Equations (2.3.9) (for pressure profile) and (2.3.10) (porosity profile based on the constitutive relationships of Fig. 2.7a,b).

The pressure profiles of latex cakes shown in Fig. 2.9a indicates that regardless of the value of p_o, the dominant pressure drop occurred in a thin region immediately adjacent to the medium. In contrast, for the two other types of cakes, pressure drop was rather uniform across the entire cake thickness. Comparing the three latex cake profiles corresponding to different p_o values, it was clear that the effect of increasing p_o merely increased the thickness of this dense layer (or skin layer). The porosity profiles of Fig. 2.9b corroborate this conclusion.

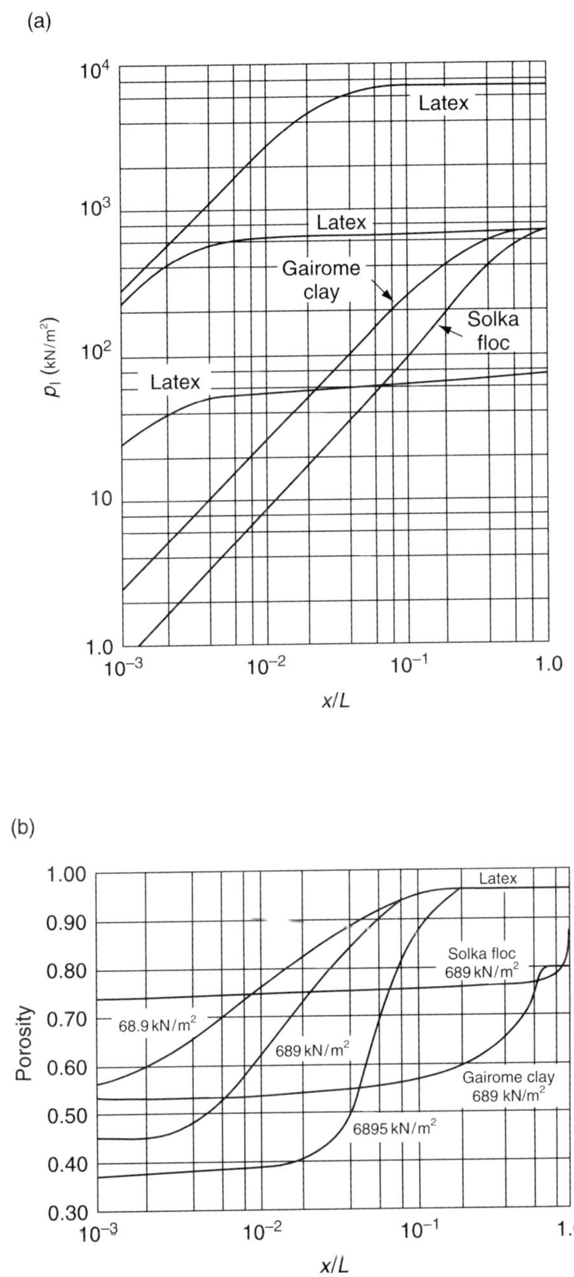

Figure 2.9a (a) Filtrate pressure profile (p_ℓ vs. x/L) of three types of cakes (Tiller and Green, 1973). (b) Porosity profile ($1-\varepsilon_s$ vs. x/L) of three types of cakes (Tiller and Green, 1973). (Tiller and Green, 1973. Reprinted by permission of The American Institute of Chemical Engineers.)

REFERENCES

Chan, S.H., Kiang, S. and Brown, M.A., *AIChE J.*, **44**, 925 (2003).

Grace, H.P., *Chem. Eng. Prog.*, **49**, 303 (1953a).

Grace, H.P., *Chem. Eng. Prog.*, **49**, 364 (1953b).

Grace, H.P., *Chem. Eng. Prog.*, **49**, 407 (1953c).

Lee, D.J., Ju, S.P., Kwan, J. and Tiller, F.M., *AIChE J.*, **46**, 110 (2000).

Okamura, S. and Shirato, M., *Kagaku Kogaku*, **19**, 104 (1955a).

Okamura, S. and Shirato, M., *Kagaku Kogaku*, **19**, 111 (1955b).

Ruth, B.F., *Ind. Eng. Chem.*, **27**, 108 (1935a).

Ruth, B.F., *Ind. Eng. Chem.*, **27**, 806 (1935b).

Ruth, B.F., *Ind. Eng. Chem.*, **38**, 567 (1946).

Ruth, B.F., Montillon, G.H. and Montonna, R.E., *Ind. Eng. Chem.*, **25**, 76 (1933a).

Ruth, B.F., Montillon, G.H. and Montonna, R.E., *Ind. Eng. Chem.*, **25**, 153 (1933b).

Sambuichi, M., Nakakura, H., Osasa, K. and Tiller, F.M., *AIChE J.*, **33**, 109 (1987).

Shirato, M. and Aragaki, T., *Kagaku Kagaku*, **33**, 205 (1969).

Shirato, M. and Okamura, S., *Kagaku Kogaku*, **20**, 670 (1956).

Shirato, M., Murase, T., Kato, H. and Fukaya, S., *Kagaka Kogaku*, **31**, 1125 (1967).

Shirato, M., Sambuichi, M., Kato, H. and Aragaki, T., *AIChE J.*, **15**, 405 (1969a).

Shirato, M., Aragaki, T., Mori, R. and Imai, K., *Kagaku Kogaku*, **39**, 576 (1969b).

Shirato, M., Murase, T., Negawa, M. and Senda, T., *J. Chem. Eng. Japan*, **3**, 105 (1970).

Shirato, M., Aragaki, T., Ichimura, K. and Ootsuji, N., *J. Chem. Eng. Japan*, **4,** 172 (1971a).

Shirato, M., Murase, T., Negawa, M. and Maridera, H., *J. Chem. Eng. Japan*, **4**, 363 (1971b).

Shirato, M., Murase, T., Iritani, E., Tiller, F.M. and Alciatore, A.F., "Filtration in the Chemical Process Industry", in *Filtration: Principle and Practices*, M.J. Matteson and C. Orr, eds, Mareel Dekker, New York (1987).

Teoh, S.K., "Studies in Filter Cake Characterization and Modeling", Ph.D. Thesis, National University of Singapore (2003).

Teoh, S.K., Tan, R.B.H. and Tien, C., "Analysis of Cake Filtration Data: A Critical Assessment of Conventional Cake Filtration Theory", to be published (2006).

Tien,C.andBai,R.,"RevisitingtheConventionalCakeFiltrationTheory:AnExaminationofItsAccuracy andUtility",*sixth World Congress of Chemical Engineering*,Melbourne,Australia(2001).

Tien, C. and Bai, R., *Chem. Eng. Sci.*, **58**, 1323 (2003).

Tien, C., Teoh, S.K. and Tan, R.B.H., *Chem. Eng. Sci.*, **56**, 5361 (2001).

Tiller, F.M., *Chem. Eng. Prog.*, **49**, 467 (1953).

Tiller, F.M., *Chem. Eng. Prog.*, **51**, 282 (1955).

Tiller, F.M., *AIChE J.*, **4**, 170 (1958).

Tiller, F.M. and Cooper, H., *AIChE J.*, **8**, 445 (1962).

Tiller, F.M. and Green, T.C., *AIChE J.*, **19**, 1266 (1973).

Tiller, F.M. and Huang, C.J., *Ind. Eng. Chem.*, **53**, 529 (1961).

Tiller, F.M. and Lu, W.M., *AIChE J.*, **18**, 569 (1972).

Tiller, F.M. and Shirato, M., *AIChE J.*, **10**, 61 (1964).

Tiller, F.M. and Yeh, C.S., *AIChE J.*, **31**, 1241 (1985).

Tiller, F.M. and Yeh, C.S., *AIChE J.*, **33**, 1241 (1987).

Tiller, F.M., Haynes, S. and Lu, W.M., *AIChE J.*, **18**, 13 (1972).

Tiller, F.M., Lu, R., Kwon, K.W. and Lee, D.J., *Water Research*, **33**, 15 (1999).

Wakeman, R.J. and Tarleton, E.S., *Filtration: Equipment Selection Modeling and Process Simulation*, Elsevier Advanced Technology, Oxford (1999).

Walker, W.H., Lewis, W.K., McAdams, W.H. and Gilliland, E.R., *Principles of Chemical Engineering* (third edn) McGraw Hill, New York (1937).

APPENDIX 2A.1: LOCATIONS OF AIR/LIQUID, LIQUID/SUSPENSION, SUSPENSION/CAKE INTERFACES

As discussed in Section 2.5.1, for centrifugal filtration, the physical domain may be considered to be composed of three regions: clear liquid, $r_\ell < r < r_s$; suspension, $r_s < r < r_c$ and cake $r_c < r < r_o$. Expressions of the locations of r_ℓ, r_s and r_c can be derived from mass balances as shown below.

For the one-dimensional case (see Fig. 2.3), first consider the suspension phase with u_ℓ and u_s denoting the liquid and particle radial velocities. Accordingly, $u_\ell = q_s/(1-\varepsilon_s)$ and $u_s = q_s/\varepsilon_s$. For centrifugal filtration, the overall mass balance gives

$$2\pi r[u_\ell(1-\varepsilon_s) + u_s\varepsilon_s] = 2\pi r_o q_{\ell_o}$$

or

$$u_\ell = \frac{r_o}{r} q_{\ell_o} \frac{1}{1-\varepsilon_s} - \frac{\varepsilon_s}{1-\varepsilon_s} u_s \tag{2A.1.1}$$

where q_{ℓ_o} is the instantaneous filtration rate.

And the solid/liquid velocity can be written as

$$(u_s - u_\ell)_c = \left(\frac{u_s}{1-\varepsilon_s}\right)_c - \frac{r_o}{r} q_{\ell_o} \frac{1}{1-\varepsilon_s} \tag{2A.1.2}$$

The quantity $(u_s - u_\ell)$ can be viewed as the sedimentation velocity in centrifugation. The subscript "c" denotes centrifugation.

In one-dimensional gravity sedimentation, if u_ℓ and u_s are the liquid and particle velocities along the direction of the gravitational force, one has

$$u_\ell(1-\varepsilon_s) + u_s\varepsilon_s = 0$$

and the sedimentation velocity, $(u_s - u_\ell)_g$, where the subscript "g" denotes gravitational sedimentation is

$$(u_s - u_\ell)_g = \left(\frac{u_s}{1-\varepsilon_s}\right)_g \tag{2A.1.3}$$

If one assumes that the ratio of $(u_s - u_\ell)_c$ to $(u_s - u_\ell)_g$ is equal to $r\omega^2/g$, the ratio of centrifugal acceleration to gravitational acceleration, $(u_s)_c$ can be found to be

$$(u_s)_c = \frac{r\omega^2}{g}(u_s)_g + q_{\ell_o}\frac{r_o}{r} \tag{2A.1.4}$$

which was first suggested by Sambuichi *et al.* (1987).

To obtain expressions of r_ℓ, r_s and r_c, the initial expressions of these quantities are

$$r_\ell = r_s = r_{\ell_o}; \quad r_c = r_o; \quad t = 0$$

where r_{ℓ_o} is the initial suspension position.

(a) *Air/liquid interface*
The overall balance gives

$$\pi r_\ell^2 - \pi r_{\ell_o}^2 = 2\pi r_o \int_o^t q_{\ell_o} \, dt$$

or

$$r_\ell^2 - r_{\ell_o}^2 = 2 r_o \int_o^t q_{\ell_o} \, dt \tag{2A.1.5}$$

(b) *Liquid/suspension interface*
By definition, dr_s/dt is the particle velocity at $r = r_s$. From Equation (2A.1.4), one has

$$\frac{dr_s}{dt} = \frac{r_s \omega^2}{g}(u_s)_g + q_{\ell_o}\frac{r_o}{r_s} \tag{2A.1.6}$$

Note that $(u_s)_g$ can be found from a number of expressions (empirical and analytical ones). Generally speaking, $(u_s)_g$ is a function of ε_s. For a number of systems, $(u_s)_g \, \varepsilon_s$ can be approximated as a linear function of ε_s.

(c) *Suspension/cake interface*
Sambuichi *et al.* (1987) obtained an expression for the cake/suspension interface in centrifugal filtration. With additional assumption that the particle volume fraction of the suspension phase remains the same as its initial value, the location of r_c becomes

$$(r_c/r_o) = \left[1 - \frac{2}{r_o(\bar{\bar{\varepsilon}}_s - \varepsilon_{s_o})} \int_o^t q_{\ell_o} \varepsilon_{s_o} \, dt \right]^{1/2} \tag{2A.1.7}$$

where ε_{s_o} is the initial particle volume fraction of the suspension and $\bar{\bar{\varepsilon}}_s$ is the average cake solidosity, defined as

$$\bar{\bar{\varepsilon}}_s = \frac{2\int_{r_c}^{r_o} r\varepsilon_s dr}{r_o^2 - r_c^2} \tag{2A.1.8}$$

APPENDIX 2A.2: PRESSURE AND COMPRESSIVE STRESS PROFILES IN CENTRIFUGAL FILTRATION

According to the multiphase flow theory, the equation of continuity of the liquid and particle phases can be written as

$$\rho_i \varepsilon_i \frac{D}{Dt}\mathbf{U}_i = -\mathbf{S}_i + \mathbf{W}_i + F_i \tag{2A.2.1}$$

$$i = \ell \text{ (liquid) or s (particle)}$$

where ρ and ε are the density and volume fractions. **U**, **S** and **W** are the velocity vector, stress vector and body force vector respectively. F_i is the force acting on phase i by phase j (and $F_i = -F_j$).

If one assumes

$$\mathbf{S}_i = \nabla \mathbf{T}_i \tag{2A.2.2a}$$

where $\mathbf{T}_i$ is the stress tensor of phase i. $\mathbf{T}_i$ may be written as

$$\mathbf{T}_i = \boldsymbol{\tau}_i + p_i \boldsymbol{\delta} \tag{2A.2.2b}$$

where p_i is the isotropic pressure of phase i, $\boldsymbol{\tau}_i$ is the shear stress tensor, and $\boldsymbol{\delta}$ the unit tensor.

For one-dimensional cake filtration, the isotropic pressure is the dominant term of the stress tensor. Equation (2A.2.2a) may be simplified to give

$$-\frac{\partial p_\ell}{\partial r} + (1-\varepsilon_s) r\omega^2 \rho + F_\ell = 0^{\dagger} \tag{2A.2.3}$$

$$\frac{\partial p_s}{\partial r} + \varepsilon_s r\omega^2 \rho_s - F_\ell = 0 \tag{2A.2.4}$$

The sum of the Equations (2A.2.3) and (2A.2.4) is

$$-\frac{\partial p_\ell}{\partial r} + (1-\varepsilon_s) r\omega^2 \rho - \frac{\partial p_s}{\partial r} + \varepsilon_s r\omega^2 \rho_s = 0 \tag{2A.2.5}$$

For the three regions, one has the following:

For $r_\ell < r < r_s$, $\varepsilon_s = 0$, and $p_s = 0$

Equation (2A.2.5) reduces to

$$-\frac{\partial p_\ell}{\partial r} + r\omega^2 \rho = 0 \tag{2A.2.6}$$

at $r = r_\ell$, $\quad p_\ell = p_o$

$^{\dagger}\rho$ is used here instead of ρ_ℓ.

Accordingly,

$$p_\ell = p_o + \frac{\rho\omega^2}{2}(r^2 - r_\ell^2) \tag{2A.2.7a}$$

and

$$(p_\ell)_{r_s} = p_o + \frac{\rho\omega^2}{2}(r_s^2 - r_\ell^2) \tag{2A.2.7b}$$

For $r_s < r < r_c$, ε_s is uniform and $p_s = 0$

Equation (2A.2.5) reduces to

$$-\frac{\partial p_\ell}{\partial r} + \rho_{SL}\omega^2 = 0 \tag{2A.2.8}$$

$$\rho_{SL} = \rho(1 - \varepsilon_s) + \rho_s\varepsilon_s \tag{2A.2.9}$$

at $r = r_s$, $\quad p_\ell = (p_\ell)_{r_s}$

Accordingly,

$$p_\ell = (p_\ell)_{r_s} + \frac{\rho_{SL}\omega^2}{2}(r^2 - r_s^2) \tag{2A.2.10a}$$

and

$$(p_\ell)_r = (p_\ell)_{r_s} + \frac{\rho_{SL}\omega^2}{2}(r_c^2 - r_s^2) \tag{2A.2.10b}$$

For $r_c < r < r_o$, ε_s varies with both r and t, and p_s does not vanish.

Equation (2A.2.5) is applicable and may be rewritten as

$$-\frac{\partial p_\ell}{\partial r} + r\omega^2\rho - \frac{\partial p_s}{\partial r} + \varepsilon_s(\rho + \rho_s)\omega^2 r = 0 \tag{2A.2.11}$$

using the same assumption as stated in 2.5.1, $q_\ell r$ is assumed to be constant across the cake region (as a consequence of neglecting q_s), by Darcy's law, one has

$$\frac{q_{\ell_o} r_o}{r} = -\frac{k}{\mu}\left[\frac{\partial p_\ell}{\partial r} - r\omega^2\rho\right] \tag{2A.2.12}$$

Substituting Equation (2A.2.12) into Equation (2A.2.11), one has

$$\frac{\mu}{k}\frac{q_{\ell_o} r_o}{r} - \frac{\partial p_s}{\partial r} + \varepsilon_s r\omega^2(\rho + \rho_s) = 0 \tag{2A.2.13}$$

with the condition

$$r = r_c, \quad p_s = 0 \quad (p_\ell)_{r_c} = p_o + \frac{\rho\omega^2}{2}(r_s^2 - r_\ell^2) + \frac{\rho_{SL}\omega^2}{2}(r_c^2 - r_s^2) \tag{2A.2.14}$$

Equations (2A.2.12) and (2A.2.13) together with the initial conditions of Equation (2A.2.14) provide a description of p_s and p_ℓ across the cake region.

APPENDIX 2A.3: A PROCEDURE FOR PREDICTING CENTRIFUGAL FILTRATION PERFORMANCE

From Equation (2.5.1.13), the rate of centrifugal filtration is given as

$$q_{\ell_o} = \frac{(\omega^2/2)[\rho(r_s^2 - r_\ell^2) + \rho_{SL}(r_c^2 - r_s^2) + \rho(r_o^2 - r_c^2)] + p_o}{\mu[w\alpha_{av} + R_m]} \tag{2A.3.1}$$

The average specific cake resistance is defined as

$$\alpha_{av} = \frac{r_o^2[(p_\ell^*)_{r_c} - (p_\ell^*)_{r_o}]}{\int_{(p_\ell^*)_{r_o}}^{(p_\ell^*)_{r_c}} \frac{r^2}{\alpha} \mathrm{d}p_\ell^*} \tag{2A.3.2}$$

and

$$w = \frac{\int_{r_c}^{r_o} (2\pi r)\varepsilon_s \rho_s \mathrm{d}r}{2\pi r_o} \tag{2A.3.3}$$

Both the cake solidosity ε_s and specific cake resistance α are functions of the effective compressive stress. w and α_{av} can be calculated if the p_s-profile (and therefore ε_s) is known. The equation relating p_ℓ, p_s and the centrifugal force and the expression of r_ℓ, r_s and r_c are given in Appendix 2A.2 and 2A.1; or

$$\frac{q_{\ell_o} r_o}{r} = -\frac{k}{\mu}\left[\frac{\partial p_\ell}{\partial x} - r\omega^2\rho\right] \tag{2A.3.4}$$

$$\frac{k}{\mu}\frac{q_{\ell_o} r_o}{r} - \frac{\partial p_s}{\partial r} + \varepsilon_s(\rho + \rho_s)\omega^2 r = 0 \tag{2A.3.5}$$

$$2r_o \int_0^t q_{\ell_o} \varepsilon_s \,\mathrm{d}t = (r_o^2 - r_c^2)(\bar{\bar{\varepsilon}}_s - \varepsilon_{s_o}) \tag{2A.3.6}$$

$$\frac{dr_s}{dt} = \frac{r_s\omega^2}{g}(u_s)_g + q_{\ell_o}\frac{r_o}{r_s} \tag{2A.3.7}$$

$$r_\ell^2 = r_{\ell_o}^2 + 2r_o \int_o^t q_{\ell_o} dt \tag{2A.3.8}$$

and

$$\bar{\bar{\varepsilon}}_s = \frac{2\int_{r_c}^{r_o} r\varepsilon_s dr}{r_o^2 - r_c^2} \tag{2A.3.9}$$

The ε_s term of Equation (2A.3.9) is the local cake solidosity, which is a function of the local compressive stress. Accordingly, $\bar{\bar{\varepsilon}}_s$ can be evaluated if the p_s-profile is known. In turn, the pressure and compressive stress profiles can be obtained from the knowledge of r_ℓ, r_s and r_c.

An iterative procedure can be used to obtain the p_ℓ- and p_s-profiles and filtration performance of V vs. t, Assuming that r_ℓ, r_s, r_c, q_{ℓ_o} and the profiles are known at t. To obtain the results at $t + \Delta t$, a new value of q_{ℓ_o} may be assumed from which the interface positions and p_ℓ- and p_s-profiles may be determined. Based on these profiles, α_{av} can be calculated according to Equation (2A.3.2) and q_{ℓ_o} according to Equation (2A.3.1). The procedure may be continued until the desired convergence is reached.

– 3 –

ANALYSIS OF CAKE FILTRATION: SOLUTIONS OF THE VOLUME-AVERAGED CONTINUITY EQUATIONS

Notation

A	defined by Equation (3.3.5) or used to denote any measurable quantity in Equation (3.7.12)
a	coefficient of Equation (3.2.1)
a_p	particle radius (m)
b	exponent of Equation (3.2.1)
c	total particle concentration (m^3/m^3)
c_1	Type 1 particle concentration (m^3/m^3)
c_2	Type 2 particle concentration (m^3/m^3)
d_p	particle diameter (m)
d_{p_1}	diameter of Type 1 particle (m)
d_{p_2}	diameter of Type 2 particle (m)
F_{ji}	interaction force vector between phase j and phase i (N/m^3)
f_{bk}	Kynch flux density function ($m\,s^{-1}$)
f'	defined by Equation (3.2.16) (-)
g	gravitational acceleration ($m\,s^{-2}$)
H_0	initial column height (m)
k	cake permeability (m^2)
k^o	cake permeability at the zero-stress state (m^2)
L	cake thickness (m)
m_i	net mass transfer rate into phase i ($kg\,m^{-3}\,s^{-1}$)
N	fine particle deposition flux ($m^3/(m^2\,s)$)
N_{Re_t}	Reynolds number defined by Equation (3.6.20)
n	total volume of particles per unit volume of suspending fluid (m^3/m^3), or the exponent of Equation (3.2.2) or (3.2.5) (-)

n_1	volume of Type 1 particle per unit volume of suspending fluid (m^3/m^3)
n_2	volume of Type 2 particle per unit volume of suspending fluid (m^3/m^3)
$\overline{n}$	exponent of Equation (3.6.18) (-)
$P_y(\varepsilon_s)$	yield stress (p_o)
p_a	parameter of the constitutive equations (Pa)
p_i	isotropic pressure of phase i (Pa)
p_ℓ	liquid pressure (Pa)
p_o	applied pressure (Pa)
p_s	compressive stress (Pa)
p_t	defined as $p_\ell + p_s$ (Pa)
p_{ℓ_m}	p_ℓ at cake/medium interface (Pa)
p_{s_m}	p_s at cake/medium interface (Pa)
q_ℓ	superficial liquid velocity ($m\,s^{-1}$)
q_s	superficial particle velocity ($m\,s^{-1}$)
q_{ℓ_m}	instantaneous filtration velocity ($m\,s^{-1}$)
R_o	defined as $\frac{1}{K^o R_m}$
R_m	medium resistance (m^{-2})
$r(\varepsilon_s)$	hindered settling factor (-)
$\mathbf{S}_i$	force vector due to stress tensor acting on phase i (N/m^3)
$\mathbf{T}_i$	stress tensor acting on phase i (Pa)
t	time (s)
t_f	time when filtration phase ends (s)
$\mathbf{U}_i$	velocity vector of phase i ($m\,s^{-1}$)
u_ℓ	liquid velocity ($m\,s^{-1}$)
u_s	particle velocity ($m\,s^{-1}$)
u_t	sedimentation velocity ($m\,s^{-1}$)
u_∞	terminal velocity
V	cumulative filtrate volume (m^3/m^2)
V_p	volume of particle (m^3)
$\mathbf{W}_i$	body force vector acting on phase i (N/m^3)
x	distance measured away from medium, or dependent variables of Equations (3.2.1) and (3.2.2)
x^0	coefficient of Equation (3.2.2)
y	independent variable of Equations (3.2.1) and (3.2.2)
y_n	parameter of Equation (3.2.2)

Greek letters

α	specific cake resistance ($m\,kg^{-1}$)
α_1, α_2	parameter characterizing the decrease in permeability due to fine retention, see Equation (3.7.11) (-)
α^0	specific cake resistance at the zero-stream state ($m\,kg^{-1}$)
β	exponent of Equation (3.2.3) (-)
Γ	defined by Equation (3.4.6b) ($m^{-2}\,Pa^{-1}\,s^{-1}$)

Δp_m	pressure drop across medium (Pa)
$\Delta \rho$	density difference, $\rho_s - \rho_\ell$ (kg m^{-3})
δ	exponent of Equation (3.2.4) (-)
ε	cake (or suspension) porosity (-)
ε_i	volume fraction of phase i (-)
ε_s	volume fraction of particle, or solidosity (-)
$\varepsilon_{s_1}, \varepsilon_{s_2}$	volume fractions of type 1 and type 2 particles, respectively (-)
ε_{s_0}	initial particle volume fraction of suspension (-)
ε_s^0	solidosity at the zero-stress state (-)
ε_{s_m}	value of ε_s at the cake/medium interface (-)
η	defined by $x/L(t)$ [Equation (3.4.7a) (-)
Λ	defined by Equation (3.4.6a) (m^2 Pa^{-1} s^{-1})
λ	filter coefficient (m^{-1}) or the Stokes law coefficient defined as $6\pi\mu a_p$
μ	fluid viscosity (Pa$-$s)
ρ_i	density of phase i (kg m^{-3})
ρ_ℓ	density of liquid (kg m^{-3})
ρ_s	density of particle (kg m^{-3})
$\boldsymbol{\tau}_i$	shear stress tensor of phase i
Φ	objective function defined by Equation (3.7.12)

In analyzing cake filtration, the problem may be considered as one concerned with the motions of a large number of particles and a fluid stream. Analysis of such a multiphase problem can be made, in principle, by satisfying the Navier-Stokes equation at each point of the fluid and equations of motion for each particle. This approach, however, is too complicated to be of practical use. As an alternative, one may derive a set of a few equations by replacing the relevant point variables with their local mean variables over a region containing many particles but smaller than the "macroscopic scale" of the intended problem. The equations obtained in the case which has two phases may be viewed as the continuity equations of two interpenetrating continua, and their solutions constitute the analysis of the problem.

The continuum approach, also known as the *multiphase flow theory*, has been applied to the analysis of a number of physical phenomena including cake filtration (Atsumi and Akiyama, 1975; Wakeman, 1978; Tosun, 1986; Stamatakis and Tien, 1991; Landman *et al.*, 1991; Landman and Russel, 1993; Landman *et al.*, 1995; Tien *et al.*, 1997; Burger *et al.*, 2001; Tien and Bai, 2003; Bai and Tien, 2005). A substantial part of the materials presented in this chapter is based on these studies

The assumptions used in the multiphase flow theory approach are fewer and less restrictive than those employed in formulating the conventional theory. Thus, the results obtained can be expected to be more accurate and complete. Furthermore, the continuity equations based on the multiphase flow theory are applicable not only to cake filtration, but to sedimentation and cake consolidation as well. These advantages are balanced by the fact that more demanding efforts are required for the solution of the continuity equation.

The basic principle of analyzing cake filtration using the continuum approach can be stated as follows. The continuity equations of mass are solved using Darcy's law for the relative particle/fluid motion, the relationship between the fluid pressure p_ℓ and the cake compressive stress p_s obtained from the continuity equations of momentum, and the constitutive relationships between cake properties (solidosity, permeability or specific cake resistance) and the cake compressive stress. The solutions give both the results of filtration performance as well as information about internal variables such as pressure, stress, and solidosity distributions and their evolutions.

It should be emphasized that the multiphase flow theory approach does have its limitations. Derivation of the continuity equations based on replacing the point variable with local mean ones itself is an assumption. More importantly, the replacement of the point variable with local mean variable "throws up a number of terms whose forms are undetermined", in the words of Anderson and Jackson (1967). There is no complete agreement on the form of some of the terms of the continuity equations. As a result, without experimental confirmation, one may have multiple results for the same problem. In cake filtration, this limitation manifests in the p_ℓ–p_s relationship obtained from the continuity equations of momentum, which will be discussed in later sections of this chapter.

3.1 FORMULATION OF THE GOVERNING EQUATIONS

The problem to be considered here is the same as that of Section 2.1 and is illustrated schematically in Fig. 3.1. The analysis presented below follows closely the previous works of Stamatakis and Tien (1991) and Bai and Tien (2005).

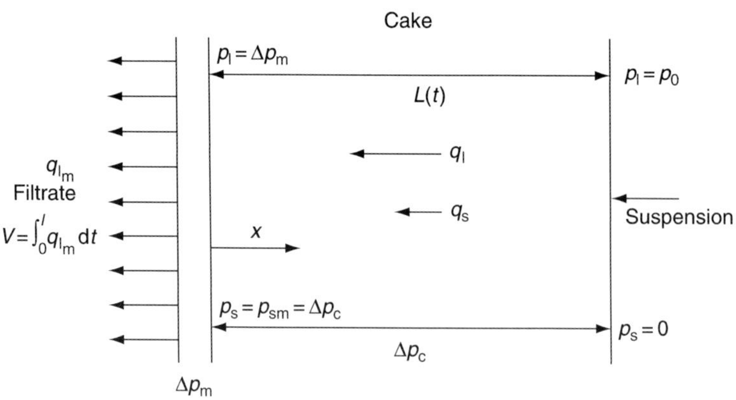

Figure 3.1 Schematic representation of cake filtration.

Equations of continuity

The continuity equations (mass and momentum) of a multiphase system can be written as (Rietema, 1982)

$$\rho_i \frac{\mathrm{d}\varepsilon_i}{\mathrm{d}t} = -\rho_i \nabla \cdot \varepsilon_i U_i + m_i \tag{3.1.1}$$

$$\rho_i \varepsilon_i \frac{\mathrm{d}}{\mathrm{d}t} \mathbf{U}i = -\mathbf{S}_i + \mathbf{W}_i + \mathbf{F}_{ji} \tag{3.1.2}$$

where i stands for $\ell =$ fluid phase and s $=$ particle phase; ε_i is the volume fraction of phase i; $\mathbf{U}_i$, the mass-averaged velocity vector of phase i; $\mathbf{S}_i$, the force vector due to stress acting on phase i; $\mathbf{W}_i$, the body force vector acting on phase i; F_{ji}, the interactions force vector which phase j acts on phase i; and m_ℓ, the net mass rate into phase i.

For the one-dimensional cake filtration problem considered here, using the same notation as before (namely, $u_\ell = q_\ell/\varepsilon$ and $u_s = q_s/\varepsilon_s$, where ε_ℓ and ε_s are the volume fractions of the void and particle phase, respectively (or porosity and solidosity)) with $m_i = 0$, Equation (3.1.1) becomes, for $0 < x < L(t)$

$$\frac{\partial \varepsilon}{\partial t} = -\frac{\partial q_\ell}{\partial x} \tag{3.1.3a}$$

$$\frac{\partial \varepsilon_s}{\partial t} = -\frac{\partial q_s}{\partial x} \tag{3.1.3b}$$

where $L(t)$ is the cake thickness.

The above equations are identical to Equations (2.1.1a) and (2.1.1b) except the signs of the respective second terms at the two equations. The differences arise from that fact that previously, the velocities are considered to move in the opposite direction of the x-coordinates which measure from the media surface. Here, the more general convention is used, that both the velocities q_ℓ and q_s are along the same direction of x (or both q_ℓ and q_s are negative).

The closing relationships which complete the description of cake filtration by Equations (3.1.3a) and (3.1.3b) are

- Darcy's law for the particle/liquid relative velocity
- The relationship between the liquid pressure p_ℓ and the particle phase compressive stress p_s.

As shown by a number of investigators previously, Darcy's law can be derived from the averaged momentum continuity equations under certain conditions. Alternatively, it may also be viewed as an empirical relationship. For the one-dimensional Cartesian case, the generalized Darcy's law gives (Shirato *et al.*, 1969)

$$\frac{q_\ell}{\varepsilon} - \frac{q_s}{\varepsilon_s} = -\frac{1}{\varepsilon}\frac{k}{\mu}\frac{\partial p_\ell}{\partial x} \tag{3.1.4}$$*

The overall continuity requirement (see Section 2.1) can be written as

$$\varepsilon + \varepsilon_s = 1 \tag{3.1.5}$$

$$q_\ell + q_s = \text{constant} = q_{\ell_m} \tag{3.1.6}$$

where q_{ℓ_m} is the value of q_ℓ at the cake/medium interface, or the instantaneous filtration velocity.

From Equation (3.1.4), one has

$$q_\ell \varepsilon_s - q_s \varepsilon = -\varepsilon_s \frac{k}{\mu}\frac{\partial p_\ell}{\partial x}$$

The left hand side of the above equation may be rewritten as

$$q_\ell - (1-\varepsilon_s)q_\ell - q_s(1-\varepsilon_s) = q_\ell - (1-\varepsilon_s)(q_\ell + q_s) = q_\ell - (1-\varepsilon_s)q_{\ell_m}$$

The following expression of q_ℓ is therefore obtained:

$$q_\ell = -\varepsilon_s \frac{k}{\mu}\frac{\partial p_\ell}{\partial x} + (1-\varepsilon_s)q_{\ell_m} = -\varepsilon_s \frac{k}{\mu}\frac{\partial p_\ell}{\partial x} + (1-\varepsilon_s)\left[-\frac{k}{\mu}\frac{\partial p_\ell}{\partial x}\right]_{x=0} \tag{3.1.7}$$

The fluid phase mass continuity equation [i.e. Equation (3.1.3a)] may be written as

$$\frac{\partial(1-\varepsilon_s)}{\partial t} + \frac{\partial}{\partial x}\left[-\varepsilon_s \frac{k}{\mu}\frac{\partial p_\ell}{\partial x} + (1-\varepsilon_s)\left(-\frac{k}{\mu}\frac{\partial p_\ell}{\partial x}\right)_{x=0}\right] = 0$$

or

$$\frac{\partial \varepsilon_s}{\partial t} = \frac{\partial}{\partial x}\left[-\varepsilon_s \frac{k}{\mu}\frac{\partial p_\ell}{\partial x}\right] - q_{\ell_m}\frac{\partial \varepsilon_s}{\partial x} \tag{3.1.8}$$

In the above equation, ε_s and k are quantities characteristic of filter cake structure. They are commonly assumed to be functions of the compressive stress of the cake phase p_s. Further, from the continuity equations of the momentum, the relationship between p_ℓ and p_s can be established. Thus the only dependent variable of Equation (3.1.8) is

---

* Equation (3.1.4) reduces to Equation (2.1.3) with $q_s = 0$. The negative sign of Equation (3.1.4) results from the fact that the direction of q_ℓ of Equation (3.1.4) is opposite to that of Equation (2.1.3) as explained previously.

p_s (or ε_s). The solution of Equation (3.1.8) corresponding to the appropriate initial, boundary and moving boundary conditions gives the complete information of the cake filtration process.

Moving boundary condition

The moving boundary condition describes cake growth and gives the information of cake thickness as a function of time. This condition may be obtained by considering mass balance at the cake surface (or cake/suspension interface). Let $q_\ell|_{L^+}$ and $q_\ell|_{L^-}$ be the fluid velocities at $x = L^+$ and $x = L^-$, respectively, and $\varepsilon_s|_{L^+}$ and $\varepsilon_s|_{L^-}$ be the corresponding cake solidosities. For a period of δt, the cake thickness increases by δL. By fluid mass balance, one has

$$[(q_\ell|_{L^-} - q_\ell|_{L^+})] \cdot \delta t = [(1 - \varepsilon_s|_{L^-}) - (1 - \varepsilon_s|_{L^+})]\delta L \tag{3.1.9a}$$

or

$$\frac{\mathrm{d}L}{\mathrm{d}t} = \frac{q_\ell|_{L^-} - q_\ell|_{L^+}}{\varepsilon_s|_{L^+} - (\varepsilon_s|_{L^-})} \tag{3.1.9b}$$

$\varepsilon_s|_{L^-}$ is the cake solidosity at the cake surface where the compressive stress is zero. $\varepsilon_s|_{L^-}$ therefore can be taken to be ε_s^0, the solidosity at the zero stress state. $\varepsilon_s|_{L^+}$ is equal to the particle volume fraction of the suspension, ε_{s_0}. Equations (3.1.9) therefore becomes

$$\frac{\mathrm{d}L}{\mathrm{d}t} = \frac{q_\ell|_{L^-} - q_\ell|_{L^+}}{\varepsilon_{s_o} - \varepsilon_s^o} \tag{3.1.10}$$

From Equation (3.1.6), one has

$$q_\ell\big|_{L^+} + q_s\big|_{L^+} = q_\ell|_{L^-} + q_s|_{L^-} = q_{\ell_m} = \left(-\frac{k}{\mu}\frac{\partial p_\ell}{\partial x}\right)_o \tag{3.1.11}$$

At $x = L^-$, by Darcy's law [Equation (3.1.4)]

$$q_\ell\Big|_{L^-} - \frac{1 - \varepsilon_s^o}{\varepsilon_s^o} q_s\Big|_{L^-} = \left[-\frac{k}{\mu}\frac{\partial p_\ell}{\partial x}\right]_{x=L^-} \tag{3.1.12a}$$

Also from Equation (3.1.11)

$$q_\ell\big|_{L^-} = -q_s\big|_{L^-} - \left(\frac{k}{\mu}\frac{\partial p_\ell}{\partial x}\right)_o \tag{3.1.12b}$$

Solving for $q_\ell|_{L^-}$ from the above two equations, one has

$$q_\ell|_{L^-} = -\varepsilon_s^o\left(\frac{k}{\mu}\frac{\partial p_\ell}{\partial x}\right)_{L^-} - (1 - \varepsilon_s^o)\left(\frac{k}{\mu}\frac{\partial p_\ell}{\partial x}\right)_o \tag{3.1.13}$$

At $x = L^+$, the suspended particles move at the same velocity as the suspending liquid, or

$$\frac{q_s|_{L^+}}{\varepsilon_{s_o}} = \frac{q_\ell|_{L^+}}{1-\varepsilon_{s_o}}$$

or

$$q_s|_{L^+} = \frac{\varepsilon_{s_o}}{1-\varepsilon_{s_o}} q_\ell|_{L^+}$$

Substituting the above expression into Equation (3.1.11), $q_\ell|_{L^+}$ is found to be

$$q_\ell\Big|_{L^+} = (1-\varepsilon_{s_o})\left(-\frac{k}{\mu}\frac{\partial p_\ell}{\partial x}\right)_o \tag{3.1.14}$$

Substituting Equations (3.1.13) and (3.1.14) into (3.1.10), the moving boundary condition of the cake/suspension interface is found to be

$$\frac{dL}{dt} = \frac{\varepsilon_s^o}{\varepsilon_s^o - \varepsilon_{s_o}}\left[\frac{k}{\mu}\frac{\partial p_\ell}{\partial x}\right]_{x=L} + q_{\ell_m} \tag{3.1.15}$$

with the initial condition

$$L = 0, \quad t = 0 \tag{3.1.16}$$

Boundary conditions

The boundary conditions vary with the mode of operation. Three specific cases may be considered. They are:

(a) *Constant pressure filtration.* The boundary conditions are

$$p_\ell = p_o, \quad p_s = 0, \quad \varepsilon_s = \varepsilon_s^o \quad \text{at } x = L \tag{3.1.17a}$$

$$-\frac{k}{\mu}\frac{\partial p_\ell}{\partial x} = \frac{-p_\ell}{R_m \mu} \quad \text{at } x = 0 \tag{3.1.17b}$$

(b) *Constant rate filtration*

$$p_s = 0, \quad \varepsilon_s = \varepsilon_s^o \quad \text{at } x = L \tag{3.1.18a}$$

$$-\frac{k}{\mu}\frac{\partial p_\ell}{\partial x} = -\frac{-p_\ell}{R_m \mu} = \text{constant} \quad \text{at } x = 0 \tag{3.1.18b}$$

(c) *Variable pressure filtration*

$$p_\ell = p_o(t), \quad p_s = 0, \quad \varepsilon_s = \varepsilon_s^o \quad \text{at } x = L \tag{3.1.19a}$$

$$-\frac{k}{\mu}\frac{\partial p_\ell}{\partial x} = -\frac{p_\ell}{R_m \mu} \quad \text{at } x = 0 \tag{3.1.19b}$$

For both cases (a) and (c), p_o is specified. One is interested in obtaining $q_{\ell_m}\left(=\left[-\frac{k}{\mu}\frac{\partial p_\ell}{\partial x}\right]_o\right)$ as a function of time. The reverse is true for case (b). The condition imposed on $x=0$ (namely, Equation (3.1.19b)) is the continuity of filtrate permeation; namely, the rate of permeation on the upstream side and that on the downstream side of the medium surface are the same. As defined before, R_m is the medium resistance.

3.2 CLOSING RELATIONSHIPS

Before attempting to find the solution of Equation (3.1.8) with appropriate boundary and moving boundary conditions, certain closing relationships must be introduced to express the various dependent quantities (ε_s, k, p_ℓ) into a common dependent variable. For this purpose, two kinds of relationships are required; namely, the constitutive relationships between cake characteristics (ε_s, k and/or α) and cake compressive stress, and the relationship between the pore liquid pressure p_ℓ and the cake compressive stress p_s during the course of filtration. They will be discussed separately as below.

Constitutive relationships

Filter cakes, in general, are compressible, i.e. they become more compact as the extent of their compression increases. The data required for establishing these relationships can be obtained from filtration experiments or independent measurements. The experimentally determined results can be fitted into empirical expressions of various forms. Two of the most widely used expressions are:

$$x = ay^b \tag{3.2.1}$$

or

$$x = x^o\left[1+\frac{y}{y_n}\right]^n \tag{3.2.2}$$

where x denotes cake properties [ε_s, k or $\alpha=(k\rho_s\varepsilon_s)^{-1}$] and y the variable responsible for cake compression (e.g. p_s). Equation (3.2.1) was used by earlier investigators and has two fitting parameters, a and b. The second expression [Equation (3.2.2)] has three parameters, x^o, y_n, and n, and is often used in more recent studies.

The main advantage of these two expressions is their simplicity. With these expressions, it is easy to explicitly express the various variables of Equation (3.1.8) into each other. The disadvantage of Equation (3.2.1) is that it has a lower limit of its applicability and does not give the zero-stress state value of x necessary for the boundary condition of Equation (3.1.17a), (3.1.18a) and (3.1.19a). This problem, however, does not arise with Equation (3.2.2) and x^o can be viewed as the value of x at the zero-stress state [$x \to x^o$ as $y \to 0$].

If the format of Equation (3.2.2) is used, the various constitutive relationships become

$$\varepsilon_s = \varepsilon_s^o \left(1 + \frac{p_s}{p_a}\right)^{\beta} \tag{3.2.3}$$

$$k = k^0 \left(1 + \frac{p_s}{p_a}\right)^{-\delta} \tag{3.2.4}$$

$$\alpha = \frac{1}{\varepsilon_s \rho_s k} = \alpha^o \left(1 + \frac{p_s}{p_a}\right)^{n} \tag{3.2.5}$$

and

$$\alpha^o = (\varepsilon_s^o \rho_s k^o)^{-1}, \quad n = -\beta + \delta \tag{3.2.6}$$

Relationship between p_ℓ *and* p_s

The relationship between p_ℓ and p_s can be obtained from Equation (3.1.2). If one ignores the inertial effect and the presence of the body force, by adding the equations of the two phase and noting that $F_{ji} = F_{ij}$, one has

$$\mathbf{S}_\ell + \mathbf{S}_s = 0 \tag{3.2.7}$$

$\mathbf{S}_i$ $(i = \ell, s)$, the force acting on phase i results from the stress tensor $\mathbf{T}_i$ acting on the same phase. $\mathbf{T}_i$ can be written as

$$\mathbf{T}_i = p_i \boldsymbol{\delta} + \boldsymbol{\tau}_i \tag{3.2.8}$$

where p_i is the isotropic pressure of phase i and $\boldsymbol{\delta}$, the unit tensor. $\boldsymbol{\tau}_i$ is the shear stress tensor of phase i.

In deriving the momentum continuity equations, different relationships between $\mathbf{S}_i$ and $\mathbf{T}_i$ have been proposed such as (Rietema, 1982):

$$\mathbf{S}_i = \nabla \mathbf{T}_i, \quad \nabla \varepsilon_i \mathbf{T}_i \quad \text{or} \quad \varepsilon_i \nabla \mathbf{T}_i \quad i = \ell \text{ or } s \tag{3.2.9}$$

In addition, for the dispersed phase (i.e. particle phase), the following relationship has also been proposed:

$$\mathbf{S}_s = \varepsilon_s (\nabla \cdot \mathbf{T}_\ell + \nabla \cdot \mathbf{T}_s) \quad \text{or} \quad \varepsilon_s \nabla \cdot \mathbf{T}_\ell + \nabla \cdot \mathbf{T}_s \tag{3.2.10}$$

Thus, by choosing different definitions as shown above, a variety of results can be determined from Equation (3.2.7). For one-dimensional cake filtration, the isotropic pressures (i.e. p_ℓ and p_s) are the dominant terms of the stress tensor. Among a number of possible p_ℓ–p_s relationships, some of the simplest ones are (Tien *et al.*, 2001)

Type (1) $\quad \mathrm{d}p_\ell + \mathrm{d}p_s = 0$ (3.2.11a)

Type (2) $\quad (1 - \varepsilon_s)\mathrm{d}p_\ell + \mathrm{d}p_s = 0$ (3.2.11b)

Type (3) $\quad (1 - \varepsilon_s)\mathrm{d}p_\ell + \varepsilon_s \mathrm{d}p_s = 0$ (3.2.11c)

Type (4) $\quad \mathrm{d}[(1 - \varepsilon_s)p_\ell] + \mathrm{d}[\varepsilon_s p_s] = 0$ (3.2.11d)

For a filter cake with $p_\ell = p_o$ at $x = L$, $p_\ell = \Delta p_m$ at $x = 0$, where Δp_m is the pressure drop across the medium and $p_s = 0$ at $x = L$. The relationship between p_ℓ and p_s across the cake and the value of p_s at the cake/medium interface p_{s_m} can be obtained by integrating the above equations. They are

Case (1)

$$p_\ell = p_o - p_s \tag{3.2.12a}$$

$$p_{s_m} = p_o - \Delta p_m \tag{3.2.12b}$$

Case (2)

$$p_\ell = p_o - \int_o^{p_s} \frac{dp_s}{1-\varepsilon} \tag{3.2.13a}$$

$$\int_o^{p_{s_m}} \frac{dp_s}{1-\varepsilon_s} = p_o - \Delta p_m \tag{3.2.13b}$$

Case (3)

$$p_\ell = p_o - \int_o^{p_s} \frac{\varepsilon_s}{1-\varepsilon_s} dp_s \tag{3.2.14a}$$

$$\int_o^{p_{s_m}} \frac{\varepsilon_s}{1-\varepsilon_s} dp_s = p_o - \Delta p_m \tag{3.2.14b}$$

Case (4)

$$p_\ell = \frac{1-\varepsilon_s^o}{1-\varepsilon_s} p_o - \frac{\varepsilon_s}{1-\varepsilon_s} p_s \tag{3.2.15a}$$

$$p_{s_m} = \frac{1-\varepsilon_s^o}{\varepsilon_{s_m}} p_o - \frac{1-\varepsilon_{s_m}}{\varepsilon_{s_m}} \Delta p_m \tag{3.2.15b}$$

where ε_{s_m} is the value of ε_s at the cake/medium interface.

Thus for a given system, with the constitutive relationship known [i.e. Equation (3.2.3)], the above expressions can be used to determine p_s corresponding to a given p_ℓ and vice versa. In Fig. 3.2, the results of p_ℓ vs. p_s for $CaCO_3$ cakes and Kaolin cakes are shown. For a fixed p_ℓ, the magnitude of the corresponding p_s increased in the order of Type (2) < Type (1) < Type (3) < Type (4). Similar behavior was found in other kinds of cakes as well (Teoh, 2003).

To complete the closing of Equation (3.1.8), the relationship of dp_ℓ/dp_s is required. If one denotes

$$\frac{\partial p_\ell}{\partial p_s} = f' \tag{3.2.16}$$

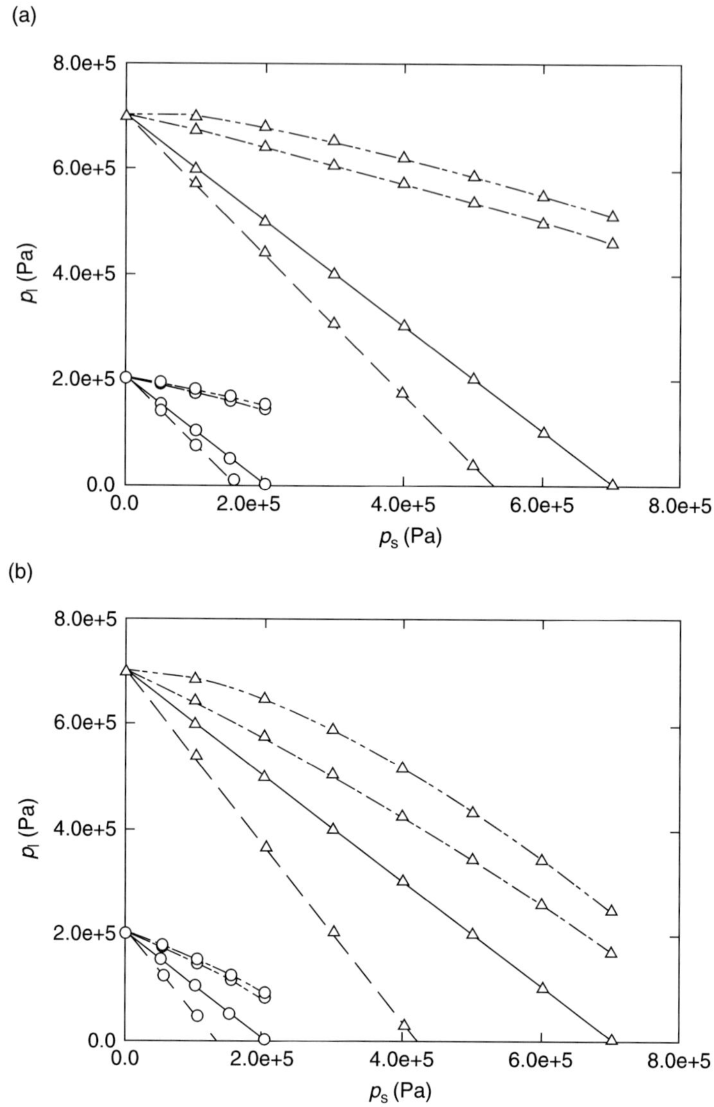

Figure 3.2 Results of p_ℓ vs. p_s: (a) $CaCO_3$ cakes, (b) Kaolin cakes; o, $p_o = 2 \times 10^5$ Pa; $\Delta p_o = 7 \times 10^5$ Pa; – Equation (3.2.11a), Case (1); – –Equation (3.2.11b), Case (2); – . – Equation (3.2.11c), Case (3); –..– Equation (3.2.11d), Case (4). (Teoh, 2003. Reprinted by permission of Dr S.-K. Teoh.)

for the four cases, f' is found to be

$$\text{Case (1)} \quad f' = -1 \tag{3.2.17a}$$

$$\text{Case (2)} \quad f' = \frac{-1}{1 - \varepsilon_s} \tag{3.2.17b}$$

Case (3) $$f' = \frac{-\varepsilon_s}{1-\varepsilon_s} \tag{3.2.17c}$$

Case (4) $$f' = -\frac{(1-\varepsilon_s^o)p_o - p_s}{(1-\varepsilon_s)^2}\frac{d\varepsilon_s}{dp_s} + \frac{\varepsilon_s}{1-\varepsilon_s} \tag{3.2.17d}$$

3.3 COMPARISONS WITH PREVIOUS STUDIES

As stated before, analysis of cake filtration based on the solution of the volume-averaged continuity equations have been attempted by a number of investigators during the past three decades. It is therefore pertinent to compare these studies with the formulation given above and identify the differences, if any, of the respective approaches used.

First of all, one major difference between what is described in Sections 3.1 and 3.2 and the other studies is the inclusion of the effect of the p_ℓ–p_s relationship in the analysis. All the studies reported before assume the validity of Equation (3.2.13a) or case (1) relationship.

In addition to this major difference, the present formulation differs from these earlier studies in several other ways. A brief comment on these differences is given below.

(1) The study of Atsumi and Akiyama (1975), which is perhaps the first rigorous analysis of cake filtration, described cake filtration as a moving boundary diffusion problem. However, one of the boundary conditions they used assumed that the void ratio [defined as $(1-\varepsilon_s)/\varepsilon_s$] at the cake/medium interface to be constant. For constant-pressure filtration, this is tantamount to the condition of negligible medium resistance which is approximately correct for large time and certainly incorrect initially.

(2) The governing equation derived by Wakeman (1978) [his Equation (8)] is the same as Equation (3.1.8), but the derivation was made in an ad hoc manner. In deriving the moving boundary condition, several omissions were made. First, his continuity requirement at the cake/suspension interface [his Equation (12)] when written with the present notation is

$$(q_\ell|_{L^+} - q_\ell|_{L^-})(\delta t) = (1-\varepsilon_s^o)\delta L \tag{3.3.1}$$

instead of Equation (3.1.9a). Furthermore, in expressing $q_\ell|_{L^-}$ from Darcy's law, the particle velocity term was ignored [This can be seen by comparing his Equation (13) with Equation (3.1.12)]. Comparing his Equation (15) with Equation (3.1.15) for dL/dt, the term q_{ℓ_m}, the instantaneous filtration velocity, was missing.

(3) There were also questions about the moving boundary conditions used in Tosun's work (1986). In obtaining his Equation (19), the average cake porosity (defined as $(1/2)\int_o^L \varepsilon dx$) was assumed to be constant. As shown in latter sections, this assumption is valid only under certain conditions.

(4) The analysis by Abboud and Corapcioglu (1993) is also based on the solution of the volume-averaged equation of continuity. The constitutive relationship expressions they used relate ε_s (and k) with their initial values and filtration pressure. It is, however, difficult to understand what is meant by the initial values of ε_s (and k) and how these values can be determined.

(5) Davis and Russel's solution was given as an extension of their solution of sedimentation. However, these authors apparently overlooked the difference between sedimentation and cake filtration. Their boundary conditions at the cake/medium interface [their Equation (126) for $x = 0$] was incorrect. More important, the physical domain (namely, the sum of the liquid, suspension and cake phases) was considered constant and not time dependent.

(6) The present formulation follows closely those of Stamatakis and Tien (1991) and Tien *et al.* (1997). But these earlier studies only used Type (1) relationship between p_ℓ and p_s.

(7) Cake filtration analyses can also be considered as a diffusion problem using the methodology developed by soil scientist (see Smiles, 1986). A more detailed discussion on treating of cake filtration by the diffusion theory is given in Chapter 5.

(8) Landman and coworkers (Landman *et al.*, 1991; Landman and Russel, 1993; Landman *et al.*, 1995) introduced the so-called "compressional rheological model" of cake filtration. The two characterizing quantities of their model are the compressive yield stress $P_y(\varepsilon_s)$ and the hindered settling factor $r(\varepsilon_s)$

The liquid/solid relative velocity is assumed to be

$$\frac{q_\ell}{1-\varepsilon_s} - \frac{q_s}{\varepsilon_s} = -\frac{V_p}{\lambda}\frac{(1-\varepsilon_s)^2}{\varepsilon_s[r]_{\ell we}}\frac{\partial p_\ell}{\partial x} \tag{3.3.2}$$

where V_p is the particle volume, $\lambda = 6\pi\mu a_p$, a_p being the particle radius and $[r]_{\ell we}$, the hindered settling factor defined by Landman *et al.* (1995).

Comparing Equations (3.3.2) with Equation (3.1.4), it is clear that the two equations are equivalent. The relationship between k and $[r]_{\ell we}$ is simply

$$k = \left(\frac{V_p}{\lambda}\mu\right)\frac{(1-\varepsilon_s)^2}{\varepsilon_s[r(\varepsilon_s)]_{\ell we}} \tag{3.3.3*}$$

The yield stress function p_y is assumed to be a function of ε_s. Since P_y is also assumed to be the same as p_s, $P_y(\varepsilon_s)$ is nothing but the inverse function of the constitutive relationship of $\varepsilon_s(p_s)$. In particular, if ε_s is given by Equation (3.2.3), $p_s = (P_y(\varepsilon_s))$ can be found to be $P_y(\varepsilon_s) = p_s(\varepsilon_s) = p_a[(\varepsilon_s/\varepsilon_s^o)^{1/\beta} - 1]$, which is precisely the yield stress function used by Landman and coworkers with ε_s^o, the solidosity at the zero-stress state, being the same as the particle volume fraction at the so-called "gel point". The equivalence between the rheological model of Landman and coworkers and the model based on the permeabilty–solidosity concept is further discussed in Chapter 7.

All the studies by Landman and coworkers were concerned with batch expression (sedimentation/filtration/consolidation). The results of Landman and Russel consider the presence of a dense cake layer ($\varepsilon_s \simeq 0.62$) at the cake/medium interface.* The so-called "similar solution" of Landman *et al.* (1995) was obtained by ignoring the medium resistance and using the assumption that both the filtrate volume and cake thickness are proportional to $t^{1/2}$, validation of which will be discussed later. One should also add that the procedure of evaluating the so-called "filtration diffusivity" proposed by Landman *et al.* (1999) based on the similar solution of Landman *et al.* (1999) is identical to the conventional t/V vs. V plot of constant pressure filtration data, which is discussed in Chapter 7.

(9) The more recent work of Burger *et al.* (2001) gives an analysis of batch expression with the formation and growth of filter cake followed by cake consolidation once the suspension is depleted. The sedimentation effect was included. Taking advantage of the analogy between the generalized Darcy law [i.e. Equation (3.1.4)] and the Kynch flux density function and the fact that the volume-averaged continuity equations [i.e. Equations (3.1.1) and (3.1.2)] are equally applicable to both the cake and suspension phases, the dynamic behavior of the process can be described by the following equation

$$\frac{\partial \varepsilon_s}{\partial t} + \frac{\partial}{\partial x}[q_{\ell_m} \varepsilon_s + f_{bk}] = \frac{\partial^2 A}{\partial x^2} \tag{3.3.4}$$

where A is defined as

$$A = \int_0^{\varepsilon_s} \frac{f_{bk}(dp_s/d\varepsilon_s)}{(\Delta\rho)g\varepsilon_s} \cdot d\varepsilon_s \tag{3.3.5}$$

where f_{bk} is the so-called "Kynch flux density function" and $\Delta\rho$ is the density difference, $\rho_s - \rho_\ell$. If Darcy's law includes the gravitational effect, f_{bk} can be related to the cake permeability by the following expressions

$$\begin{aligned} f_{bk} &= -(\Delta\rho)g\varepsilon_s^2\left(\frac{k}{\mu}\right) \\ &= -\left(\frac{9}{2}\right)\varepsilon_s^2 u_\infty \left(\frac{k}{a_p^2}\right) \end{aligned} \tag{3.3.6}$$

where u_∞ is the particle terminal velocity according to Stoke's law, or

$$u_\infty = \frac{2}{9}\frac{a_p^2(\Delta\rho)g}{\mu} \tag{3.3.7}$$

The function f_{bk} has the unit of velocity ($m\,s^{-1}$) and $-(f_{bk}/u_\infty)$ can be expected to be a function of ε_s.

---

* Landman and Russel justified the condition by citing the experimental data of Shen (see their Fig. 3.2(b)). However, they failed to recognize that even at relatively high compression (greater than 100 psi), cake solidosity rarely reaches values greater than 0.6 (for example, see Fig. 7.2 of Chapter 7).

As shown in Section 3.6, Equation (3.3.4) is the same as Equation (3.1.8) if the gravitational effect is introduced into the volume-averaged momentum equations with the assumption that $\mathbf{S}_i = \nabla \cdot \mathbf{T}_i$ [see Equation (3.2.9)].*

Burger *et al.* developed an algorithm for the numerical solution of Equation (3.3.4) with certain initial and boundary conditions and the specification of the spatial domain. The numerical results obtained provide detailed information of filtration performance and the evolution of cake structure during filtration and cake consolidation.

The formulation of Burger *et al.* may seem to have the advantage of conceptual simplicity as the same equation is used to describe the three different processes involved; sedimentation and filtration occurring simultaneously followed by cake consolidation. However, there are certain problems with the approach including:

(1) Solutions of Equation (3.3.4) requires the knowledge of f_{bk} which is not always available. Experimental data obtained from C-P measurements and particle settling experiments provide only segments of the required data. In coping with this difficulty, Burger *et al.* filled the gap by connecting segments of experimental data with certain physical argument and the assumption that f_{bk} is a continuous function of ε_{s}. As a result, the possibility that ε_{s} may display a discontinuity at the cake/suspension interface is ignored.

(2) Over the range of $\varepsilon_{\mathrm{s_o}} < \varepsilon_{\mathrm{s}} < \varepsilon_{\mathrm{s_m}}$, the governing equation of Burger *et al.* [i.e. Equation (3.2.5)] is a strongly degenerate (mixed hyperbolic-parabolic) partial differential equation with A vanishing in the suspension region, since $p_{\mathrm{s}} = 0$ throughout the region. As a result, the algorithm they developed is rather involved and complicated. From computational point of view, it may be better to apply different algorithms for the solutions of the cake and suspension regions separately.

(3) The "agreement between experiments and predictions" achieved by Burger *et al.* was realized by arbitrarily adjusting certain constitutive relationship data (k vs. p_{s}). The observed agreement may not necessarily constitute a legitimate validation of their model.

3.4 NUMERICAL ANALYSIS OF CAKE FILTRATION

We now return to the numerical solution of Equation (3.1.8) together with the interface moving boundary condition and boundary conditions using the closing relationships of Section 3.2. Equation (3.1.8) is given as

$$\frac{\partial \varepsilon_{\mathrm{s}}}{\partial t} = \frac{\partial}{\partial x}\left[-\varepsilon_{\mathrm{s}}\frac{k}{\mu}\frac{\partial p_{\ell}}{\partial x}\right] - q_{\ell_{\mathrm{m}}}\frac{\partial \varepsilon_{\mathrm{s}}}{\partial x} \tag{3.1.8}$$

* Under these conditions, Equation (3.2.11) becomes $(\mathrm{d}p_{\ell} + \rho_{\ell} g) + (\mathrm{d}p_{\mathrm{s}} + \Delta\rho g \varepsilon_{\mathrm{s}}) = 0$.

Using the relationship given by Equation (3.2.16), one has

$$\frac{\partial p_\ell}{\partial x} = \frac{\partial p_\ell}{\partial p_s}\frac{\partial p_s}{\partial x} = (f')\frac{\partial p_s}{\partial x} \tag{3.4.1}$$

From the constitutive relationship of Equations (3.2.3) and (3.2.4), one has

$$k = k_o\left(\frac{\varepsilon_s}{\varepsilon_s^o}\right)^{-\delta/\beta} \tag{3.4.2}$$

Equation (3.1.8) becomes

$$\frac{\partial \varepsilon_s}{\partial t} = \frac{\partial}{\partial x}\left[(-f')\frac{k^o}{\mu}\varepsilon_s\left(\frac{\varepsilon_s}{\varepsilon_s^o}\right)^{-\delta/\beta}\frac{\partial p_s}{\partial x}\right] - q_{\ell_m}\frac{\partial \varepsilon_s}{\partial x} \tag{3.4.3}$$

By definition

$$-q_{\ell_m} = \frac{p_{\ell_m}}{\mu R_m} \tag{3.4.4}$$

where p_{ℓ_m} is the liquid pressure at the cake/membrane interface. Equation (3.4.3) now becomes

$$\frac{\partial \varepsilon_s}{\partial t} = \Lambda\frac{\partial}{\partial x}\left[(-f')\varepsilon_s^{-n/\beta}\frac{\partial p_s}{\partial x}\right] + \Gamma(p_{\ell_m})\frac{\partial \varepsilon_s}{\partial x} \tag{3.4.5}$$

where

$$\Lambda = \frac{k^0}{\mu}(\varepsilon_s^o)^{\delta/\beta} \tag{3.4.6a}$$

$$\Gamma = \frac{1}{R_m \mu} \tag{3.4.6b}$$

$$n = \delta - \beta \tag{3.4.6c}$$

For numerical calculations, it is more convenient to immobilize the cake/suspension interface by introducing a set of new independent variables η and $\bar{t}$ defined as

$$\eta = x/L(t) \tag{3.4.7a}$$

$$\bar{t} = t \tag{3.4.7b}$$

In terms of the new variables and by virtue of Equation (3.2.3), p_s may be written as

$$p_s = p_a\left[\left(\frac{\varepsilon_s}{\varepsilon_s^o}\right)^{1/\beta} - 1\right] \tag{3.4.8}$$

Equation (3.4.5) may be rewritten as

$$\frac{\partial \varepsilon_s}{\partial \bar{t}} = \frac{1}{L^2}\frac{k^o}{\mu\beta}\frac{p_a}{\partial \eta}\frac{\partial}{\partial \eta}\left[(f')\left(\frac{\varepsilon_s}{\varepsilon_s^o}\right)^{\frac{1-\delta}{\beta}}\frac{\partial \varepsilon_s}{\partial \eta}\right] + \frac{1}{L}\left[\eta\frac{dL}{d\bar{t}} + \frac{1}{\mu R_m}p_{\ell_m}\right]\frac{\partial \varepsilon_s}{\partial \eta} \tag{3.4.9}$$

For convenience, the bar over t is omitted in later discussions.

The moving boundary conditions of Equation (3.1.15) now becomes

$$\frac{dL}{dt} = \frac{\varepsilon_s^o}{\varepsilon_s^o - \varepsilon_{s_o}}\left(\frac{k^o}{\mu}\right)\left(\frac{1}{L}\right)(-f')_{\varepsilon_s^o}\left(\frac{\partial p_s}{\partial \eta}\right)_{\eta=1} + q_{\ell_m} \tag{3.4.10}$$

$$\frac{\partial p_s}{\partial \eta} = \frac{p_a}{\beta\varepsilon_s^o}\left(\frac{\varepsilon_s}{\varepsilon_s^o}\right)^{(1/\beta)-1}\frac{\partial \varepsilon_s}{\partial \eta} \tag{3.4.11}$$

$$q_{\ell_m} = -\left(\frac{k^o}{\mu}\right)\left(1+\frac{p_{s_m}}{p_a}\right)^{-\delta}(-f')_{\varepsilon_{sm}}\left(\frac{\partial p_s}{\partial \eta}\right)_{\eta=0}\frac{1}{L} = \frac{-p_{\ell_m}}{\mu R_m} \tag{3.4.12}$$

where $(-f')_{\varepsilon_s^o}$ and $(-f')_{\varepsilon_{sm}}$ are values of $-f'$ evaluated at $\varepsilon_s = \varepsilon_s^o$ and $\varepsilon_s = \varepsilon_{s_m}$ ($\varepsilon_{s_{mi}}$ values of ε_s with $p_s = p_{s_m}$ or $p_\ell = p_{s_m}$), respectively, and the initial conditions are

at $t = 0$, (3.4.13a)

$$\frac{dL}{dt} = \frac{\varepsilon_{s_o}}{\varepsilon_s^o - \varepsilon_{s_o}}(-q_{\ell_m}) \tag{3.4.13b}$$

$L = 0$

$$q_{\ell_m} = \frac{-p_o}{\mu R_m} \tag{3.4.13c}$$

The boundary conditions given by Equations (3.1.17a)–(3.1.19b) now become

(a) *Constant rate filtration*, q_{ℓ_m} = constant

$$k^o\left(1+\frac{p_s}{p_a}\right)^{-\delta}(R_m/L)(-f')\frac{\partial p_s}{\partial \eta} = p_\ell = \frac{q_{\ell_m}}{\mu R_m} \quad \text{at} \quad \eta = 0 \tag{3.4.14a}$$

$$\varepsilon_s = \varepsilon_s^o \text{ (or } p_s = 0) \quad \text{at} \quad \eta = 1 \tag{3.4.14b}$$

(b) *Constant pressure filtration*, p_o = constant

$$k^o\left(1+\frac{p_s}{p_a}\right)^{-\delta}(-f')\left(\frac{\partial p_s}{\partial \eta}\right)\left(\frac{1}{L}\right) = \frac{p_o}{R_m} \quad \text{at} \quad \eta = 0 \tag{3.4.15a}$$

$$\varepsilon_s = \varepsilon_s^o \text{ (or } p_s = 0) \quad \text{at} \quad \eta = 1 \tag{3.4.15b}$$

(c) *Variable pressure filtration,* $p_o = p_o(t)$

$$k^o \left(1 + \frac{p_s}{p_a}\right)^{-\delta} (-f') \left(\frac{\partial p_s}{\partial \eta}\right) \left(\frac{1}{L}\right) = \frac{p_o(t)}{R_m} \quad \text{at} \quad \eta = 0 \tag{3.4.16a}$$

$$\varepsilon_s = \varepsilon_s^o \ (\text{or } p_s = 0) \quad \text{at} \quad \eta = 1 \tag{3.4.16b}$$

The term p_ℓ of Equation (3.4.15a) can be written as $p_o - p_s$ if the relationship of $p_\ell - p_s$ is that of case (1). For the other cases, p_ℓ can be replaced by Equations (3.2.13b)–(3.2.14b) or (3.2.15b). In sum, the governing equation of cake filtration are Equations (3.4.9)–(3.4.13c) and the boundary conditions are given by Equations (3.4.14a)–(3.4.16b) according to the operation mode. In Equation (3.4.9), the cake solidosity ε_s is the dependent variable although the compressive stress p_s can also be considered as such. For a specific set of operating conditions and the knowledge of the constitutive relationships (namely, ε_s vs. p_s and k vs. p_s) and the relationship between p_ℓ and p_s, this system of equations can be solved numerically as was done before (Stamatakis and Tien, 1991; Tien *et al.*, 1997; Burger *et al.*, 2001). The results yield filtration performance, L vs. t and q_{ℓ_m} vs. t (for the case when the operating pressure, p_o is specified) or p_o vs. t (for the case of constant rate filtration). The cumulative filtrate volume V is simply

$$V = \int_0^t q_{\ell_m} \mathrm{d}t \tag{3.4.17}$$

Besides filtration performance, the solution also gives information about the internal structure of the cake and its evolution with time – namely, profiles of ε_s, p_s and p_ℓ as functions of time as well as the local liquid and particle velocities. A brief presentation and discussion of the results obtained for a few cases is given in the next section.

3.5 EXAMPLES OF NUMERICAL RESULTS

A set of numerical results obtained from the solutions of Equations (3.4.9)–(3.4.13c) for the constant-pressure filtration of Hong Kong Kaolin suspensions ($\varepsilon_{s_o} = 0.15$) is presented here to demonstrate the utility and scope of the numerical analysis of cake filtration discussed above. The conditions used in the calculation are listed in Table 3.1. The results presented are those with p_ℓ–p_s relationships of Types (1), (2) and (3) [see Equations (3.2.13a)–(3.2.13c)].

The results obtained are of two kinds: filtration performance and the internal information of the cake formed. In Fig. 3.3a–c, the results of the cake thickness L, the cumulative filtrate volume V and the instantaneous filtration rate q_{ℓ_m} are shown as functions of time. Both L and V increase with time while q_{ℓ_m} decreases as expected. The effects of applying different p_ℓ–p_s relationships in obtaining these results are consistent; namely, Type (2) relationship gives the highest value followed by Types (1) and (3).

Table 3.1

Conditions and parameter values used in sample calculations and comparisons with experiments

System	Hong Kong Kaolin (Figs 3.3–3.8)	$CaCO_3$ (Figs 3.9, 3.10)
ρ_s (kg m^{-3})	1500	2655
ε_{s_0} (-)	0.15	0.0076
μ (Pa s)	0.001	0.001
p_o (Pa)	100,000	100,000, 500,000, 800,000
$(1/k^o)(1/R_m)$ (-)	100	186, 108, 89
ε_s^o (-)	0.27	0.2
k^o (m^2)	3.22×10^{-5}	4.892×10^{-14}
β (-)	0.09	0.13
δ (-)	0.465	0.57
p_a (Pa)	1370	4.4×10^4

The average cake solidosity $\bar{\bar{\varepsilon}}$ [defined by Equation (2.1.8)] and the cake solidosity profiles (ε_s vs. x/L) at various times are shown in Figs 3.4 and 3.5. $\bar{\bar{\varepsilon}}$ is found to show rapid increase with time initially but approaches to nearly constant value. Type (3) p_ℓ–p_s relationship gives the highest value of $\bar{\bar{\varepsilon}}_s$ at any given time followed by Types (1) and (2). This trend is opposite to that shown in Fig. 3.3a–c but physically consistent, as cake thickness increases with the decrease of cake solidosity.

The compressive stress profile shown in Fig. 3.6a–b suggests a few things. The effects of applying different p_ℓ–p_s relationships are significant. Second, similar to what is shown in Fig. 3.5, Type (3) relationship gives the highest p_s values followed by Types (1) and (2). This is expected since higher compression yields greater values of ε_s. It is also clear from these results that p_s/p_a vs. x/L (p_a is one of the parameters of the constitutive relationships given by Equations (3.2.3)–(3.2.5)] is not a unique function.

In Figure 3.7a–c, the results of the pressure profile, p_ℓ vs. x/L at various times are shown. The effect of the p_ℓ–p_s relationship is less significant. Furthermore as time increases, the results according to the different relationships seem to merge into a single curve. The results of filtration performance shown in Fig. 3.3b seem to confirm this. These figures, especially Fig. 3.7a, show that while the pressure drop across the filter cake increases with time, the pressure drop across the filter medium even at $t = 1000$ s remains significant and still accounts for 20% of the total pressure applied.

The liquid velocity profiles are shown in Fig. 3.8a–c. The results are presented in the form of q_ℓ/q_{ℓ_m} vs. x/L at $t = 1{,}500$ and 1500 s with various medium resistance (or $1/k^o R_m = 10{,}100$ and 1000) and Type (1) p_ℓ–p_s relationship. The deviation for the constant liquid velocity assumption used in the conventional theory, in all cases, is less than 10%.

Comparisons of the numerical results with experimental data are given in Figs 3.9–3.10. The experimental results of L vs. t and V vs. t obtained by Teoh (2003)

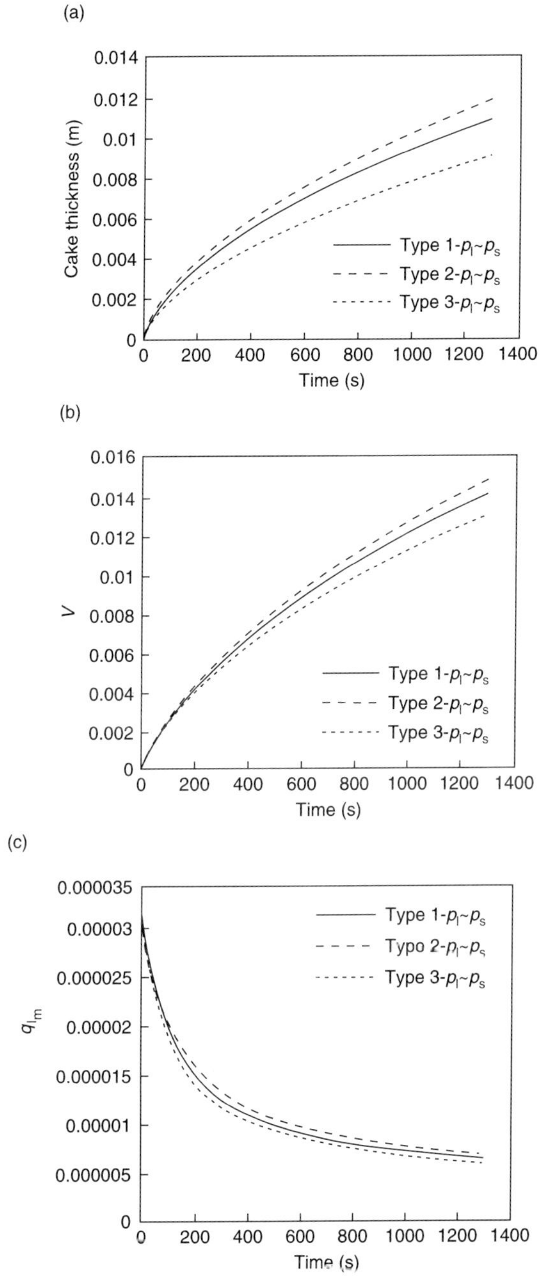

Figure 3.3 Constant pressure cake filtration performance; Hong Kong Kaolin suspension; (a) cake thickness vs. time; (b) cumulative filtration volume vs. time; (c) filtration velocity vs. time. Conditions used in calculation given in Table 3.1. (Bai and Tien, 2005. Reprinted by permission of Elsevier.)

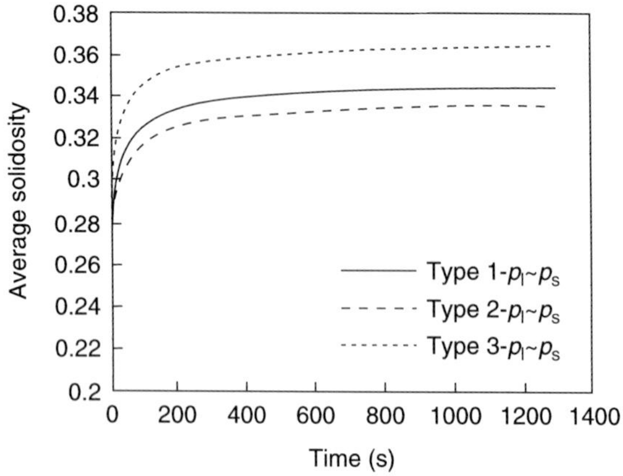

Figure 3.4 Average cake solidosity vs. time, constant pressure filtrate of Hong Kong Kaolin suspension. Conditions used in calculation given in Table 3.1. (Bai and Tien, 2005. Reprinted by permission of Elsevier.)

on the filtration of $CaCO_3$ suspension under constant pressure $p_o = 1,58$ bars were used. The numerical results are those corresponding to the three types of p_ℓ–p_s relationships. The effect of the p_ℓ–p_s relationship is more pronounced in the case of L vs. t, which was also shown previously. Considering all the results, Type (1) relationship seems to give the best agreement between experiments and predictions. This point will be discussed further later in Chapter 7.

3.6 SEDIMENTATION EFFECT

Bockstal *et al.* (1985), Tiller *et al.* (1995) and, more recently, Benesch *et al.* (2004) examined the effect of particle sedimentation on cake filtration performance. In these studies, the sedimentation effect was considered only in terms of the presence of particle flux due to sedimentation at the cake surface and the subsequent modification of the conventional cake filtration theory. The possible effect of sedimentation within filter cakes was not investigated.

Burger *et al.* (2001) presented a more exact analysis of batch cake filtration/consolidation as mentioned before. Although cake formation and growth was part of the analysis, it is not clear how the study may be adopted to cake filtration in continuous operation without first specifying the moving boundary condition at cake surface. The analysis given below is intended to outline a procedure which can be applied to include the sedimentation effect for cake filtration of both batch and continuous operations.

The analysis is made as an extension of that given earlier in this chapter. The mass (particle) continuity equation is given by Equation (3.1.3b) or

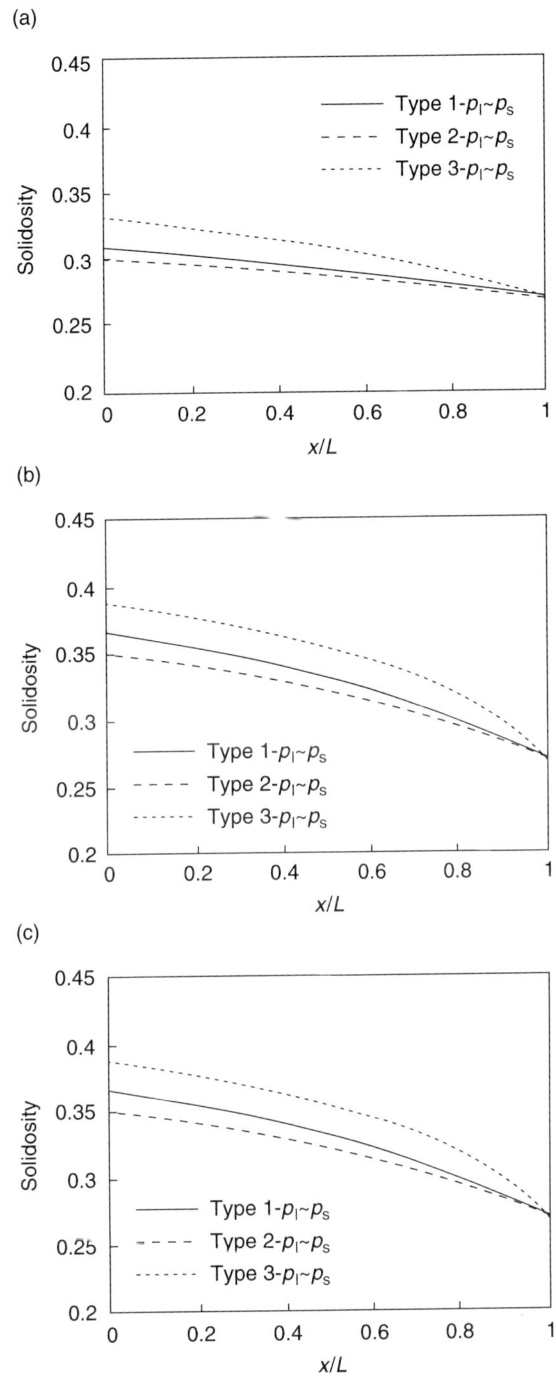

Figure 3.5 Cake solidosity profiles at different time. Constant pressure filtration of Hong Kong Kaolin suspension: (a) 10 s; (b) 100 s; (c) 1000 s. Conditions used in calculation given in Table 3.1.

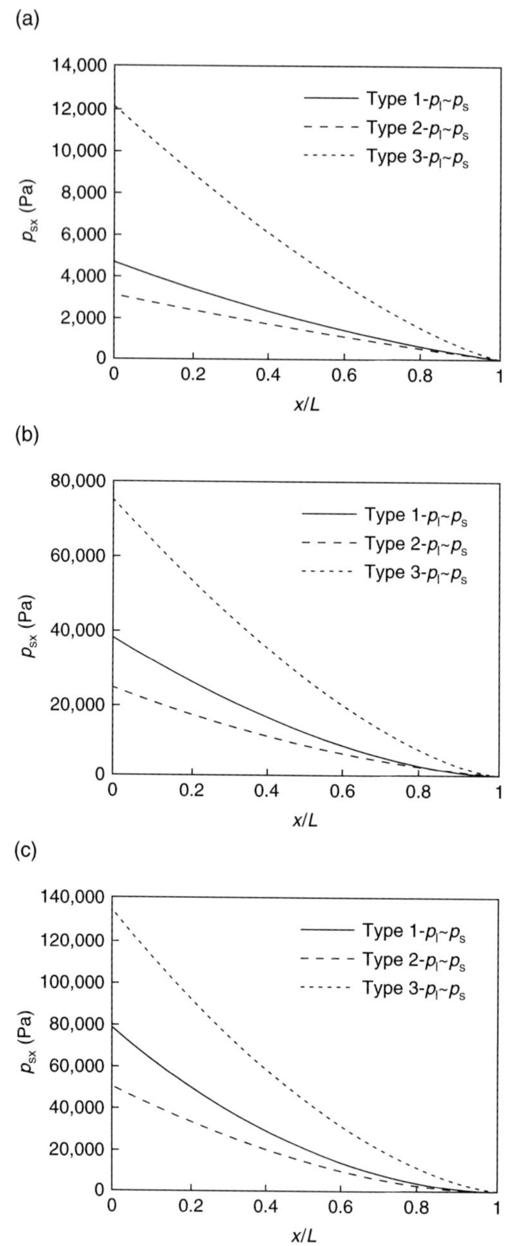

Figure 3.6 Cake compressive stress profiles at different time: Constant pressure filtration of Hong Kong Kaolin suspensions: (a) 10 s; (b) 100 s; (c) 1000 s. Conditions used in calculation given in Table 3.1.

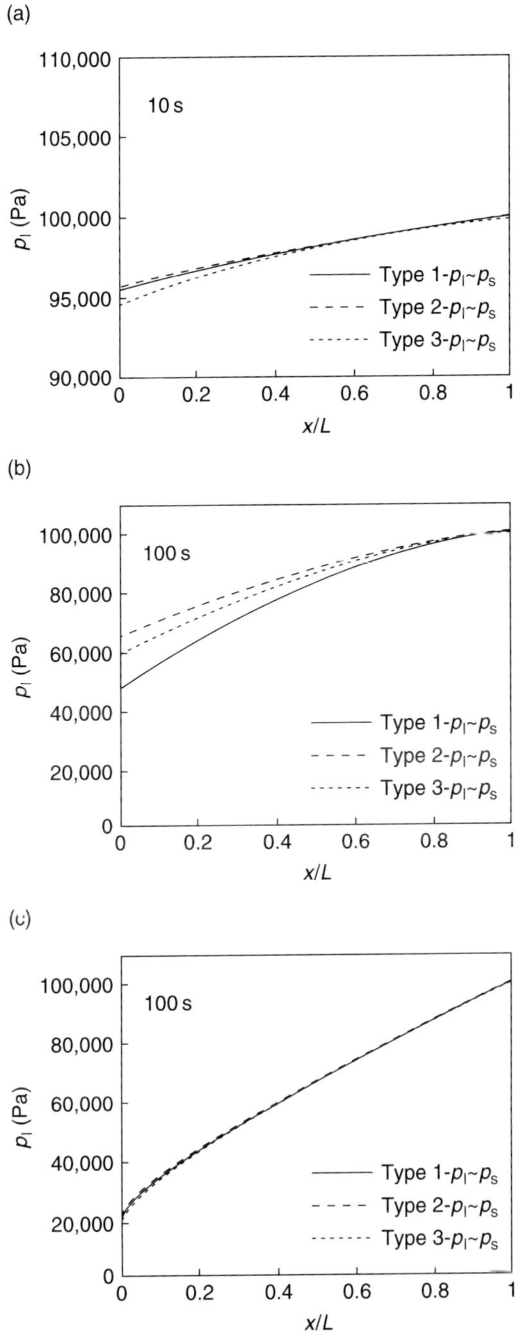

Figure 3.7 Liquid pressure profile at different time. Constant pressure filtration of Hong Kong Kaolin suspension. Conditions used for calculations given in Table 3.1.

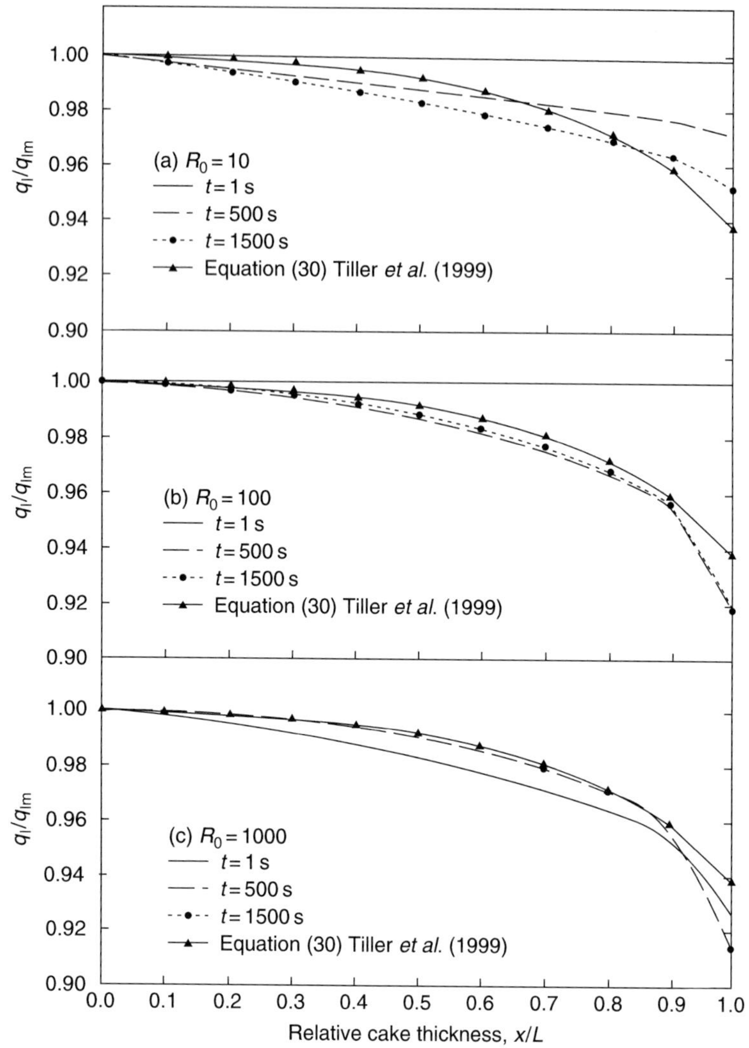

Figure 3.8 Liquid velocity profile across filter cake.

$$\frac{\partial \varepsilon_s}{\partial t} + \frac{\partial q_s}{\partial x} = 0 \tag{3.6.1}$$

With the inclusion of the body force terms, the momentum continuity equation now becomes

$$\frac{\partial p_\ell}{\partial x} + \frac{\partial p_s}{\partial x} + \rho g + (\Delta\rho) g \varepsilon_s = 0 \tag{3.6.2}$$

where $\Delta\rho$ is the density difference $\rho_s - \rho$.

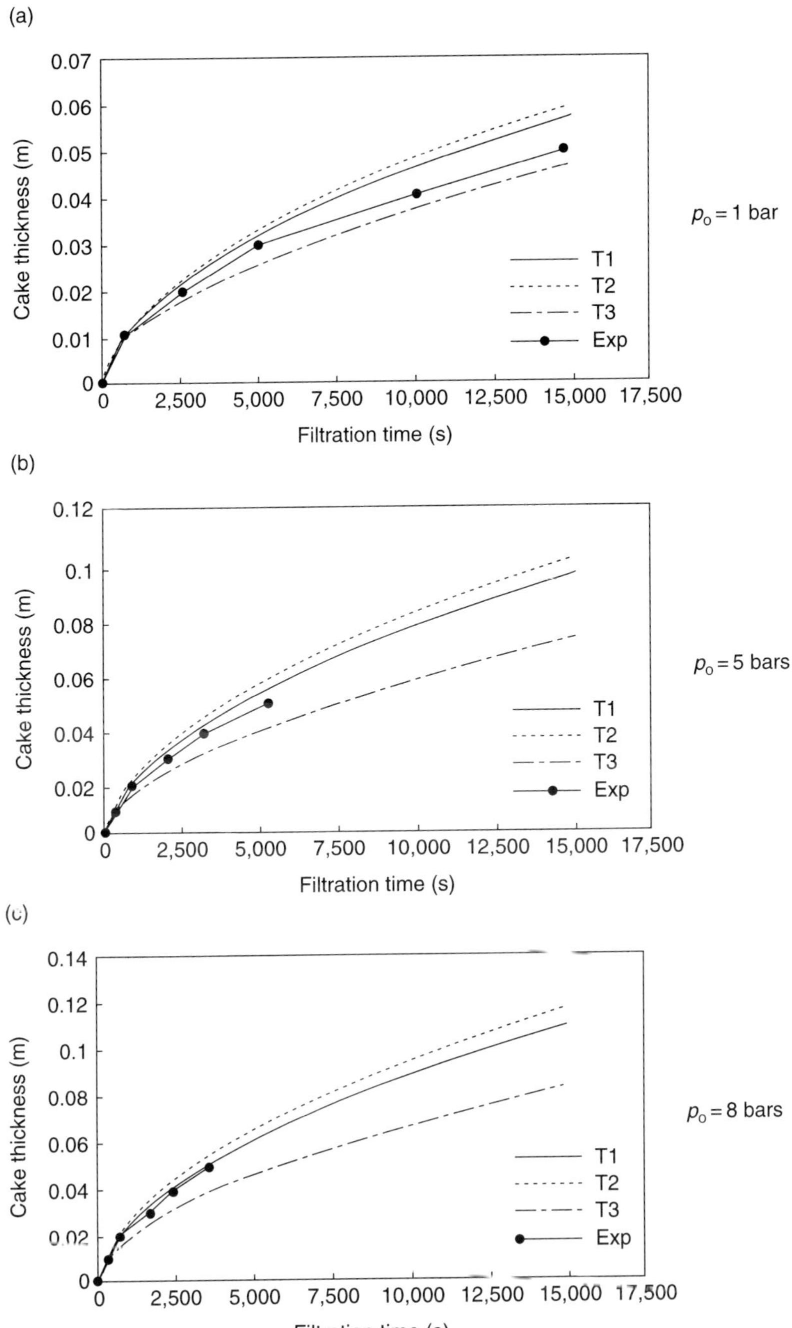

Figure 3.9 Comparison between calculated performance results (L vs. t) with experiments. Constant pressure filtration of $CaCO_3$ suspension conditions used for calculations given in Table 3.1.

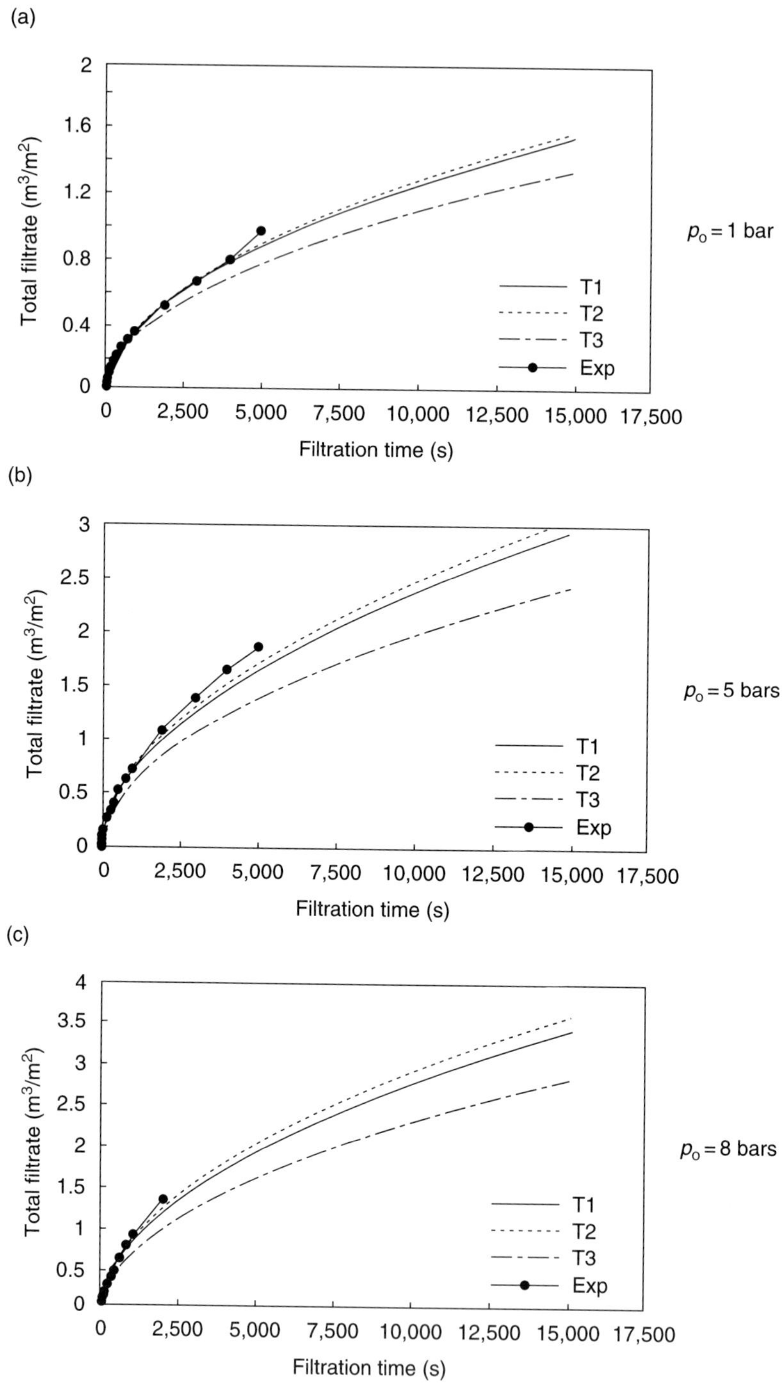

Figure 3.10 Comparison between calculated V vs. t with experiments. Constant pressure filtration of $CaCO_3$ suspensions. Conditions used for calculations given in Table 3.1.

Equation (3.6.2) was obtained under the following conditions, $\mathbf{S}_i = \nabla \mathbf{T}_i$ [Equation (3.2.9)] and $W_\ell = (1-\varepsilon_s)\rho g$, $W_s = \varepsilon_s \rho_s g$ [Equation (3.1.2)]. Thus Equation (3.6.2) can be considered as equivalent to Equation (3.2.11a) since it may be rewritten as

$$\frac{\partial p_\ell^*}{\partial x} + \frac{\partial}{\partial x}\left[p_s + \int_0^x \Delta\rho g \varepsilon_s \mathrm{d}x\right] = 0 \tag{3.6.3}$$

where

$$p_\ell^* = p_\ell + \rho g z \tag{3.6.4}$$

Assuming that Darcy's law [Equation (3.1.4)] may be modified as

$$\frac{q_\ell}{\varepsilon} - \frac{q_s}{\varepsilon_s} = -\frac{1}{\varepsilon}\frac{k}{\mu}\frac{\partial p_\ell^*}{\partial x} \tag{3.6.5}$$

following the procedure used in deriving Equation (3.1.7), one now has

$$q_s = \varepsilon_s q_{\ell_m} + \varepsilon_s\left[\frac{k}{\varepsilon\mu}\frac{\partial p_\ell^*}{\partial x}\right] \tag{3.6.6}$$

where

$$q_{\ell m} = \left[-\frac{k}{\mu}\frac{\partial p_\ell^*}{\partial x}\right]_{x=0} \tag{3.6.7}$$

Substituting Equations (3.6.6) and (3.6.7) into (3.6.1) yields

$$\frac{\partial \varepsilon_s}{\partial t} + q_{\ell m}\frac{\partial \varepsilon_s}{\partial x} = \frac{\partial}{\partial x}\left[-\varepsilon_s \frac{k}{\varepsilon\mu}\frac{\partial p_\ell{}^*}{\partial x}\right] \tag{3.6.8}$$

Combining Equations (3.6.8) and (3.6.3), one has

$$\frac{\partial \varepsilon_s}{\partial t} + q_{\ell m}\frac{\partial \varepsilon_s}{\partial x} - \frac{\partial}{\partial x}\left[\varepsilon_s^2 \frac{k}{\mu}\Delta\rho g\right] = \frac{\partial}{\partial x}\left[\varepsilon_s \frac{k}{\mu}\frac{\partial p_s}{\partial x}\right] = \frac{\partial}{\partial x}\left[\varepsilon_s \frac{k}{\mu}\frac{\partial p_s}{\partial \varepsilon_s}\frac{\partial \varepsilon}{\partial x}\right] \tag{3.6.9}$$

Equation (3.6.9) is the same as Equation (9) of Burger *et al.* (2001) since

$$\frac{k}{\mu} = \frac{-f_{bk}}{(\Delta\rho) g \varepsilon_s^2} \tag{3.6.10}$$

where f_{bk} is the so-called "Kynch batch flux density function" used by Burger *et al.* in their formulation.

Equation (3.6.9) applies over the region $0 < x < L(t)$. The boundary condition imposed depends upon the mode of operation and are those given by Equation (3.1.17a)–(3.1.19b). The moving boundary condition given by Equation (3.1.15), however, requires modification as shown below.

The basic expression of $\mathrm{d}L/\mathrm{d}t$ given by Equation (3.1.9b) remains valid, or

$$\frac{\mathrm{d}L}{\mathrm{d}t} = \frac{q_\ell|_{L^-} - q_\ell|_{L^+}}{\varepsilon_s|_{L^+} - \varepsilon_{s|_{L^-}}} \tag{3.6.11}$$

The values of $\varepsilon_s|_{L^+}$, $\varepsilon_s|_{L^-}$ and $q_\ell|_{L^-}$ used before remain the same, or

$$\varepsilon_s|_{L^+} = \varepsilon_{s_o} \tag{3.6.12}$$

$$\varepsilon_s|_{L^-} = \varepsilon_s^o \tag{3.6.13}$$

$$q_\ell|_{L^-} = -\varepsilon_s^o\left(\frac{k}{\mu}\frac{\partial p_\ell^*}{\partial x}\right)_{L^-} + (1-\varepsilon_s^*)q_{\ell_m} \tag{3.6.14}$$

For $q_\ell|_{L^+}$, at the cake surface due to particle sedimentation, one has

$$-\frac{q_s|_{L^+}}{\varepsilon_{s_o}} - \frac{-q_\ell|_{L^+}}{1-\varepsilon_{s_o}} = u_t \tag{3.6.15}$$

where u_t is the particle sedimentation velocity.

Solving Equations (3.6.15) and (3.1.11) simultaneously, $q_\ell|_{L^-}$ is found to be

$$\frac{\mathrm{d}L}{\mathrm{d}t} = q_{\ell_m} + \frac{1-\varepsilon_{s_o}}{\varepsilon_s^o - \varepsilon_{s_o}}\varepsilon_{s_o}u_t + \frac{1}{\varepsilon_s^o - \varepsilon_{s_o}}\left(-\frac{k}{\mu}\frac{\partial p_\ell}{\partial x}\right)_{L^-} \tag{3.6.16}$$

and

$$L = 0, \quad t = 0 \tag{3.6.17}$$

Equations (3.6.9) together with Equation (3.6.3) with the same boundary conditions given in Section 3.1 and the moving boundary conditions of (3.6.16) and (3.6.17) constitute a complete description of cake filtration including the sedimentation effect. To obtain the solution, the constitutive relationships of ε_s vs. p_s and k vs. p_s are required. The particle settling velocity at the cake surface must also be known. As an approximation, u_t may be taken to be the terminal velocity given by the Stokes law.

In comparing the above formulation with that of Burger *et al.* (2001), although the equations considered [namely, Equation (3.6.9) vs. Equation (9) of Burger *et al.*] are the same, the systems treated are different, namely continuous vs. batch. The range of ε_s covered in the batch expression case extends from ε_{s_o} to ε_{s_m} while in the continuous filtration case, ε_s ranges from ε_s^o to ε_{s_m}. The problem encountered by Burger *et al.*, namely, the incompleteness of the constitutive relationship information, is not present in the present case. There are, however, limitations in the present formulation. Equation (3.6.3) represents just one of the possible relationships between p_ℓ, p_s and the gravitational force. Using other types of relationships, as was done previously, would significantly complicate the required computation and may even render it impractical.

Intuitively, one may argue that the sedimentation effect is manifested mainly in the presence of the u_t term in Equation (3.6.16). Accordingly, an approximate assessment of the sedimentation effect can be made as follows. First, one may relate the sedimentation velocity u_t with the particle concentration of the suspension ε_{s_o} by the classical Richardson–Zaki correlation (1954), or

$$\frac{(-u_t)}{u_\infty} = (1 - \varepsilon_{s_o})^{\bar{n}} \tag{3.6.18}$$

where u_∞ is the terminal velocity according to the Stokes law or

$$u_\infty = \frac{d_p^2 g(\rho_s - \rho_\ell)}{18\mu} \tag{3.6.19}$$

According to Rowe (1987), $\bar{n}$ can be estimated from the correlation given below:

$$\frac{4.7 - \bar{n}}{\bar{n} - 2.35} = 0.175 N_{Re_t}^{0.75} \tag{3.6.20}$$

where

$$N_{Re_t} = \frac{d_p u_\infty \rho_\ell}{\mu} \tag{3.6.21}$$

Substituting Equation (3.6.18) into (3.6.16), one has

$$\frac{dL}{dt} = \frac{\varepsilon_s^o}{\varepsilon_s^o - \varepsilon_{s_o}} \left(\frac{k}{\mu} \frac{\partial p_\ell}{\partial x} \right)_{L^-} + q_{\ell_m} + \frac{\varepsilon_{s_o}}{\varepsilon_s^o - \varepsilon_{s_o}} u_\infty (1 - \varepsilon_{s_o})^{\bar{n}+1} \tag{3.6.22}$$

An estimate of the effect of the sedimentation can, therefore, be made by comparing the magnitude of the last term of the dL/dt expression with the sum of the first two terms. In estimating the latter, if one ignores the particle velocity, the liquid velocity across the cake at any instant is constant, or

$$\left(\frac{k}{\mu} \frac{\partial p_\ell}{\partial x} \right)_{L^-} = \left(\frac{k}{\mu} \frac{\partial p_\ell}{\partial x} \right)_{o} = -q_{\ell_m}$$

and

$$\frac{\varepsilon_s^o}{\varepsilon_s^o - \varepsilon_{s_o}} \left(\frac{k}{\mu} \frac{\partial p_\ell}{\partial x} \right)_{L^-} + q_{\ell_m} = \frac{\varepsilon_{s_o}}{\varepsilon_s^o - \varepsilon_{s_o}} (-q_{\ell_m})$$

The approximate expression of dL/dt is

$$\begin{aligned} \frac{dL}{dt} &\simeq \frac{\varepsilon_{s_o}}{\varepsilon_s^o - \varepsilon_{s_o}} (-q_{\ell_m}) + \frac{\varepsilon_{s_o}}{\varepsilon_s^o - \varepsilon_{s_o}} u_\infty (1 - \varepsilon_{s_o})^{\bar{n}+1} \\ &= \frac{\varepsilon_{s_o}}{\varepsilon_s^o - \varepsilon_{s_o}} \left[(-q_{\ell_m}) + u_\infty (1 - \varepsilon_{s_o})^{\bar{n}+1} \right] \end{aligned} \tag{3.6.23}$$

The value of $\bar{n}$ ranges from 2.35 to 4.7 [see Equation (3.6.20)]. $(1 - \varepsilon_{s_o})^{\bar{n}+1}$ may be close to or a fraction of unity depending upon the value of ε_{s_o} [for example with

$\varepsilon_{s_0} = 0.1$, and assuming $\overline{n} = 3.5$, an approximate arithmetic average of 2.35 and 4.7, $(1-\varepsilon_{s_0})^{\overline{n}+1}$ is 0.692; however, with $\varepsilon_{s_0} = 0.02$, $(1-\varepsilon_{s_0})^{\overline{n}+1}$ is nearly 0.92]. Thus the effect of sedimentation can be seen by comparing the value of the terminal velocity u_∞ and the filtration velocity. For the data shown in Fig. 3.9, at an operating pressure of 1 bar, the rate of filtration of 2% $CaCO_3$ suspension was approximately $3.825 \times 10^{-4}\,ms^{-1}$ during the first 1000 s and reduced to $8.37 \times 10^{-5}\,ms^{-1}$ between 5000 and 6000 s. The value of u_∞ for $CaCO_3$ particles with diameter of 3.85 μm is about $1.3716 \times 10^{-5}\,ms^{-1}$. Taking $(1-\varepsilon_{s_0})^n$ to be 0.92, neglecting the sedimentation gives an error of 3% during the first 1000 s but increases to 16% during the period of 5000–6000 s. On the other hand, for filtration carried out at 8 bars, the errors were found to be approximately 1.4% during the first 1000 s and 5% during the period of 4000–5000 s suggesting that the relative importance of the sedimentation effect is determined by the magnitude of the operating pressure in pressure filtration.

3.7 EFFECT OF FINE PARTICLE RETENTION

Most of the cake filtration analyses were made with the assumption that particles of suspension to be treated are of uniform size. In practice, the sizes of particles present in a suspension often cover a wide range. While the majority of the particles may be retained to form a cake, the finer particles together with filtrate flow may penetrate into and become retained within the cake formed. The permeability of the cake (and therefore the specific cake resistance) depends not only on the extent of compression but also on the degree of fine particle retention.

The effect of fine retention can be seen in two aspects. In predicting filtration performance, ignoring fine retention may lead to an overestimation of filtration rate and therefore the cumulative filtrate volume. On the other hand, the constitutive relationship obtained from constant pressure filtration data may be in serious error unless the effect of fine retention is properly accounted for.

Tien *et al.* (1997) presented a rigorous analysis of the effect of fine particle retention on cake filtration. Their analysis is a generalization of that presented in Sections 3.1–3.3. An outline of the work is given below.

The basic assumptions used in formulating the analysis are:

(a) The suspended particles can be considered as two types. Type 1 particles with diameter d_{p_1} which are completely retained at the cake/suspension interface, and Type 2 particles with diameter d_{p_2} that penetrates into the cake. These particles may be retained within the cake in the manner of deep-bed filtration. Based on particle deposition considerations (Tien, 1989), $d_{p_1}/d_{p_2} < 10^{-1}$.
(b) The coordinate system is the same as shown in Figure 3.1.
(c) There is no sedimentation effect in the suspension.
(d) The media resistance remains constant, i.e. there is no medium clogging by particles.

The notations used are essentially the same as those of previous sections with modifications to account for the fact that Type 2 particles are present in the filtrate as it flows through the cake. If c, c_1 and c_2 are volume concentrations of total particles, particles of Type 1 and particles of Type 2, respectively, and n, n_1 and n_2 are the corresponding values per unit volume of suspending liquid, one has

$$c = c_1 + c_2 \tag{3.7.1}$$

$$n = n_1 + n_2 \tag{3.7.2}$$

$$c_1 = \frac{n_1}{1+n_1+n_2} \simeq \frac{n_1}{1+n_1} \tag{3.7.3}$$

$$c_2 = \frac{n_2}{1+n_2+n_2} \simeq \frac{n_2}{1+n_2} \tag{3.7.4}$$

since $n_2 \ll n_1$. Within the cake, $c_1 = 0$ and $c_2 = n_2$.

As the cake is composed of two types of particles, its solidosity ε_s can be written as

$$\varepsilon_s = \varepsilon_{s_1} + \varepsilon_{s_2}, \quad \varepsilon_s + \varepsilon = 1 \tag{3.7.5}$$

Using the same procedure as outlined in Section 3.1, the volume-averaged continuity equations are

$$\frac{\partial \varepsilon_s}{\partial t} = -\frac{\partial}{\partial x}\left[\varepsilon_s \frac{k}{\mu}\frac{\partial p_\ell}{\partial x}\right] - \left(q_{\ell_m} + \int_o^x N\,dx\right)\frac{\partial \varepsilon_s}{\partial x} + (1-\varepsilon_s)N \tag{3.7.6a}$$

$$\frac{\partial \varepsilon_{s_2}}{\partial t} = -\frac{\partial}{\partial x}\left[q_s \frac{\varepsilon_{s_2}}{\varepsilon_s}\right] + N \tag{3.7.6b}$$

$$\frac{\partial[(1-\varepsilon_s)n_2]}{\partial t} = -\frac{\partial}{\partial x}[q_\ell n_2] - N \tag{3.7.6c}$$

where N is the rate of fine particle deposition within the cake. Equation (3.7.6a) is equivalent to Equation (3.1.8) (Note that in the absence of Type 2 particles and if $\varepsilon_{s_o} = 0$, $n_2 = 0$, $N = 0$, these two equations become the same). Equations (3.7.6b) and (3.7.6a) are the equations for the fine particles retained in the cake and present in the filtrate passing through the cake.

The expressions of q_ℓ, q_s and q_{ℓ_m} are

$$q_\ell + q_s = q_{\ell_m} + \int_o^L N\mathrm{d}x \tag{3.7.7a}$$

$$q_\ell = -\varepsilon_s \frac{k}{\mu}\frac{\partial p_\ell}{\partial x} + (1-\varepsilon_s)q_{\ell_m} + (1-\varepsilon_s)\int_o^x N\mathrm{d}x \tag{3.7.7b}$$

$$q_s = -q_\ell + q_{\ell_m} + \int_o^x N\mathrm{d}x \tag{3.7.7c}$$

$$q_{\ell_m} = \left[-\frac{k}{\mu}\frac{\partial p_\ell}{\partial x}\right]_o \tag{3.7.7d}$$

The rate of retention of Type 2 particles N is (Tien, 1989)

$$N = \lambda(1-\varepsilon_s)\left|\frac{q_\ell}{1-\varepsilon_s} - \frac{q_s}{\varepsilon_s}\right| n_2 \tag{3.7.8}$$

The moving boundary condition of the cake/suspension interface is

$$\frac{dL}{dt} = \frac{\varepsilon_s^o}{\varepsilon_s^o - c_o}\left(\frac{k}{\mu}\frac{\partial p_\ell}{\partial x}\right)_{L^-} + q_{\ell_m} + \int_o^L N dx \tag{3.7.9}$$

and

$$L = 0, \quad t = 0 \tag{3.7.10}$$

The relationship between p_ℓ and p_s, as before, is given by Equation (3.2.12). The constitutive relationship between ε_s and p_s is that of Equation (3.2.3). On the other hand, cake permeability can be significantly reduced with fine particle retention. To account for it, Equation (3.2.4) may be modified to give (Tien *et al.*, 1997)

$$k = k^o\left(1 + \frac{p_s}{p_a}\right)^{-\delta}(1 + \alpha_1 \varepsilon_{s_2}^{\alpha_2})^{-1} \tag{3.7.11}$$

where α_1 and α_2 are empirical constants.

The boundary condition given in Section 3.4 for different operating modes are readily applicable.

Numerical solutions of the above systems of equations were obtained by Tien *et al.* Details of the method used (including the procedure of initialization of the numerical solutions) can be found in the work of Tien *et al.* (1997).

To demonstrate the effect of fine particle retention in filtration performance, some of the results obtained previously are shown in Figs 3.11 and 3.12. These results were obtained using conditions listed in Table 3.2 and with $f' = -1$. Fig. 3.11a–c give the results of L vs. t, q_{ℓ_m} vs. t and V vs. t for the case of constant pressure filtration ($p_o = 900{,}000$ Pa). The increase in the extent of fine particle retention is shown to affect adversely on filtration performance. The results of constant rate filtration ($|q_{\ell_m}| = 2 \times 10^{-5}$ m^2/s-m^2) are shown in Figs. 3.12a (L vs. t) and 3.12b (p_o vs. t). With fine particle retention, the total amount of particle deposition increases. Thus, the cake thickness can be expected to increase with the increase of the filter coefficient. At the same time, with the increase of fine particle retention, the cake permeability decreases according to Equation (3.7.11). The operating pressure required to maintain a fixed filtration rate can be expected to increase with the increase of λ. This is confirmed by the results shown in Fig. 3.12b.

In practice, filtration data are often used to determine the constitutive relationship (or, more precisely, the determination of the constitutive relationship parameters). Similar to any parameter search problem, one may define an objective function Φ to be

$$\Phi = \sum_{i=1}^{m}[(A)_j^E - (A)_j^P]^2 \tag{3.7.12}$$

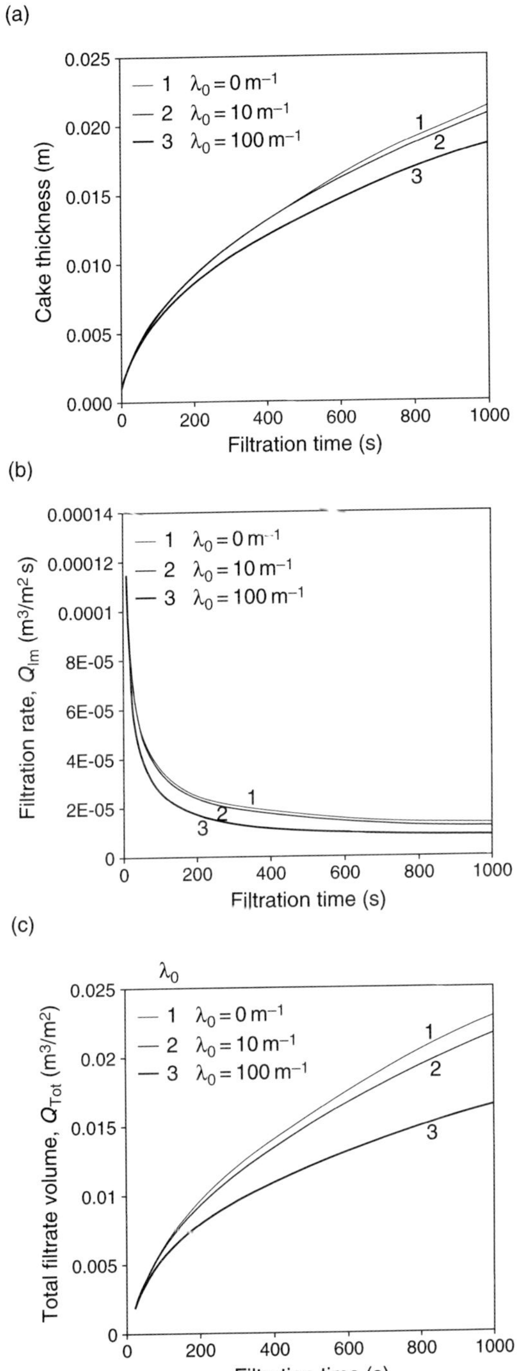

Figure 3.11 Effect of fine retention on filtration performance: (a) L vs. t; (b) q_{ℓ_m} vs. t; (c) V vs. t. Constant Pressure Filtration. Conditions used for calculations given in Table 3.2. (Tien, Bai and Ramarao, 1997. Reprinted by permission of The American Institute of Chemical Engineers.)

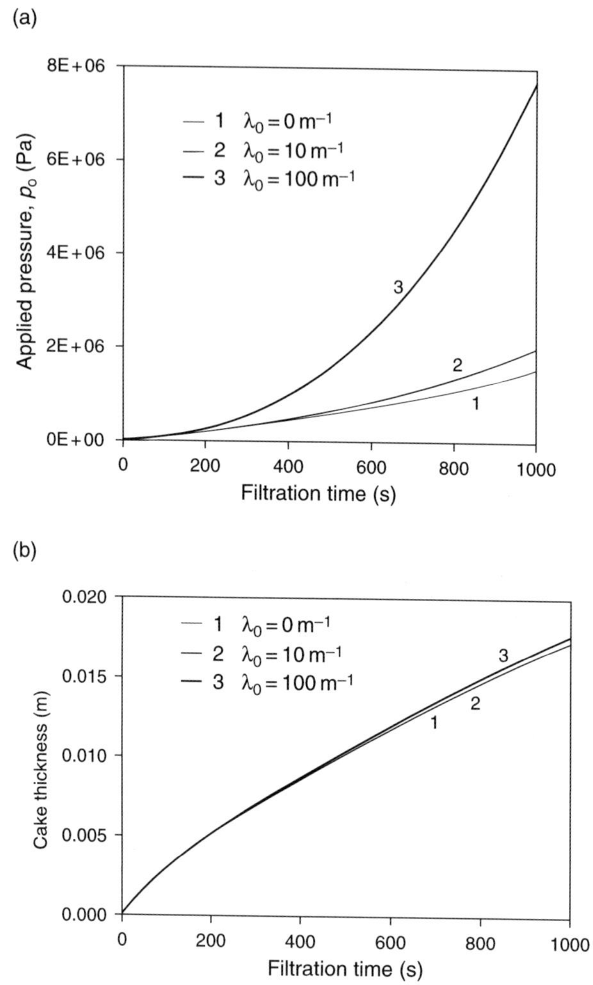

Figure 3.12 Effect of fine retention on filtration performance: (a) p_o vs. t; (b) L vs. t. Constant rate filtration. Conditions used for calculations given in Table 3.2. (Tien, Bai and Ramarao, 1997. Reprinted by permission of The American Institute of Chemical Engineers.)

where A is a measureable quantity (A may be taken to be V or L for constant pressure filtration). The superscripts "E" and "P" denote, respectively, the experimental and predicted values. The subscript j refers to the j-th data point, and there are m data points. The search is made by minimizing the objective function Φ.

The importance of fine particle retention in the determination of constitutive relationship parameters can be seen from some sample calculation results presented below. If one uses the numerical results of Figs 3.11c and 3.12c as experimental data, the constitutive parameters can be determined from the minimization of Φ with A being V or p_o. The results obtained from the three sets of data of V vs. t and those of p_o vs. t of Fig. 3.11c are presented in Table 3.3. For the case of no retention (i.e. $\lambda_o = 0$), the parameter values

Table 3.2

Conditions and parameter values used to obtain results shown in Figs 3.11 and 3.12

ε_{s_o} (-)	0.2	ε_s^o (-)	0.268
n_{2_o} (m^3/m^3)	0.05, 0.01, 0.0005	k^o (m^2)	3.4965×10^{-15}
μ (Pa s)	0.001	β (-)	0.09
p_o (Pa)	900,000	δ (-)	0.49
(constant pressure case)			
q_{ℓ_m} ($m^3\,m^{-2}\,s^{-1}$)	2.0×10^{-5}	$p_a(p_o)$	1200
(constant rate case)			
$(1/k^o)(1/R_m)$ (-)	100	α_1 (-)	30
λ (m^{-1})	0, 10, 100	α_2 (-)	1.0

Table 3.3

Parameters of constitutive relationships obtained from search optimization of results of Figs 3.11c and 3.12c[a]

Filtration mode		p_a	δ	β	ε_s^o	k^o
Constant pressure	$\lambda_o = 0$	1200	0.4900	0.0900	0.269	3.5×10^{-15}
(Fig. 3.11c)	$\lambda_o = 10$	1188	0.5292	0.0900	0.269	3.5×10^{-15}
	$\lambda_o = 100$	1925	0.6746	0.0926	0.278	3.5×10^{-15}
Constant rate	$\lambda_o = 0$	1200	0.4900	0.0900	0.269	3.5×10^{-15}
(Fig. 3.12c)	$\lambda_o = 10$	2943	0.6010	0.0902	0.291	3.5×10^{-15}
	$\lambda_o = 100$	2780	0.6978	0.0900	0.290	3.5×10^{-15}

[a] Units of quantities are the same as given in Table 3.2.

obtained are nearly the same as those used in the numerical analysis (see Table 3.2). For the other two cases, the values of β, δ and ε_s^o obtained are significantly different from those given in Table 3.2. The difference is understandable since their various parameters must be adjusted to account for the decrease in V vs. t because of cake clogging by fine particles.

3.8 BATCH FILTRATION/CONSOLIDATION

As discussed in Section 2.5.2, batch filtration/consolidation or expression is an operation aimed at removing water (or liquid) from a finite volume of solid/liquid mixtures. The first step of the operation is filtration or filtration with sedimentation, taking place in the suspension phase. Once the suspension is depleted and the cake formed contains all the particles present in the suspension initially, consolidation takes place until the compressive stress throughout the cake becomes uniform (if the wall friction effect is negligible).

The formulation given by Burger *et al.* (2001), in principle, can be applied to analyze the problem. However, because of the problems stated before, a simple procedure which treats filtration and consolidation separately is proposed and described as below.

(1) If the sedimentation effect can be ignored, the procedure described in Section 3.4 can be applied to the filtration step of the process extending from $t = 0$ to $t = t_f$. t_f is reached if the cake mass per unit medium reaches the value equal to the particle mass present in the suspension initially or

$$\int_0^{L(t_f)} \varepsilon_s dz = H_o \varepsilon_{s_o} \tag{3.8.1}$$

where H_o is the initial suspension height and ε_{s_o} the initial particle volume fraction.

For the analysis of the consolidation step which extends from $t > t_f$, it is more convenient to describe the progress of consolidation in terms of the compressive stress distribution. The governing equations can be obtained directly from Equation (3.4.9) since ε_s and p_s are related by the constitutive relationship (e.g. p_s may be expressed as $p_a[(\varepsilon_s/\varepsilon_s^o)^{1/\beta} - 1]$ if Equation (3.2.3) is used. dL/dt now is given as

$$\frac{dL}{dt} = q_{\ell_m} \tag{3.8.2}$$

since the change of cake thickness is directly related to the amount of water expelled from the cake.

The compressive stress distribution can then be found from the solution of Equation (3.4.9) (or its equivalent with p_s as the dependent variable) and (3.7.12) with the initial condition

$$L = L(t_f) \qquad \text{at} \quad t = t_f \tag{3.8.3}$$

$$p_s = p_s(x, \ t_f) \qquad 0 < x < L_f \tag{3.8.4}$$

where $p_s(x, t_f)$ is the compressive stress profile obtained from the filtration step calculation at $t = t_f$.

The boundary conditions at $x = L$ now become

$$p_t = p_\ell + p_s = p_o \tag{3.8.5}$$

where p_o is the applied pressure.

The numerical results should show an increase of the compressive stress and decrease of the liquid pressure throughout the cake. When the compressive stress throughout the cake becomes uniform, consolidation may be considered to be complete.

(2) If the sedimentation effect is to be considered, one may follow the approximate method outlined in Section 3.6 for the first step (i.e. filtration) calculation, namely applying Equation (3.6.16) instead of Equation (3.1.15) for the moving cake/suspension interface condition. The criterion of the completion of the filtration step [i.e. Equation (3.2.1)] and the calculation of cake consolidation are the same as described in the preceding paragraphs.

(3) If a more exact account of the sedimentation effect is desired, it may be more convenient to divide the physical domain during the first step into two parts, $0 < x < L(t)$ for the cake phase and $L(t) < x < H(t)$ for the suspension phase (including the top layer of clear liquid), and applying appropriate algorithms accordingly; for example, a combination of the method described in Section 3.4 and the rather simple method of Stamatakis and Tien (1992) for sedimentation may be applied. Equation (3.1.15) [and Equation (3.6.16)] can no longer be applied as the moving cake/suspension interface condition and new expressions based on continuity requirement need to be derived.

REFERENCES

Abboud, W.M. and Corapcioglu, M.Y., *J. Colloid Interface Sci.*, **160**, 304 (1993).
Anderson, T.B. and Jackson, R., *Ind. Eng. Chem. Fundamentals*, **6**, 527 (1967).
Atsumi, K. and Akiyama, T., *J. Chem. Eng. Japan*, **8**, 487 (1975).
Bai, R. and Tien, C., *Chem. Eng. Sci.*, **60**, 301 (2005).
Benesch, T., Meier, U. and Schutz, W., *Sep. Pur. Tech.*, **35**, 37 (2004).
Bockstal, F., Fourarge, L., Hermia, J. and Rahier, G., *Filtrat. Separat.*, **25**, 255 (1985).
Burger, R., Concha, F. and Karlsen, K.H., *Chem. Eng. Sci.*, **56**, 4537 (2001).
Landman, K.A. and Russel, W.B., *Phys. Fluids*, A, **5**, 530 (1993).
Landman, K.A., Sirakoff, C. and White, L.B., *Phys. Fluids*, A, **3**, 1495 (1991).
Landman, K.A., Stankovich, J.M. and White, L.R., *AIChE J.*, **45**, 1875 (1999).
Landman, K.A., White, L.R. and Eberl, M., *AIChE J.*, **41**, 1687 (1995).
Richardson, J.F. and Zaki, W.N., *Trans. Inst. Chem. Engrs.*, **32**, 35 (1954).
Rietema, K., *Chem. Eng. Sci.*, **37**, 1125 (1982).
Rowe, P.N., *Chem. Eng. Sci.*, **43**, 2795 (1987).
Shirato, M., Sambuichi, M., Kato, H. and Aragaki, T., *AIChE J.*, **15**, 405 (1969).
Smiles, D.E., "Principles of Constant Pressure Filtration", in *Encyclopedia of Fluid Mechanics*, Vol. 5, N.P. Cheremisirouff, ed., Gulf Publishing, Houston, Tx (1986).
Stamatakis, K. and Tien, C., *Chem. Eng. Sci.*, **46**, 1917 (1991).
Teoh, S.II., "Studies in Filter Cake Characterization and Modeling", Ph.D. Thesis, National University of Singapore (2003).
Tien, C., *Granular Filtration of Aerosols and Hydrosols*, Butterworths, Stoneham, MA (1989).
Tien, C. and Bai, R., *Chem. Eng. Sci.*, **58**, 1323 (2003).
Tien, C., Bai, R. and Ramarao, B.V., *AIChE J.*, **43**, 33 (1997).
Tien, C., Teoh, S.K. and Tan, B.H.R., *Chem. Eng. Sci.*, **56**, 5361 (2001).
Tiller, F.M., Hsyung, N.B. and Cong, D.Z., *AIChE J.*, **41**, 1153 (1995).
Tosun, I., *Chem. Eng. Sci.*, **41**, 2563 (1986).
Wakeman, R.J., *Trans. Inst. Chem. Engrs.*, **56**, 258 (1978).

–4–

CAKE FILTRATION AS A DIFFUSION PROCESS

Notation

b	exponent of Equation (4.3.2.3) (-)
$D(e)$	filtration diffusivity ($m^2\,s^{-1}$)
D_o	value of $D(e)$ at $e = e^o (m^2\,s^{-1})$
$E(m, o)$	function defined by Equation (4.4.2.5a) (-)
$E(o, t')$	function defined by Equation (4.4.2.5b) (-)
e	void ratio (-)
e^o	value of e corresponding to ε_s^o (-)
e_o	value of e corresponding to ε_{s_o} (-)
$e(p_o)$	value of e at $p_s = p_o$ (-)
e^*	defined as $[e - e(p_o)]/[e_o - e(p_o)]$ (-)
e_{av}	defined by Equation (4.4.1.5) (-)
e_∞	ultimate value of e_{av}
$f(m)$	e-profile at the beginning of consolidation
f'	defined by Equation (4.1.13) (-)
g	gravitational acceleration (ms^{-2})
k	permeability (m^2)
k^o	values of k at $p_s = 0$ (m^2)
$k_m{}^*$	a constant present in Equation (4.3.1.14)
L	height of suspension undergoing expression (m)
L_o	initial value of L (m)
L_1	value of L upon the completion of filtration (m)
L_∞	ultimate value of L obtained in consolidation (m)
M	total volume of particles present per unit medium area (m^3/m^2)
m	defined by Equation (4.1.7a) (m)
m_L	value of m at the cake surface (m)
$\overline{m}$	average wet to dry cake mass ratio (-)
m^*	defined as m/M (-)

p_a a parameter present in Equations (4.3.2.1) and (4.3.2.2) (Pa)
p_ℓ liquid pressure (Pa)
p_o applied (operating) pressure
p_s compressive stress (p_g)
q_ℓ superficial liquid velocity (ms^{-1})
q_s superficial particle velocity (ms^{-1})
$q_{\ell s}$ liquid/particle relative velocity (ms^{-1})
R_m medium resistance (m^{-2})
S defined by Equation (4.3.1.9) ($mt^{-1/2}$)
S_n defined as $D\left[\frac{(2n-1)\pi}{2M}\right]^2$ (s^{-1})
s particle mass fraction of suspension (-)
t time (s)
$\bar{t}$ time (s)
t^* defined as $D_o t/M^2$ (-)
u_c extent of consolidation defined by Equation (4.4.5) or (4.4.6)
V cumulative filtrate volume (m^3/m^2)
w_o particle mass per unit medium surface area ($kg\,m^{-2}$)
x distance away from filter medium (m)

Greek letters

β exponent of Equation (4.3.2.1) (-)
δ exponent of Equation (4.3.2.2) (-)
ε porosity (-)
ε_s solidosity (-)
ε_s^o value of ε_s at $p_s = 0$ (-)
ε_{s_o} initial particle volume fraction (-)
$(\varepsilon_s)_{av}$ average value of ε_s
λ similarity variable defined as $m/t^{1/2}$ ($m\,s^{-1/2}$)
μ viscosity (Pa s)
ν kinetic viscosity ($m^2\,s^{-1}$)
ν^* value of ν at $T = 293\,K$
ρ density of filtrate ($kg\,m^{-3}$)
ρ_s density of particle ($kg\,m^{-3}$)

From the presentations given in the preceding chapters, it is clear that the central problem of cake filtration is the flow of filtrate through porous media (i.e. filter cake) undergoing growth and compaction simultaneously, a phenomenon which can also be found in a number of natural or engineered systems. In certain ways, cake filtration is similar to the seepage of water through soil, which changes volume as the water content changes, a topic of study which has commanded the interests and efforts of soil scientists for decades. As a result of these efforts, a body of well-developed information on the topic is readily available in the literature.

Smiles (1970, 1986) and Smiles *et al.* (1982), among other investigators, presented cake filtration analyses using the soil science approach. The major differences between Smiles' work and those of the previous chapters reside mainly in the use of the material domain instead of the physical domain and the void ratio e, defined as $\varepsilon_s/(1-\varepsilon_s)$, instead of ε_s as the dependent variable in describing cake filtration. As a result, the continuity equation can be transformed to a diffusion equation–like expression which can then be manipulated to yield results (either in terms of predicting filtration performance or for data interpretation) with certain advantages. In the following sections, we will present details of this type of analysis based largely on the works of Smiles (1986) and Ramarao *et al.* (2002), but using terminologies and notations as closely as possible to those of previous chapters.

4.1 FORMULATION OF CAKE FILTRATION AS A DIFFUSION PROBLEM

The starting point of the analysis is the same as before, namely, Equations (3.1.3a) and (3.1.3b) or

$$\frac{\partial \varepsilon}{\partial t} = -\frac{\partial q_\ell}{\partial x} \tag{4.1.1a}$$

$$\frac{\partial \varepsilon_s}{\partial t} = -\frac{\partial q_s}{\partial t} \tag{4.1.1b}$$

with the meanings of the symbols as before.

The liquid/solid relative velocity, $q_{\ell s}$, is given by the generalized Darcy's law as

$$\frac{q_{\ell s}}{\varepsilon} = \frac{q_\ell}{\varepsilon} - \frac{q_s}{\varepsilon_s} = \frac{-1}{\varepsilon}\frac{k}{\mu}\frac{\partial p_\ell}{\partial x} \tag{4.1.2}$$

The void ratio e is defined as

$$e = \frac{\varepsilon}{\varepsilon_s} \tag{4.1.3}$$

It is simple to show that

$$\varepsilon_s = \frac{1}{1+e}, \qquad \varepsilon = \frac{e}{1+e} \tag{4.1.4}$$

From Equation (4.1.2), one has

$$q_\ell = q_{\ell s} + e q_s \tag{4.1.5a}$$

$$q_{\ell s} = -\frac{k}{\mu}\frac{\partial p_\ell}{\partial x} \tag{4.1.5b}$$

Substituting Equations (4.1.5a), (4.1.5b) into (4.1.1a), and after rearrangement, one has

$$\frac{\partial e}{\partial t}+\frac{q_s}{\varepsilon_s}\frac{\partial e}{\partial x}=-\frac{1}{\varepsilon_s}\frac{\partial}{\partial x}\left[-\frac{k}{\mu}\frac{\partial p_\ell}{\partial x}\right] \tag{4.1.6}$$

If a new set of independent variables $(m, \bar{t})$ defined as

$$\mathrm{d}m=\varepsilon_s\,\mathrm{d}x-q_s\mathrm{d}t \tag{4.1.7a}$$

$$\mathrm{d}\bar{t}=\mathrm{d}t \tag{4.1.7b}$$

is used, Equation (4.1.6) becomes

$$\frac{\partial e}{\partial \bar{t}}+\frac{\partial}{\partial m}\left[-\frac{k}{\mu}\frac{1}{1+e}\frac{\partial p_\ell}{\partial m}\right]=0 \tag{4.1.8}$$

The relationship between the material coordinate m and the physical coordinate x can be found from the integration of Equation (4.1.7a), or

$$m=\int_{\text{along } x=0}(-q_s)\,\mathrm{d}t+\int_{\text{along } t=t}\varepsilon_s\,\mathrm{d}x=\int_0^x\varepsilon_s\,\mathrm{d}x=\int_0^x\frac{\mathrm{d}x}{1+e} \tag{4.1.9}$$

since q_s vanishes along $x=0$ (cake-membrane interface).

The quantity $\partial p_\ell/\partial m$ of Equation (4.1.8) may be written as

$$\frac{\partial p_\ell}{\partial m}=\frac{\partial p_\ell}{\partial p_s}\frac{\partial p_s}{\partial e}\frac{\partial e}{\partial m} \tag{4.1.10}$$

This means that there exists a relationship between p_ℓ and p_s as stated before and e is a unique function of p_s. The last condition is the same as assuming ε_s being a unique function of p_s as was done previously.

Substituting Equations (4.1.10) into (4.1.8) and dropping the bar over $\bar{t}$ yield

$$\frac{\partial e}{\partial t}=\frac{\partial}{\partial m}\left[D(e)\frac{\partial e}{\partial m}\right] \tag{4.1.11}$$

where

$$D(e)=(f')\left[\frac{k}{\mu(1+e)(\mathrm{d}e/\mathrm{d}p_s)}\right] \tag{4.1.12}$$ [1]

[1] Smiles assumed $-f'=1$ or Type 1 relationship between p_ℓ and p_s. Accordingly, $D(e)=\dfrac{k}{\mu(1+e)(-\mathrm{d}e/\mathrm{d}p_s)}$

with

$$f' = \frac{\partial p_\ell}{\partial p_s} \tag{4.1.13}$$

The liquid/solid relative velocity, in terms of the new variable, is

$$q_{\ell s} = -\frac{k}{\mu}\frac{1}{1+e}\frac{\partial p_\ell}{\partial m} = -D\frac{\partial e}{\partial m} \tag{4.1.14}$$

At cake/membrane interface, or $m = 0$, $q_s = 0$:

$$(q_{\ell s})_{m=0} = (q_\ell)_{m=0} = \left(-D\frac{\partial e}{\partial m}\right)_{m=0} \tag{4.1.15}$$

$(q_\ell)_{m=0}$ is the instantaneous filtration velocity. If V denotes the cumulative filtrate volume, and noticing that q_ℓ is inherently negative, one has

$$|(q_\ell)|_{m=0} = \frac{\mathrm{d}V}{\mathrm{d}t} = \left(D\frac{\mathrm{d}e}{\mathrm{d}m}\right)_{m=0} \tag{4.1.16}$$

In Equation (4.1.14), the filtrate flux (relative to solid) is found to be proportional to the gradient of the void ratio. $D(e)$ is therefore termed the *filtration diffusivity*.

To complete the analysis, one must first specify the material domain to which Equation (4.1.11) applies as well as the initial and boundary conditions. This will be discussed in the following section.

4.2 MATERIAL DOMAIN AND INITIAL AND BOUNDARY CONDITIONS

For one-dimensional rectilinear case, Equation (4.1.11) may be applied to any one of the following three cases.

Case (1)

System confined to the cake region, or $0 < m < m_L(t)$ and

$$m_L = 0, \quad t = 0 \tag{4.2.1a}$$

$$\frac{\mathrm{d}m_\mathrm{L}}{\mathrm{d}t} = \frac{1}{e_0}|(q_\ell)_{m=0}| = \frac{1}{e_0}\left[D\frac{\partial e}{\partial m}\right]_{m=0} \tag{4.2.1b}$$

with

$$e_0 = \frac{1-\varepsilon_{s_0}}{\varepsilon_{s_0}} \tag{4.2.2}$$

and ε_{s_0} is the particle volume fraction of the suspension to be filtered.

Case (2)

System of infinite extent, or $m > 0$

Case (3)

System of Finite Extent, or $0 < m < M$. M is given as

$$M = L_0 \varepsilon_{s_0} = \frac{L_0}{1 + e_0} \tag{4.2.3}$$

where L_0 is the initial height of the system.

Case (3) represents batch filtration or expression. Schematic representations of the three cases are shown in Fig. 4.1. The moving boundary, initial and boundary conditions of these three cases are:

Case (1)

$0 < m < m_L(t)$, or the cake region.

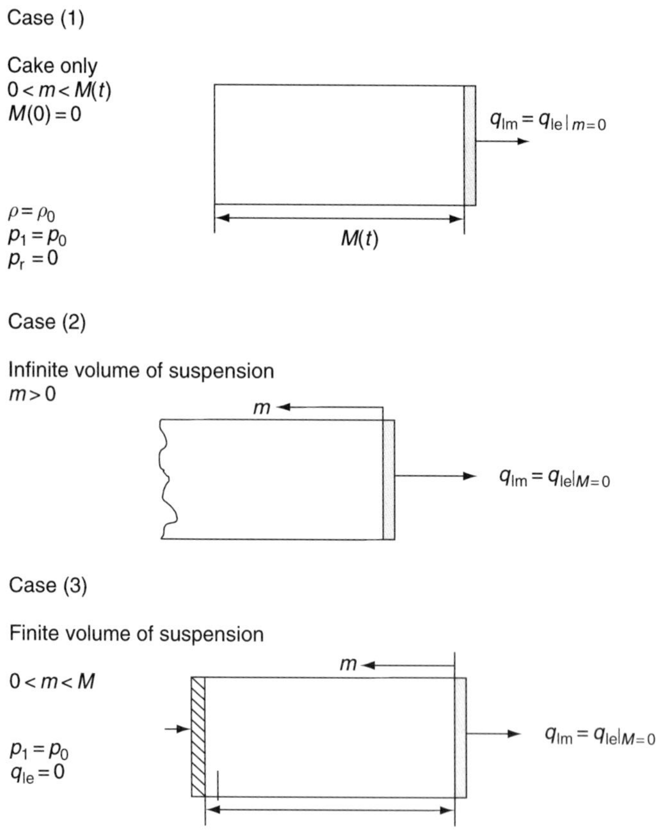

Figure 4.1 Schematic representation of the three domains discussed in 4.2.

The moving boundary condition is given by Equation (4.2.1b) and the initial condition is Equation (4.2.1a). The boundary conditions are:

At $m = M$

$$D\frac{\partial e}{\partial m} = \frac{p_\ell}{\mu R_m} \quad \text{if } p_o \text{ is specified} \tag{4.2.4a}$$

or

$$D\frac{\partial e}{\partial m} = |(q_\ell)_{m=0}| \quad \text{if } (q_\ell)_{m=0} \text{ is specified} \tag{4.2.4b}$$

At $m = m_L(t)$

$$e = e^0 = \frac{1-\varepsilon_s^0}{\varepsilon_s^0}, \qquad p_s = 0, \qquad p_\ell = p_0 \tag{4.2.5}$$

Case (2)

$m > 0$, system of infinite extent.

The initial condition is:

$$e = e_{o,} \qquad m > 0, \qquad t \leq 0 \tag{4.2.6}$$

The moving boundary conditions are those of Equations (4.2.1a) and (4.2.1b). The boundary conditions are

$$D\frac{\partial e}{\partial m} = \frac{p_\ell}{\mu R_m} \quad \text{if } p_o \text{ is specified} \tag{4.2.7a}$$

at $m = 0$

$$D\frac{\partial e}{\partial m} = |(q_\ell)_{m=0}| \quad \text{if } (q_\ell)_{m=0} \text{ is specified} \tag{4.2.7b}$$

$$e \to e_o \quad \text{as} \quad m \to \infty \tag{4.2.8}$$

Case (3)

$0 < m < M$, system of finite extent.[2]

The initial condition is

$$e = e_o, \qquad 0 < m < M, \qquad t \leq 0 \tag{4.2.9}$$

---

[2] For the solution of Case (3), D as a function of e with e ranging from $e(p_o)$ to e_o is assumed to be known. The cake thickness can be found from the value of m corresponding to $e = e^o$ (see Section 4.3).

The boundary conditions are

At $m = M$

$$D\frac{\partial e}{\partial m} = \frac{p_\ell}{\mu R_m} \quad \text{if } p_o \text{ is specified} \tag{4.2.10a}$$

$$D\frac{\partial e}{\partial m} = |(q_\ell)_{m=0}| \quad \text{if } (q_\ell)_{m=0} \text{ is specified} \tag{4.2.10b}$$

and

$$\frac{\partial e}{\partial m} = 0 \tag{4.2.11a}$$

$$p_\ell = p_o \tag{4.2.11b}$$

4.3 REMARKS CONCERNING THE SOLUTION OF THE DIFFUSION EQUATION

The solution of Equation (4.1.11) with the appropriate initial, boundary and moving boundary conditions given in Section 4.2 provides a complete description of the dynamics of cake filtration. The solution can be obtained if the relationship of e vs. p_s and k vs. e and the relationship between p_ℓ and p_s are known [see the definition of $D(e)$ of Equation (4.1.12)]. Since the relationships of e vs. p_s and k vs. e can be readily obtained from the constitutive relationships of ε_s vs. p_s and k vs. p_s, it is clear that there exists an equivalence between the analysis given in Chapter 3 and what is shown above. However, the equivalence is not complete. The differences are:

(1) If Equation (4.1.11) is applied only to the cake region, the equivalence between the two approaches is exact. The cake constitutive relationships and the p_ℓ–p_s relationship can be used to establish the relationships of e vs. p_s, k vs. e, and $D(e)$. The results obtained from the solution of Equation (4.1.12) in terms of m can be readily converted into the physical coordinate by means of Equation (4.1.9).

(2) If one considers a large body of suspension undergoing cake filtration, as its limit, the material domain may be assumed to be semi-infinite, and the cake formed at the membrane surface is a part of the domain. On the other hand, Equation (4.1.11) applies to both the cake region and the suspension region. In fact, the solution of Equation (4.1.11) does not give any information about cake thickness as a function of time unless extraneous assumptions are introduced.

Another problem one faces is the availability of the filtration diffusivity data over a sufficiently wide range of e values of interest. The value of e corresponding to a relatively dilute suspension is large (for $\varepsilon_{s_o} = 0.02$, the corresponding e is of the order of 50). In contrast, the solidosity of a well-compressed filter cake may be about 0.3–0.5 (depending upon the applied pressure), and the corresponding value of e ranges from 1.0 to 2.3. Neither the – C-P measurement nor the method based on the t/V vs. V plot of constant pressure filtration data is able to give data over such a wide range of ε_s (or e).[3]

(3) For the estimation of the cake thickness history, one may assume that there exists a threshold value of e (or ε_s) which distinguishes the cake from the suspension. The simplest assumption which can be made is to consider the threshold value to be ε_s^o [see Equation (3.2.3)], or at the cake/suspension interface, one has

$$e = \frac{1}{1+\varepsilon_s^o} = e^o \quad \text{at} \quad m = m_L \tag{4.3.1}$$

And the cake thickness history can be found implicitly from the expression

$$\int_0^L \frac{dx}{1+e} = m_L \tag{4.3.2}$$

(4) For cases of finite m-domain, initially, the behavior is similar to that of case (2) (semi-infinite domain). The problem regarding cake thickness history is the same as case (2).

Once the suspension is depleted (i.e. $e = e^o$ at $m = M$), dewatering of the cake begins, and Equation (4.1.11) applies as well if the appropriate $D(e)$ data are available. The entire problem considered is that of filtration-consolidation as mentioned earlier (see Section 2.5.2). The use of the diffusion equation in describing expressions will be discussed later.

It is not the purpose of this monograph to provide a complete list of the solutions of Equation (4.1.11). Instead, two example solutions are shown in order to illustrate the essential features of cake filtration analysis based on the soil science approach and to demonstrate its utility.

4.3.1 Constant pressure filtration of suspension of infinite volume

The problem of constant pressure filtration of suspensions of semi-infinite extent with negligible medium resistance and type-1 p_ℓ–p_s relationship will be first considered. If the medium resistance is negligible, the pressure drop across the medium vanishes and that across the cake is the same as p_o, the applied pressure. With type-1 p_ℓ–p_s relationship, p_s at membrane/cake interface ($m = 0$) is p_o. Replacing Equation (4.2.7a), the boundary condition at $m = 0$ is

$$e = e(p_o) \tag{4.3.1.1}$$

where $e(p_o)$ is the value of e at $p_s = p_o$.

[3] For more on this point, see Chapter 7.

Since the domain is semi-infinite, and the conditions of Equations (4.2.6) and (4.2.8) are consistent, a similar solution of Equation (4.1.11) can be obtained. The similarity variable λ is defined as

$$\lambda = \frac{m}{t^{1/2}} \tag{4.3.1.2}$$

Equation (4.1.11) becomes

$$\frac{\mathrm{d}}{\mathrm{d}\lambda}\left[D(e)\frac{\mathrm{d}e}{\mathrm{d}\lambda}\right] + \frac{\lambda}{2}\frac{\mathrm{d}e}{\mathrm{d}\lambda} = 0 \tag{4.3.1.3}$$

The initial and boundary conditions of Equation (4.2.6), (4.2.8) and (4.3.1.1) become

$$\mathrm{e} \to \mathrm{e_o} \quad \text{as} \quad \lambda \to \infty \tag{4.3.1.4a}$$

$$\mathrm{e} = \mathrm{e(p_o)} \quad \text{at} \quad \lambda = 0 \tag{4.3.1.4b}$$

Before presenting some typical results of the solution of Equations (4.3.1.3)–(4.3.1.4b), certain manipulations of Equation (4.3.1.3) which yield practically useful information will be discussed.

Integrating Equation (4.3.1.3) from $\lambda = \lambda$ to $\lambda = \infty$, if one assumes $\mathrm{d}e/\mathrm{d}\lambda \to 0$ as $\lambda \to \infty$ from Equation (4.3.1.4a), one has

$$-D(e)\frac{\mathrm{d}e}{\mathrm{d}\lambda} = -\frac{1}{2}\int_{e}^{e_o} \lambda \, \mathrm{d}e \tag{4.3.1.5}$$

If the void ratio profiles at different times can be determined experimentally, the values of $\mathrm{d}e/\mathrm{d}\lambda$ and the integral of Equation (4.3.1.5) at various times can be readily determined. This, in turn, allows the determination of $D(e)$ from the knowledge of the e-profiles. The method developed by Matano (1932) was based on this principle.

From the definition of λ [Equation (4.3.1.2)], one has

$$\frac{\partial e}{\partial m} = \frac{\mathrm{d}e}{\mathrm{d}\lambda}\frac{\partial \lambda}{\partial m} = t^{-1/2}\frac{\mathrm{d}e}{\mathrm{d}\lambda} \tag{4.3.1.6}$$

Combining Equations (4.3.1.5) and (4.3.1.6), one has

$$\frac{1}{2}\int_{e}^{e_o} \lambda \, \mathrm{d}e = D(e)\frac{\mathrm{d}e}{\mathrm{d}\lambda} = t^{1/2}D(e)\frac{\partial e}{\partial m}$$

or

$$D(e)\frac{\partial e}{\partial m} = t^{-1/2}\left[\frac{1}{2}\int_{e}^{e_o} \lambda \, \mathrm{d}e\right] \tag{4.3.1.7}$$

Applying the above expression at $\lambda = 0$ $[e = e(p_o)]$, and by virtue of Equation (4.1.16), one has

$$\frac{dV}{dt} = \left[D(e)\frac{\partial e}{\partial m}\right]_{m=0} = t^{-1/2}\left[\frac{1}{2}\int_{e(p_o)}^{e_o} \lambda \, de\right] \tag{4.3.1.8}$$

Since e is a unique function of λ, the inverse relationship λ vs. e is also unique. Accordingly, the integral of the above equation is dependent only upon e_o and $e(p_o)$. Thus for a given filtration run (or expression), one has

$$\frac{dV}{dt} = St^{-1/2}$$

with

$$S = \frac{1}{2}\int_{e(p_o)}^{e_o} \lambda \, de \tag{4.3.1.9}$$

The cumulative filtrate volume, V, then becomes

$$V = 2\ St^{1/2} \tag{4.3.1.10}$$

Recalling the V vs. t relationship of the conventional theory [Equations (2.2.2) and (2.2.11)], if the medium resistance is omitted, one has

$$\mu s\ \rho[(1 - \overline{m}s)_{\Delta p_c = p_o}][\alpha_{av}]_{\Delta p_c = p_o}\frac{V^2}{2} = t \tag{4.3.1.11}$$

Comparing Equations (4.3.1.11) with (4.3.1.10), S is found to be

$$S^2 = \left[\frac{p_0(1 - \bar{m}s)}{2\mu s\rho[\alpha_{av}]_{p_o}}\right] \tag{4.3.1.12}$$[4]

In Fig. 4.2, the results of constant pressure filtration of $CaCO_3$ suspension ($\varepsilon_{s_0} = 0.02$, $p_o = 100\,\text{kPa}$) and three different cake thickness settings obtained by Teoh (2003) are shown in the form of V vs. $\sqrt{t}$. Also shown in the curve is the linear relationship according to Equation (2.3.1.10) with S estimated from Equation (4.3.1.12). The reasonable good agreement between the data and Equation (4.3.1.10) with S estimated according to Equation (4.3.1.12) for $t \geq 100\,\text{s}$ provides a degree of verification of Equation (4.3.1.10).

Further comparison between experiments and data is shown in Fig. 4.3. Figure 4.3 is taken from Smiles (1986) and gives constant pressure filtration data of bentonite

[4] See Chapter 7 (Section 7.3.5) for more detailed discussion.

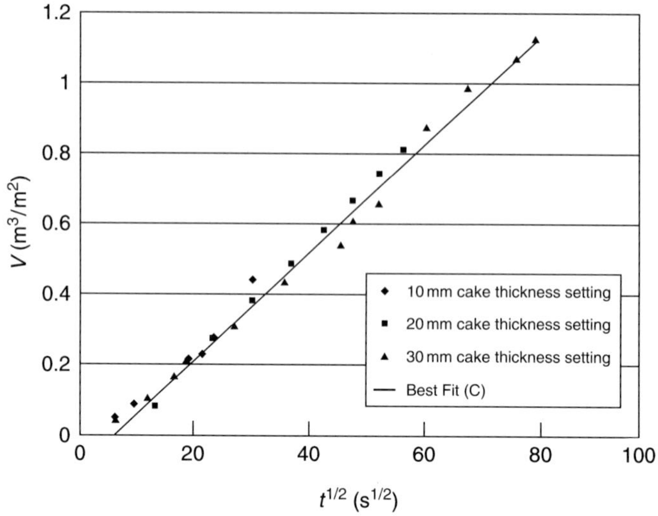

Figure 4.2 Comparison of Teoh's data of constant pressure filtration with Equations (4.3.1.10) and (4.3.1.12); 2% $CaCO_3$ suspension with $p_o = 100\,\text{kPa}$.

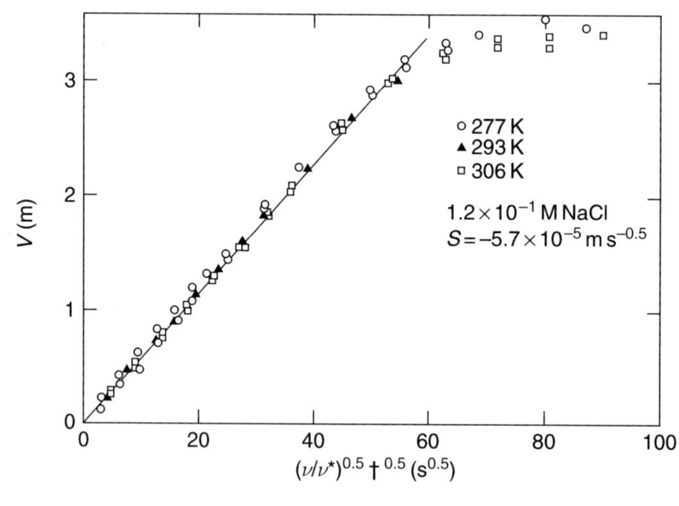

Figure 4.3 Further test of the validity of Equation (4.3.1.10); Constant pressure filtration of bentonite suspension ($\varepsilon_{s_o} = 0.0294$) with $p_o = 63.7\,\text{kPa}$ at various temperatures. (Smiles, 1986. Reprinted by permission of Elsevier.)

suspension ($\varepsilon_{s_o} = 0.0294$, $p_o = 63.7\,\text{kPa}$) in the form of V vs. $(\nu/\nu^*)^{0.5}\,t^{1/2}$ with ν being the kinematic viscosity of the filtrate. The term $(\nu/\nu^*)^{0.5}$ with ν^* being the value of ν at 293 K was introduced in order to account for the temperature effect as the data were obtained at different temperatures. It is clear that Equation (4.3.1.10) remains valid up to certain time. Although Smiles offered no explanation about the failure at longer times (or higher degree of cake compaction), it is conceivable that the constitutive relationship of e vs. p_s used (which was determined by rather crude C-P cell measurements) may not cover the entire range of p_s of interest.

Returning to the solutions of Equations (4.3.1.3)–(4.3.1.4b), the results of Smiles (1970) will be discussed below. These results were obtained using an iterative procedure (Philip, 1955) and the relationship between e vs. p_s is assumed to be

$$e = -8 \log\left(\frac{p_s}{\rho g}\right) + 40 \tag{4.3.1.13}$$

where $(p_s/\rho g)$ has the unit of cm.

The filtration diffusivity $D(e)$ was given as

$$D(e) = \frac{2.3\ k_m^*}{8} \frac{e^3}{(1+e)^2} \exp\left[\frac{2.3}{8}(40-e)\right] \tag{4.3.1.14}$$

with

$D(e)$ in $cm^2\,s^{-1}$ and $k_m^* = 3 \times 10^{-11}$.

The solution of Equation (4.3.1.3) gives the results of e vs. $mt^{-1/2}$ for specified values of e_o and $e(p_o)$. Using Equation (4.1.9), the results can be converted to e vs. $xt^{-1/2}$, which are shown in Fig. 4.4.

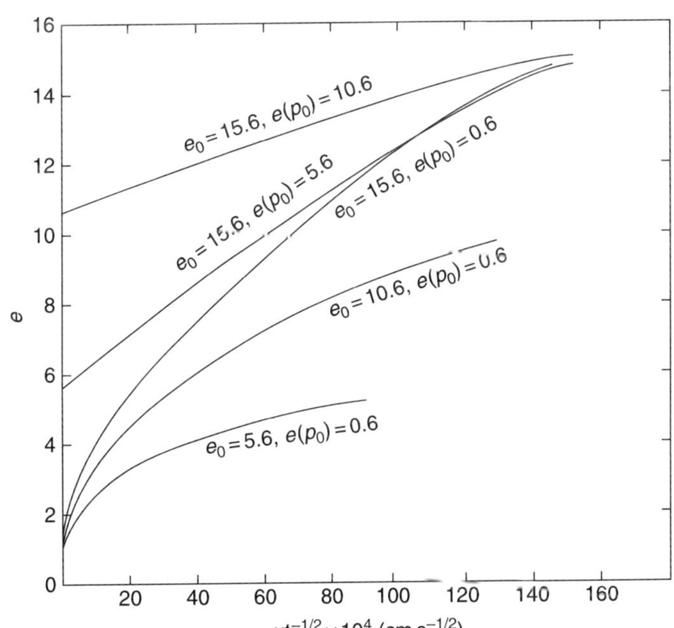

Figure 4.4 Void ratio profiles corresponding to various combinations of e_o and $e(p_o)$ from the solution of Equation (4.1.11) with $D(e)$ given by Equation (4.3.1.14). (Smiles, 1970. Reprinted by permission of Elsevier.)

From Equations (4.3.1.9) and (4.3.1.10), the cumulative filtrate volume can be shown to be

$$V = t^{1/2}\left[\int_{e(p_o)}^{e_o} \lambda \, de\right] \tag{4.3.1.15}$$

Based on the results of Fig. 4.4, values of $Vt^{-1/2}$ vs. $e(p_o)$ for specified values of e_o [or $Vt^{-1/2}$ vs. e_o for specified values of $e(p_o)$] are given in Fig. 4.5. These results confirm what is expected: namely, the greater the difference between e_o and $e(p_o)$, the greater the filtrate volume collected.

The value of $e(p_o)$ is directly related to the applied pressure. Using the relationship of Equation (4.3.1.13), the results corresponding to one of the curves (for $e_o = 15.6$) of Fig. 4.5 can be converted into $Vt^{-1/2}$ vs. p_o, which is shown in Fig. 4.6.

4.3.2 Constant pressure filtration of suspension of finite volume

As stated before, solution of the filtration diffusion equation [namely, Equation (4.1.11)] requires the knowledge of the filtration diffusivity as a function of the void ratio. For constant pressure filtration of suspension of finite volume, the range of the void ratio

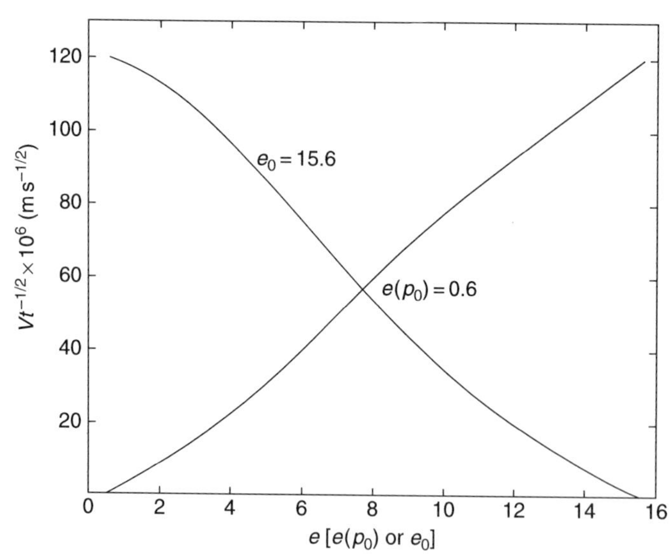

Figure 4.5 Plot of $Vt^{-1/2}$ vs. e_o [or $e(p_o)$] from the solution of Equation (4.1.11) with $D(e)$ given by Equation (4.3.1.14). (Smiles, 1970. Reprinted by permission of Elsevier.)

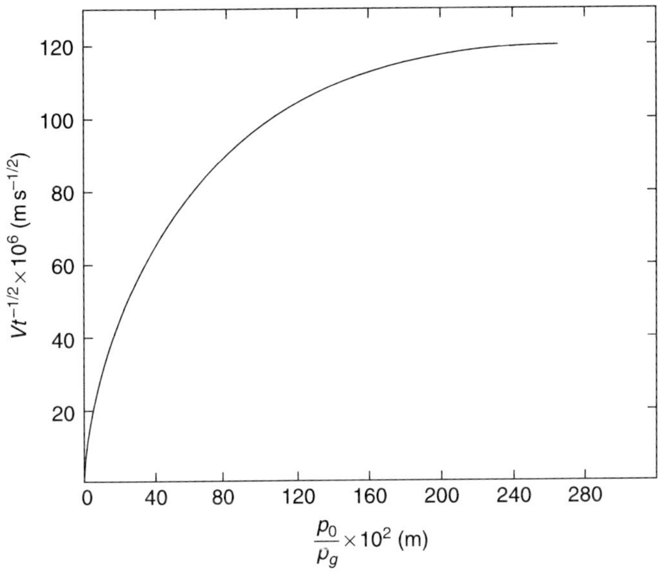

Figure 4.6 Replot of the results of Fig. 4.5 in the form of $Vt^{1/2}$ vs. p_o. (Smiles, 1970. Reprinted by permission of Elsevier.)

extends from $e = e(p_o)$ to $e = e_o$. Generally speaking, information of $D(e)$ over such a range of e cannot be obtained from a single set of independent measurements. For example, the commonly applied C-P measurement yields results over a range of e values corresponding to filter cakes but not suspensions. Often, one is forced to extend the constitutive relationship (or D vs. e) beyond the range of e over which the measurements were made. To illustrate the kind of extrapolation one may make and the possible error caused, the results of an example problem worked out by Ramarao *et al.* (2002) are presented below.

The problem considered by Ramarao *et al.* (2002) was constant pressure filtration of Hong Kong Pink Kaolin suspension, under conditions given in Table 4.1.

The particle volume fraction of the suspension was assumed to be $\varepsilon_s = 0.2$ or $e_o = 4$. The volume of the suspension was $10^{-1}\ \mathrm{m^3/m^2}$, and the corresponding M according to Equation (2.4.2.3) was $2 \times 10^{-2}\ \mathrm{m^3/m^2}$. The constitutive relationships of Hong Kong Pink Kaolin cakes (ε_s vs. p_s, k vs. p_s may be expressed as:

$$\varepsilon_s = \varepsilon_s^o \left(1 + \frac{p_s}{p_a}\right)^{\beta} \tag{4.3.2.1}$$

$$k = k^o \left(1 + \frac{p_s}{p_a}\right)^{-\delta} \tag{4.3.2.2}$$

Table 4.1

Values of parameters and conditions used in obtaining results of Figs 4.7–4.11

ε_s^o (-)	0.269
k^o (m^2)	3.4965×10^{-15}
p_a (kPa)	1.2
β (-)	0.09
δ (-)	0.49
ρ_s (kg m^{-3})	2658
μ (Pa s)	0.001
ε_{s_o} (-)	0.2
p_o (kPa)	900
$R = \dfrac{1}{k^o R_m}$ (m^{-1})	100
D_o (m^2 s^{-1})	7.95×10^{-5}
b (-)	−7.667

with the values of ε_s^o, k^o and p_a given in Table 4.1. Using these relationships and with $f' = -1$, values of $D(e)$ vs. e were were obtained from Equation (4.1.12), which can be approximated by the following expression

$$D = D_0(1+e)^{-b} \tag{4.3.2.3}$$

Values of D_o and b are given in Table 4.1.

The D vs. e relationship so established is applicable for $e < 0.731/0.269 = 2.7174$. On the other hand, for the present example problem, e extends up to $e = 8$. A procedure estimating D beyond $e = 2.7174$ must be devised before the solution can be obtained.

There are two obvious possibilities. First, one may assume that $D = 0$ for $e > 2.7174$. This is tantamount to assigning Equation (4.1.11) only to the cake region. The other possibility is to extend the established expression of Equation (4.3.2.3) beyond $e = 2.7174$. This assumption was employed by Ramarao *et al.* (2002).

Equation (4.1.11), in its dimensionless form, was solved numerically. The dimensionless void ratio e^* was defined as $[e - e(p_o)]/[e_o - e(p_o)]$ and the dimensionless independent variables were $m^* = m/M$ and $t^* = D_o t/M^2$. The e^*-profile at various times are shown in Fig. 4.7. In Fig. 4.8, the variation of e^* at $m^+ = 1$ (or at the piston surface) with time is displayed.

For the results shown in Fig. 4.7, the initial particle volume fraction of the suspension and the cake solidosity at $p_o = 900$ kPa are 0.2 and 0.4882 respectively. The corresponding e^* values are zero and unity. Initially, the e^*-profile coincides with the m^*-axis. As shown in Fig. 4.7, cake formation occurred almost instantaneously. However, after 600 s of filtration, more than half of the original suspension remained unfiltered. Even at $t \simeq 3000$ s, 20% of the initial particles remained in their initial state if one

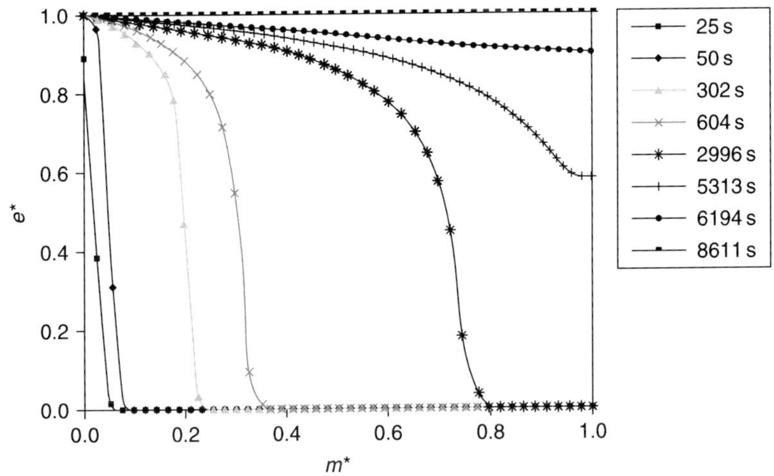

Figure 4.7 Dimensionless void rate profiles at various times, filtration of Hong Kong Pink Clay suspension under conditions listed in Table 4.1. (Ramarao, Tien and Satyadev, 2002. Reprinted by permission of Taylor & Francis.)

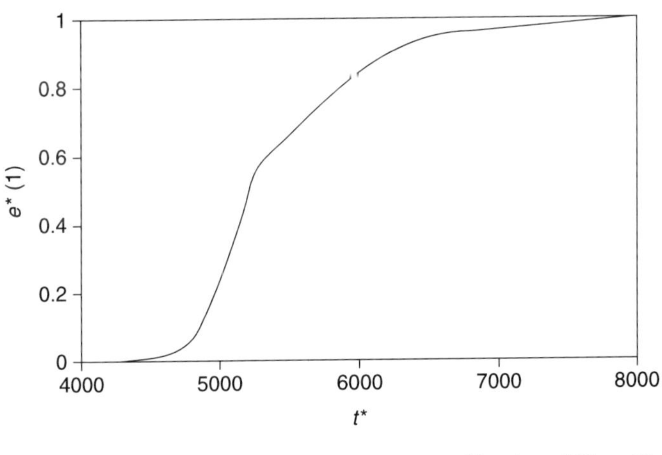

Figure 4.8 Values of e^* at $m^* = 1$ vs. t^*; Constant pressure filtration of Hong Kong Pink Clay suspension under conditions listed in Table 4.1. (Ramarao, Tien and Satyadev, 2002. Reprinted by permission of Taylor & Francis.)

assumes that cake formation takes place at $\varepsilon_s = \varepsilon_s^o$ (or $e^* = 0.6643$). Furthermore, at $t = 5300$ s, some of the particles were still in the suspension phase. The e^*-profile did not have any discontinuity. This is expected since D was assumed to be a continuous function of e.

The results of e^* at $m^* = 1$ vs. t^* (see Fig. 4.8) show that $e^* |_{m^*=1}$ remained at its initial value ($e^* = 0$) initially. At $t^* = 4000$, the value of e^* begins to increase. Thus for $t^* < 4000$, the filtration behavior was the same as what can be expected for suspensions of infinite extent.

With the assumption that the cake is characterized by a threshold value of the particle volume fraction, namely the cake solidosity at the zero stress state, ε_s^o, from the e^*-profile at various times, the cake thickness history L vs. t can be readily obtained. The results are shown in Fig. 4.9. Also included are the results obtained from the numerical solution of the volume-averaged continuity equation and those according to the conventional theory (Stamatakis and Tien, 1991). The results based on the diffusion equation give higher cake thickness values. This is not surprising since the solution of Equation (4.1.11) was obtained by ignoring the medium resistance while the medium resistance was considered by Stamatakis and Tien (1991).

The results of the cumulative filtrate volume vs. time are shown in Figs 4.10 and 4.11. The results are presented in the form of V vs. $t^{1/2}$. The initial part of the results is shown in Fig. 4.10. Also included in the figures are the results according to the conventional theory and those from the solution of the volume-averaged continuity equations (Stamatakis and Tien, 1991). Similar to Fig. 4.9, the results given by the solution of the diffusion equation are higher than those of Stamatakis and Tien (1991) for the reason already stated. However, the linearity of V vs. $t^{1/2}$ is not obeyed for larger time as shown in Fig. 4.11.

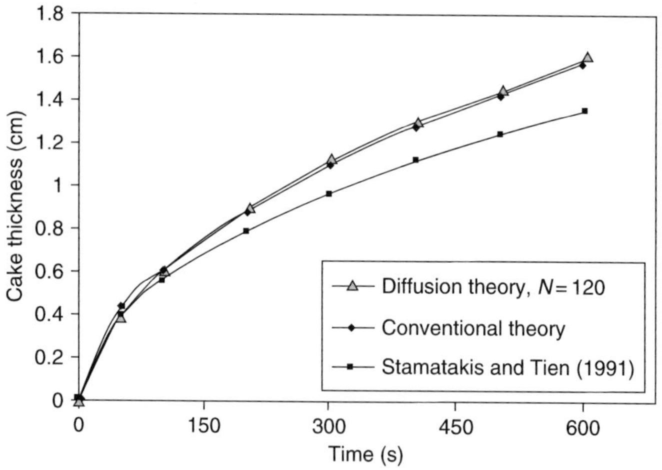

Figure 4.9 Cake thickness vs. time, filtration of Hong Kong Pink Clay suspension under conditions listed in Table 4.1. (Ramarao, Tien and Satyadev, 2002. Reprinted by permission of Taylor & Francis.)

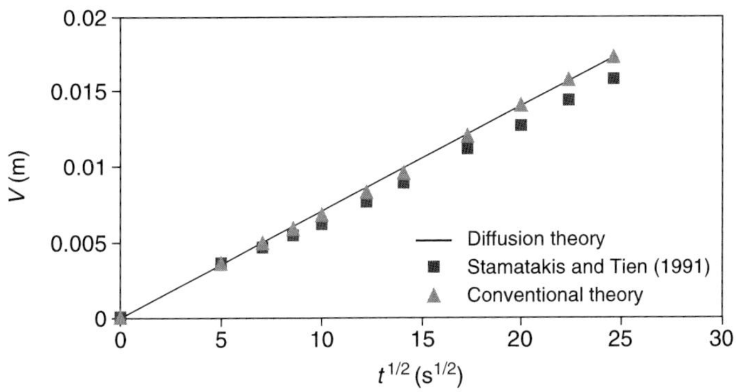

Figure 4.10 Cumulative filtration volume vs. time. Initial period of constant pressure filtration of Hong Kong Red Clay suspension under Conditions listed in Table 4.1. (Ramarao, Tien and Satyadev, 2002. Reprinted by permission of Taylor & Francis.)

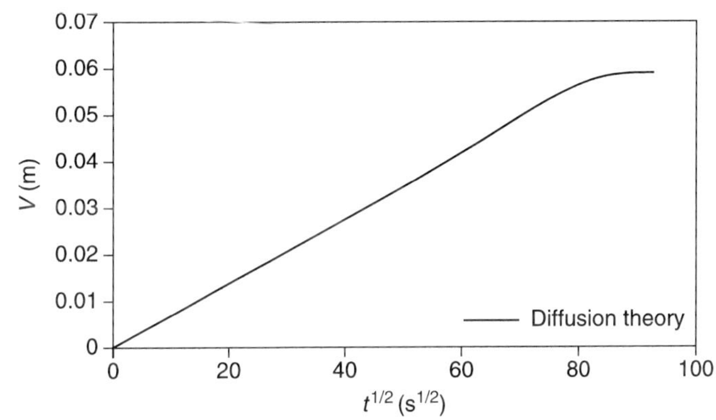

Figure 4.11 Cumulative filtrate volume vs. time. Constant pressure filtration of Hong Kong Red Clay suspension under conditions listed in Table 4.1. (Ramarao, Tien and Satyadev, 2002. Reprinted by permission of Taylor & Francis.)

4.4 EXPRESSION OF SOLID/LIQUID MIXTURES

A brief discussion of expression operation is given in Section 2.5.2. As stated there, Shirato and coworkers (1967, 1970a,b, 1987) defined *expression* as the removal of water from solid/liquid mixtures through the application of mechanical force. In the more general case of mixtures with dilute solid concentrations, *expression* operation consists of two phases: *filtration* followed by *consolidation*. If the solid concentration of the mixture to be treated is sufficiently high (or the mixture is a semi-solid, using Shirato's words), the expression consists only of consolidation. A brief account of analyzing the filtration phase of expression by the conventional cake filtration theory is given in

Chapter 2 (see Section 2.5.2). Presently, application of the diffusion equation for the description of the consolidation phase is discussed.

For expression of a finite volume of solid/liquid mixture (L_o, m^3/m^2) with a particle mass fraction, s, such that

$$\frac{s}{s+(1-s)(\rho_s/\rho)} < \varepsilon_s^o \tag{4.4.1}$$

where ρ_s and ρ are the particle and liquid densities, respectively, and ε_s^o is the cake solidosity at the zero stress state, the process proceeds in two stages, filtration followed by consolidation. At the end of the filtration phase, the volume of the mixture is L_1, the relationship between L_o and L_1 is given by Equation (2.5.2.4) or

$$L_0 - L_1 = w_0 \frac{1-\overline{m}s}{s\rho} \tag{4.4.2}$$

where $\overline{m}$ is the average wet to dry cake mass ratio upon completion of the filtration phase. w_o is the total mass of particles present in the mixture (kg/m^2) and is given as

$$w_0 = (L_0)(\varepsilon_{s_0})\rho_s = M\rho_s \tag{4.4.3}$$

where ε_{s_0} is the particle volume fraction of the mixture initially.

As the amount of particle present is constant during expression, one has

$$w_0 = (L)(\varepsilon_s)_{av}\, \rho_s$$

or

$$L = \frac{w_0}{(\varepsilon_s)_{av}\, \rho_s} = \frac{w_0(1+e_{av})}{\rho_s} \tag{4.4.4}$$

with $(\varepsilon_s)_{av}$ and e_{av} being the average solidosity and the average void ratio at any instant. The extent of consolidation u_c may be defined as

$$u_c = \frac{L_1 - L}{L_1 - L_\infty} = \frac{(e_{av})_{t=0} - (e_{av})}{(e_{av})_{t=0} - (e_\infty)_\infty} \tag{4.4.5}$$

where L_∞ is the limiting value of the mixture volume, or the volume achieved as $t \to \infty$ and e_∞, the corresponding e value. If the initial solid concentration is sufficiently high (or Equation (4.4.1) is not obeyed), the extent of consolidation u_c then becomes

$$u_c = \frac{L_0 - L}{L_0 - L_\infty} = \frac{e_0 - e_{av}}{e_0 - (e_{av})_\infty} \tag{4.4.6}$$

where e_o is given as $(1+\varepsilon_{s_0})^{-1}$.

The value of e_{av} can be evaluated if the void ratio profile throughout the mixture is known. In turn, the profiles are part of the solution of the filtration diffusion equation [i.e. Equation (4.1.11)] corresponding to appropriate initial and boundary condition. Two such specific solutions will be presented below.

4.4.1 Consolidation of solid/liquid mixture under constant pressure, p_o

Equation (4.1.11), with simplification, will be used as the starting point. If an average value of $D(e)$ can be used as an approximation, one has

$$\frac{\partial e}{\partial t} = D(e)\frac{\partial^2 e}{\partial m^2} \qquad 0 < m < M \tag{4.4.1.1}$$

with M being w_o/ρ_s.

Assuming that the particle concentration is uniform throughout the mixture initially, the initial condition is given as

$$e = e_o \quad \text{for} \quad 0 < m < w_o/\rho_s \quad \text{for} \quad t \le 0 \tag{4.4.1.2}$$

The boundary conditions are

$$\frac{\partial e}{\partial m} = 0 \quad \text{at} \quad m = M = w_o/\rho_s \tag{4.4.1.3a}$$

$$e = e(p_o) \quad \text{at} \quad m = 0 \tag{4.4.1.3b}$$

The boundary condition of Equation (4.4.1.3a) states that at the top of the mixture (see case 3, Fig. 4.1) where the mechanical force is applied, both the liquid and particles move down at the same velocity and there is no relative motion between them. Equation (4.4.1.3b) is based on the assumption that the medium resistance is negligible. With $f' = -1$ (see Equation 4.1.13), the cake compressive stress at the medium surface is p_o, the applied pressure at $m = w_o/\rho_s$. With p_o = constant, the solution of Equation (4.4.1.1) with the initial and boundary conditions of Equations (4.4.1.2), (4.4.1.3a) and (4.4.1.3b) is given as (Carlaw and Jaeger, 1959)

$$\frac{e - e(p_0)}{e_0 - e(p_0)} = \frac{4}{\pi}\sum_{n=l}^{\infty}\frac{1}{(2n+1)}\exp\left[-D\frac{(2n+1)^2\pi^2}{M^2}t\right]\sin\frac{(2n+1)\pi m}{M} \tag{4.4.1.4}$$

where $e(p_o)$ is the ultimate value of e of the mixture. Using the same notation introduced previously, $e(p_o) = e_\infty$.

The average e, or e_{av}, by definition is

$$e_{av} = \frac{\int_0^{w_0} e\, dm}{M} \tag{4.4.1.5}$$

Substituting Equation (4.4.1.4) into (4.4.1.5), one has

$$\frac{e_{av} - e_\infty}{e_0 - e_\infty} = \frac{8}{\pi^2}\sum_{n=l}^{\infty}\frac{1}{(2n+1)}\exp\left[-D\frac{(2n+1)^2\pi^2}{M^2}t\right] \tag{4.4.1.6}$$

The extent of consolidation as a function of time can be found by substituting Equation (4.4.1.6) into (4.4.6), or

$$u_c = \frac{e_0 - e_{av}}{e_0 - e_\infty} = \left[1 - \frac{e_{av} - (e_{av})_\infty}{e_0 - (e_{av})_\infty}\right]$$
$$= 1 - \sum_{n=l}^{\infty} \frac{8}{(2n+1)^2\pi^2} \exp\left[-D\frac{(2n+1)^2\pi^2}{M^2}t\right] \quad (4.4.1.7)$$

4.4.2 Expression of dilute suspension of finite volume

For a more general case of expression, one may begin with a dilute suspension of particles such that Equation (4.4.1) is obeyed. Expressions take place under an operating pressure which may vary with time in some specified manner. For example, variation of p_o may go on during filtration and continue into the consolidation phase until the pressure reaches a maximum value p_{max}, which is then maintained until the consolidation is completed.

The filtration phase can be easily analyzed by the conventional cake filtration theory (see Section 2.2). The consolidation phase is described by Equation (4.1.11). The initial and boundary conditions are

$$e = e(m, 0) = f(m) \quad t \leq 0 \quad (4.4.2.1)$$

where $f(m)$ is the e-profile across the cake at the end of the filtration phase (or $t = 0$), $f(m)$ can be obtained from the corresponding ε_s vs. x profile by noting that $\varepsilon_s = (1+e)^{-1}$ and from the relationship between m and x given by Equation (4.1.9). The corresponding solidosity profile can be obtained from either the conventional theory (see Section 2.3) or the solution of the appropriate continuity equations (see Chapter 3).

The boundary conditions are:

$$\frac{\partial e}{\partial m} = 0 \quad \text{at} \quad m = M = w_o/\rho_s \quad (4.4.2.2a)$$

$$e = e(p_o) \quad \text{at} \quad m = 0, \quad p_o = p_o(t) \quad t < t_2 \quad (4.4.2.2b)$$
$$= e(p_{max}) = e_\infty \quad \text{at} \quad m = 0, \quad p_o = p_{max} \quad t > t_2$$

The rationale of using Equation (4.4.2.2a) as the boundary condition at $m = M$ is similar to that of Equation (4.4.1.3a). The same assumption of negligible medium resistance is used for formulating Equation (4.4.2.2b). Unlike the problem considered in Section 4.4, p_o varies with time until it reaches a maximum value (p_{max}) (or $t = t_2$) and is kept constant afterwards. The time variable begins with the commencement of the consolidation phase.

The solution of Equation (4.1.1.1) together with Equations (4.4.2.2a) and (4.4.2.2b) is (Carslaw and Jaeger, 1959)

$$e - e_\infty = 2 - \sum_{1}^{\infty} \exp(S_n t) \sin \frac{(2n-1)\ \pi m}{2M} \left[\int_0^{2M} E(m,\ 0) \sin \frac{(2n-1)\pi m}{2M} \frac{\mathrm{d}m}{2M} + \frac{2S_n}{(2n-1)\pi} \int_0^{t} E(0, t'), \exp(S_n t')\, \mathrm{d}t' \right] \quad t < t_2 \tag{4.4.2.3a}$$

$$e - e_\infty = 2 \sum_{n=1}^{\infty} \exp(-S_n t) \sin \frac{(2n-1)\pi}{2M} m \left[\int_0^{2M} E(m,\ 0) \sin \frac{(2n-1)\pi m}{2M} \mathrm{d}\left(\frac{m}{2M}\right) + \frac{2S_n}{(2n-1)\pi} \int_0^{t_2} E(0,\ t') \exp\ (S_n t') \mathrm{d}t' \right] \quad t > t_2 \tag{4.4.2.3b}$$

with

$$S_n = D\left[\frac{(2n-1)\pi}{2M}\right]^2 \tag{4.4.2.4}$$

$$E(m,\ 0) = f(m) - e_\infty \tag{4.4.2.5a}$$

and

$$E(m,\ t') = e(p_0) - e_\infty \tag{4.4.2.5b}$$

The average void ratio, from Equations (4.4.1.5) and (4.4.2.3a) [or (4.4.2.3b)], is found to be

$$\begin{aligned} e_{av} - e_\infty &= 2 \sum_{1}^{\infty} \exp(-S_n t) \left[\int_0^{2M} E(m,\ 0) \sin \frac{(2n-1)\pi m}{2M} \mathrm{d}\left(\frac{m}{2M}\right) + \frac{2S_n}{(2n-1)\pi} \int_0^{t} E(0,\ t') \exp(S_n t') \mathrm{d}t \right] \int_0^1 \sin (2n-1)\pi \left(\frac{m}{2M}\right) \mathrm{d}\left(\frac{m}{2M}\right) \\ &= \frac{4}{\pi} \sum_{n=1}^{\infty} \frac{\exp(-S_n t)}{(2n-1)} \left[\int_0^1 E(m,\ 0) \sin((2n-1)\pi m^*)\, \mathrm{d}m^* + \frac{2S_n}{(2n-1)\pi} \int_0^{t} E(0,\ t) \exp(S_n t') \mathrm{d}t' \right] \quad t < t_2 \end{aligned} \tag{4.4.2.6a}$$

and

$$e_{av} - e_{\infty} = \frac{4}{\pi}\sum_{n=1}^{\infty}\frac{\exp(-S_n t)}{(2n-1)}\left[\int_0^1 E(m,0)\sin[(2n-1)\pi m^2]\mathrm{d}m' + \frac{2S_n}{(2n-1)\pi}\int_0^{t_2} E(0,t')\exp(S_n t')\mathrm{d}t'\right] \quad t > t_2 \tag{4.4.2.6b}$$

From Equation (4.4.2.6a), one has

$$(e_{av})_{t=0} - (e_{\infty}) = \frac{4}{\pi}\sum_{n=1}^{\infty}\frac{1}{(2n-1)}\int_0^1 E(m,\ 0)\sin((2n-1)\pi m^*)\mathrm{d}m^* \tag{4.4.2.7}$$

According to Equation (4.4.5), the extent of consolidation, u_c, is

$$u_c = \frac{(e_{av})_{t=0} - (e_{av})}{(e_{av})_{t=0} - e_{\infty}} = 1 - \frac{(e_{av}) - e_{\infty}}{(e_{av})_{t=0} - e_{\infty}}$$

$$= 1 - \frac{\sum_{n=1}^{\infty}\frac{\exp(-S_n t)}{(2n-1)}\left[\int_0^1 E(m,0)\sin((2n-1)\pi m^*)\mathrm{d}m^* + \frac{2S_n}{(2n-1)\pi}\int_0^t E(0,t')\exp(S_n t')\mathrm{d}t'\right]}{\sum_{n=1}^{\infty}\frac{1}{2n-1}\int_0^1 E(m,0)\sin((2n-1)\pi m^*)\mathrm{d}m^*}$$

$$\text{for } t < t_2 \tag{4.4.2.8a}$$

$$u_c = 1 - \frac{\sum_{n=1}^{\infty}\frac{\exp(-S_n t)}{(2n-1)}\left[\int_0^1 E(m,0)\sin\{(2n-1)\pi m^*\}\mathrm{d}m^* + \frac{2S_n}{(2n-1)\pi}\int_0^{t_2} E(0,t')\exp(S_n t')\mathrm{d}t'\right]}{\sum_{n=1}^{\infty}\frac{1}{2n-1}\int_0^1 E(m,0)\sin\{(2n-1)\pi m^*\}\mathrm{d}m^*}$$

$$\text{for } t > t_2 \tag{4.4.2.8b}$$

The results given above [i.e. Equations (4.4.1.7), (4.4.2.8a) and (4.4.2.8b)], in principle, can be used for predicting expression performance. More likely, they can be applied for data interpretation and to provide a rational basis of scale-up. These applications are, of course, subject to the inherent limitation of using the diffusion equations in describing consolidation; namely, the system studied must not display time-dependent rheological behavior. Shirato *et al.* (1970a,b) presented consolidation experimental data of a number of systems and compared them with the appropriate equations. As an example, in Fig. 4.12, consolidation data of Hara Gairome Clay/Solka Floc mixtures are shown in the form of u_c vs. $\sqrt{t}/w_0$. Theoretical predictions refer to those from Equation (4.4.1.7) with $D(e)$ obtained from C-P measurements, namely, $e = 2.411 - 0.308p_s$, p_s is $\mathrm{lb_f\,in^{-2}}$ and $\alpha = (k\ \rho_s\ \varepsilon_s)^{-1} = 1.283 \times 10^{12} + 2.021 \times 10^{12}\ p_s^{0.772}$, α in $\mathrm{ft\,lb_m}^{-1}$, p_s in $\mathrm{lb_f\,in^{-2}}$ and the effective p_s taken being the average compressive stress. The agreement was indeed rather good (Shirato *et al.*, 1970b).

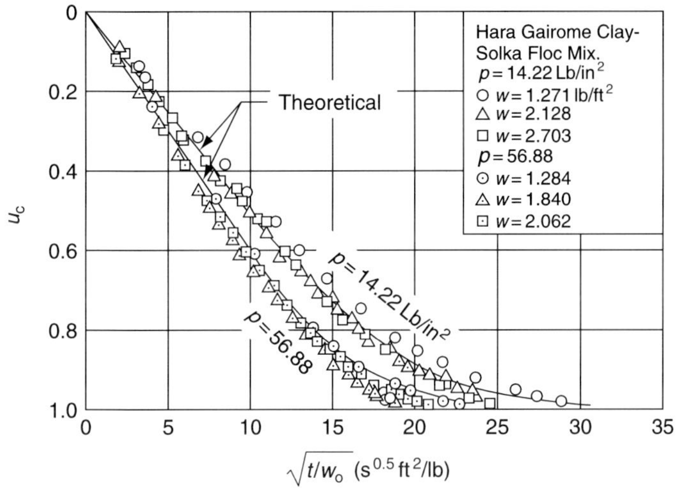

Figure 4.12 Extent of consolidation vs. time. (Shirato, Murase, Kato and Fukaya, 1970. Reprinted by permission of Elsevier.)

REFERENCES

Carslaw, H.S. and Jaeger, J.C., *Conduction of Heat in Solids*, second Edition, Oxford Press (1959).

Matano, C., *Jap. J. Phys.*, **8**, 109 (1932).

Philip, J.R., *Trans. Faraday Soc.*, **6**, 885 (1955).

Ramarao, B.V., Tien, C. and Satyadev, C.N., "Determination of the Constitutive Relationship for Filter Cakes in Cake Filtration Using the Analogy between Filtration and Diffusion", in *Transport Processes in Bubble, Drops and Particles*, second Edition, D. DeKee and R.P. Chhabra (eds), Taylor and Francis, New York (2002).

Shirato, M., Murase, T., Kato, H. and Fukaya, S., *Kagaka Kogaka*, **31**, 1125 (1967).

Shirato, M., Murase, T., Negawa, M. and Senda, T., *J. Chem. Eng. Japan*, **3**, 105 (1970a).

Shirato, M., Murase, T., Kato, H. and Fakaya, S., *Filtration and Separation*, 277 (1970b).

Shirato, M., Murase, T., Iritani, E., Tiller, F.M. and Alicicature, A.F., "Filtraiton in the Chemical Industry", in *Filtration and Principles and Practices*, M.J. Matteson and C. Orr (eds), Marcel and Dekker, New York (1987).

Smiles, D.E., *Chem. Eng. Sci.*, **25**, 485 (1970).

Smiles, D.E., "Principles of Constant Pressure Filtration", in *Encyclopedia of Fluid Mechanics*, Vol. 5, N.P. Cheremisinoff (ed.), Gulf Publishing, Houston, Tx, (1986).

Smiles, D.E., Raats, P.A.C. and Knight, J.H., *Chem. Eng. Sci.*, **37**, 707 (1982).

Stamatakis, K. and Tien, C., *Chem. Eng. Sci*, **46**, 1917 (1991).

Teoh, S.-K., "Studies in Filter Cake Characterization and Modelling", Ph.D. Thesis, National University of Singapore (2003).

– 5 –

SIMULATION OF CAKE FORMATION AND GROWTH

Notation

A	cross-sectional area of a particle segment [see Equation (5.4.1.2)] (m^2)
$A(t)$	Brownian diffusion force (N)
a_p	particle radius (m)
C_D	drag coefficient (-)
d_p	particle diameter (m)
e	charge of an electron ( )
F	time constant (s^{-1})
$(\mathbf{F}_D)$	drag force vector (N)
$(\mathbf{F}_D)_i$	drag force experienced by the i-th particle [see Equation (5.4.3.2)] (N)
$(\mathbf{F}_{DL})_i$	double layer force vector experienced by the i-th particle [see Equation (5.4.3.3)] (N)
$(\mathbf{F}_{DL})_{ij}$	double layer force between the i-th and the j-th particles
$\mathbf{F}_{DM}$	Brownian diffusion force vector (N)
$\mathbf{F}_e$	external force vector (N)
$\mathbf{F}_G$	gravitational force vector (N)
F_p	force acting along the tangential direction (N)
F_q	force acting along the normal direction (N)
F_t	total force acting on a particle (N)
$\mathbf{F}_{Lo}$	London-van der Waals force between a particle and medium surface (N)
$(\mathbf{F}_{Lo})_i$	London-van der Waals force vector experienced by the i-th particle [see Equation (5.2.11)] (N)
$(\mathbf{F}_{Lo})_{ij}$	London-van der Waals force vector of the i-j pair [see Equation (5.2.8)] (N)
f	friction coefficient (-)
$\overline{f}$	correction factor for $\mathbf{F}_D$ [see Equation (5.2.5)] (-)
f_1	compressive force (N)
f_2	cohesive force of particle (N)
$(\mathbf{f}_D)_i$	defined by Equation (5.4.3.8) (N/kg)
$(\mathbf{f}_{DL})_i$	defined by Equation (5.4.3.9) (N/kg)
$(\mathbf{f}_{Lo})_i$	defined by Equation (5.4.3.10) (N/kg)

$G(n_a)$	particle height distribution function [(see Equation (5.4.3.12)] (-)
g	gravitational acceleration vector (m/s^2)
H	Hamaker's constant (J)
h	protrusion height (m)
k	cake permeability (m^2), or the Boltzman constant (J/T)
k_1	constant of the Kozeny-Carman Equation [see Equation (5.4.1.3)] (-)
L	length of the entering line (see Fig. 5.4) (m)
M	total number of segments
m_j	concentration of the j-th ion species (number of ions/m^3)
m_i	normally distributed number (-)
m_p	particle mass (kg)
N	total number of particles (-), or number of segments (-)
N_i	a random number (-)
N_{Re}	Reynolds number defined as $d_p u_\infty \rho/\mu$ (-)
$\mathbf{n}$	unit normal vector or an arbitrary unit vector (-)
$\mathbf{n}_3$	unit vector along the z-direction (-)
n_a	number of particles with height equal to or less than z (-)
n_i	a normally distributed number or the number of particles present in the i-th segment (-)
n_o	electrolyte concentration (number of ions/m^3)
p_ℓ	fluid pressure (Pa)
p_s	compressive stress (Pa)
$(q_\ell)_i$	filtrate flow rate of the i-th segment (m/s)
q_{ℓ_m}	filtration velocity (m/s)
$\overline{q}$	a quantity defined by Equation (5.2.19a) (J/kg)
$\mathbf{R}_v$	velocity increment due to the Brownian motion (m)
$\mathbf{R}_z$	displacement increment due to Brownian motion (m)
$\mathbf{r}_i$	position vector of the i-th particle (m)
$\mathbf{r}_i^*$	defined as $\mathbf{r}_i/d_p$ (-)
S	cross-sectional area (m^2)
T	absolute temperature (K)
t	time (s)
$\mathbf{u}$	fluid velocity vector (m/s)
$\mathbf{u}_p$	particle velocity vector (m/s)
w_o	areal mass density of cake (kg/m^2)
x, y, z	coordinates (m)
x_1, y_1, z_1	dimensions of the simulation domain (m)
z	separation distance (m)
z_j	valance of the j-th ion species (-)
z_{ij}	separation distance between the i-th and the j-th particles (m)
z^*	charge per indifferent electrolyte ion or dimensionless separation distance (-)

Greek letters

α	specific cake resistance (m kg^{-1})
α_{sp}	retardation factor (-)
$\overline{\alpha}$	fraction of the force F_t sustained by a contacting particle (-)
β	friction coefficient [see Equations (5.2.19a) and (5.2.19b)] (s^{-1})
γ	surface charge constant given by Equation (5.2.13) (-)
$(\Delta p_{\ell})_i$	pressure drop across the *i*-th segment (Pa)
(Δt)	time increment (s)
δ	angle between the force F_t and the line connecting two contacting particles (see Fig. 5.3a) (-)
δ_{ij}	angle formed between the line connecting the *i-j* pair of particles and the force F_t (-)
$\overline{\delta}$	angle between the vertical axis and the line connecting two contacting particles
ε	cake porosity (-)
ε_s	cake solidosity (-)
ε_{s_o}	particle volume fraction of suspension (-)
$\hat{\varepsilon}$	dielectric constant
κ	reciprocal of the Debye layer thickness (m^{-1})
μ	fluid viscosity (Pa s)
ρ	fluid density (kg/m^2)
ρ_s	particle density (kg/m^3)
σ_{v_i}	quantity given by Equation (5.2.18a)
σ_{z_i}	quantity given by Equation (5.2.18b)
σ_{vz_i}	quantity given by Equation (5.2.18a)
ψ	surface potential (V)
ψ_c	medium surface potential (V)
ψ_p	particle surface potential (V)

The explosive growth of computing power of the past three decades has made it practical to conduct simulations of physical phenomena and engineering systems of sufficient complexities. Such simulation studies, which may be viewed as computer experiments, attempt to replicate what happens in reality based on well-developed principles and relevant information. Simulation studies are therefore potentially capable of providing a variety of useful information, both on macroscopic and microscopic scales, of the problems to be investigated.

A number of simulation studies of cake filtration have appeared in the literature in recent years (Lu and Hwang, 1993, 1995; Hoflinger *et al.*, 1994; Fu, 1996; Schmidt, 1996; Fu and Dempsey, 1998; Jeon and Jung, 2004). While these studies may only represent some initial efforts of examining cake filtration by the simulation methodology, they have clearly demonstrated the potential and capability of the simulation approach. It is conceivable that through simulations, information such as cake structure and morphology and the effect of solution variables (pH, surface potential of particles, etc.) on cake structure can be obtained. A brief account of the principle of cake filtration simulation will be presented in the following sections followed by a summary of the results and discussions of some of these studies.

5.1 GENERAL CONSIDERATIONS

On the most basic level, cake filtration involves the motions of fluids and particles and particle collisions as they move toward the filter medium. A complete understanding of cake filtration therefore requires the knowledge of the flow of filtrate in the void space of the cake region, the movement of the individual particles, their contact with the medium and each other, and the aftermath once a contact (collision) is made. For the last issue, the stability criteria, which take into account the geometry, the particle material characteristics and the various forces acting on particles must be devised.

Detailed information of filtrate flow in the void space may be obtained from the solution of the Navier-Stokes equation (assuming that the filtrate is Newtonian). However, the efforts required may be excessive since the boundaries of the void space are complex and undergo changes during the growth and compaction of filter cakes. For all the simulation studies of cake filtration conducted so far, filtrate flow was assumed to follow Darcy's law and neglecting cake particle movement, or

$$\mathbf{u} = -\frac{k}{\mu}\nabla p_\ell \tag{5.1.1}$$

where $\mathbf{u}$ is the filtrate velocity vector, ∇p_ℓ is the fluid pressure gradient, and μ, the viscosity.

The cake permeability may be assumed to be a function of the local porosity (or solidosity). A number of permeability expressions are listed in Table 5.1. Figure 5.1 shows a comparison of these expressions over the range of $0.01 < \varepsilon_s < 0.5$. On account of the large differences among these expressions, shown in Fig. 5.1, selecting an appropriate expression of k is often a practical problem one faces in applying Darcy's law in cake simulation.

The motion of a particle (or particle trajectory) can be obtained from the trajectory equation based on Newton's law. If the virtual mass force, the Bassett force and the pressure gradient force are ignored, the trajectory equation is given as

$$\frac{4\pi}{3}a_p^3\,\rho_s\frac{D}{Dt}\mathbf{u}_p = \frac{4\pi}{3}a_p^3\,\rho_s F(\mathbf{u}-\mathbf{u}_p) + \mathbf{F}_e + \frac{4}{3}\pi a_p^3 A(t) \tag{5.1.2}$$

where $\mathbf{u}$ and $\mathbf{u}_p$ are the filtrate and particle velocity vectors, ρ_s is the particle density, a_p is the particle radius, F is the time constant, $\mathbf{F}_e$ is the external force vector which, for the present discussion, includes the gravitational force and the particle–particle interaction forces and $A(t)$ is the Brownian diffusion force acting on a particle (per unit mass). The

Table 5.1

Expression of cake permeability[a]

Expression	Investigator(s)
$\frac{a_p^2(1-\varepsilon_s)^3}{k_1\varepsilon_s^2}$ $k_1 = 45$ (Kozeny) $= 180$ (Carman)	Kozeny (1927), Carman (1937)
$k_{DL}\left[1+\frac{3}{4}\varepsilon_s(1-\sqrt{(8/\varepsilon_s)-3})\right]$	Brinkman (1947)
$\frac{a_p^2}{16\varepsilon_s^{1.5}[1+56\varepsilon_s^3]}$	Davis (1952)
$k_{DL}\frac{6-9\varepsilon_s^{1/3}+9\varepsilon_s^{5/3}-6\varepsilon_s^2}{6+4\varepsilon_s^{5/3}}$	Happel (1958)
$\frac{k_{DL}}{\zeta(\alpha, \alpha/\varepsilon_s^{1/3})}$ $\alpha = \frac{\alpha_p}{\sqrt{k}}$	Neals and Nader (1974)
$k_{DL}\left[1+\frac{3}{\sqrt{2}}\varepsilon_s^{1/2}+\frac{135}{64}\varepsilon_s \ell n\varepsilon_s\right.$	Howells (1974), Hinch (1977)
$\left.+16.456\varepsilon_s+\cdots\right]$	Kim and Russel (1985)
$\frac{3a_p^2}{20\varepsilon_s}[-\ell n\varepsilon_s-0.931+\theta(\ell n\phi)^{-1}]$	Jackson and James (1986)
$k_{DL}=\frac{2a_p^2}{9\varepsilon_s}$	Happel and Brenner (1991)

[a]Reproduced from Kim and Stolzenbach (2002).

basic principle of simulation is rather straightforward. Based on the operating conditions (i.e. the applied pressure or the filtration rate) and the local particle concentration, the filtrate flow field can be determined from Equation (5.1.1). With the forces acting on particles known, particle trajectories can be determined from Equation (5.1.2). Once a particle makes contact with another particle or medium, based on the assumed stability criterion, subsequent particle placement/movement may be determined leading to the formation, growth and compaction of a cake.

Two techniques may be applied in simulating particle motion: sequential addition and collective rearrangement. For the former, particles are introduced one at a time, and each of its trajectory determined. Once a particle makes contact, stability criterion may be applied to all previously placed particles. For the latter, a large number of particles are placed in a random manner, and particle trajectories and stability criteria are then used to effect rearrangement. The collective rearrangement approach has the advantage of being

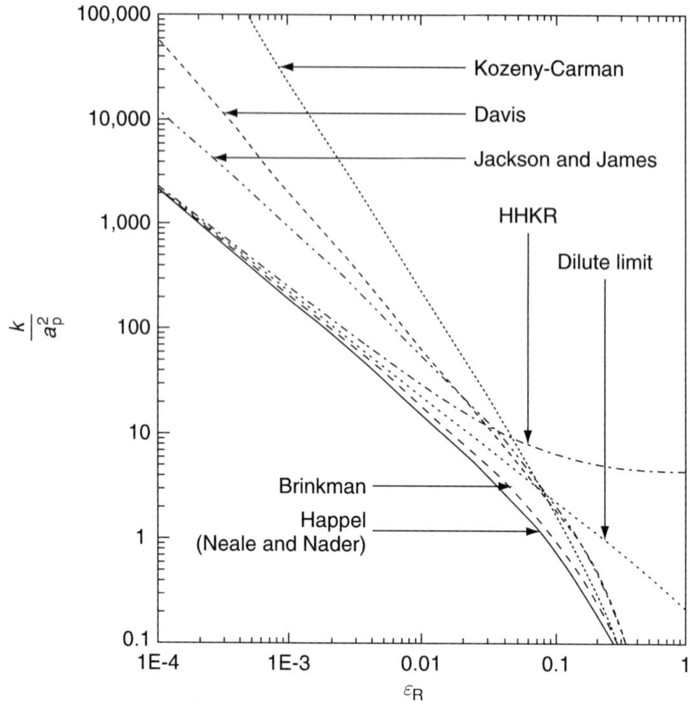

Figure 5.1 Permeability as a function of ε_s according to expressions listed in Table 5.1. (Kim and Stolzenbach, 2002. Reprinted by permission of Elsevier.)

more efficient, but the question of possible bias due to the assumed initial positions of particles must be dealt with. As a compromise, a hybrid of these two approaches may be used. Multiple additions of particles may be carried out first before a "particle by particle" check on stability is implemented.

5.2 FORCES ACTING ON PARTICLES

Various force terms of the trajectory equation [i.e. Equation (5.1.2)] are expressed as below.

(1) *Drag force imparted by filtrate flow*, $\mathbf{F}_\mathrm{D}$

$$\mathbf{F}_\mathrm{D} = \left(\frac{4}{3}\right) \pi a_\mathrm{p}^3 \rho_\mathrm{s} F(\mathbf{u} - \mathbf{u}_\mathrm{p}) \tag{5.2.1}$$

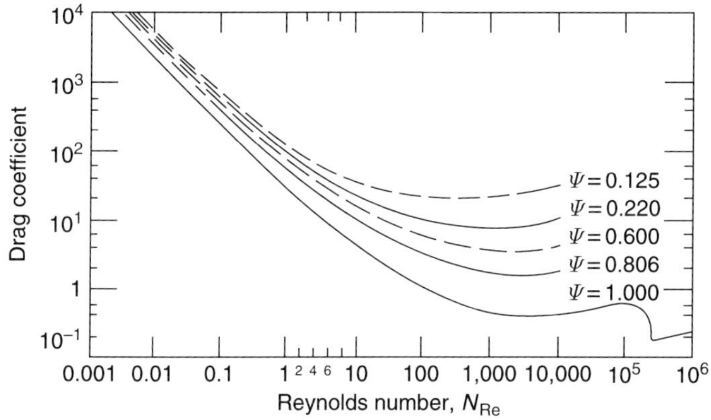

Figure 5.2 Drag coefficient as a function of N_{Re}.

The time constant is

$$F = \frac{3}{8} C_D \frac{\rho}{\rho_p} \frac{|u - u_p|}{a_p} \tag{5.2.2}$$

The drag coefficient, C_D, is a function of the particle Reynolds number, N_{Re}, defined as $d_p U_\infty \rho / \mu$ as shown in Fig. 5.2. The characteristic velocity, U_∞, in this case, is $|\mathbf{u} - \mathbf{u}_p|$. For $N_{Re} < 1.0$, C_D is given as

$$C_D = \frac{24}{N_{Re}} \tag{5.2.3}$$

Combining Equations (5.2.1), (5.2.2) and (5.2.3), one has

$$\mathbf{F}_D = 6\pi\mu a_p (\mathbf{u} - \mathbf{u}_p) \tag{5.2.4}$$

which is the Stokes law. For $N_{Re} < 1$, this expression can be used to estimate the drag force imparted by fluid to particles except in the following situations:

(a) If the particle size is comparable to the mean free path of the entraining fluid (i.e. filtrate) – for example in gas filtration under low pressure and high temperature – the Cunningham correction factor must be introduced into Equation (5.2.4). For a discussion and estimation of the Cunningham correction factors, see Davis (1945, 1973).

(b) If a particle is sufficiently close to a stationary surface (for example, filter medium), the hydrodynamic retardation effect needs to be accounted for. Discussions of the hydrodynamic retardation effect can be found in Tien (1989).

(c) If sufficient numbers of particles are present in a given region, the "hindered settling" effect must be considered. The drag force acting on the particle may be written as

$$\mathbf{F}_D = \frac{6\pi\mu a_p (\mathbf{u} - \mathbf{u}_p)}{\bar{f}} \tag{5.2.5}$$

where $(1/\overline{f})$ may be considered as the correction factor. A number of expressions of the correction factor are available in the literature. A simple expression given by Reed and Anderson (1980) is

$$\overline{f} = \frac{1 - 1.88\varepsilon_s}{1 + 5.0\varepsilon_s} \tag{5.2.6}$$

(2) *External forces*
A number of external forces may be present in cake filtration. For the discussion here, the external force will be limited to the following.
(a) *Gravitational force*, $\mathbf{F}_G$. Including the buoyancy force effect, $\mathbf{F}_G$ is given as

$$\mathbf{F}_G = \frac{4}{3}\pi a_p^3(\rho_p - \rho_s)\mathbf{g} \tag{5.2.7}$$

where $\mathbf{g}$ is the gravitational acceleration vector.
(b) *Particle–particle interaction forces.* Two types of forces, the London-van der Waals force (attractive) and the double layer force, are often considered in cake simulation studies.
(i) *London-van der Waals force.* This force arises from the instantaneous dipole moments by the temporary asymmetrical distribution of electrons around atomic nuclei. For particles of radii $(a_p)_i$ and $(a_p)_j$, the London-van der Waals force between the two particles is given as

$$(\mathbf{F}_{Lo})_{ij} = -\frac{H}{6z_{ij}^2}\frac{(a_p)_i(a_p)_j}{(a_p)_i + (a_p)_j}\frac{\mathbf{r}_i - \mathbf{r}_j}{|\mathbf{r}_i - \mathbf{r}_j|} \tag{5.2.8}$$

where H is the Hamaker constant, z_{ij} is the separation distance between the i-j particle pair, and $\mathbf{r}_i$ and $\mathbf{r}_j$ are the position vectors of the i-th and j-th particles, respectively.

If both particles are of the same size, $(a_p)_i = (a_p)_j = a_p$, the above expression reduces to

$$(\mathbf{F}_{Lo})_{ij} = -\frac{Ha_p}{12z_{ij}^2}\frac{\mathbf{r}_i - \mathbf{r}_j}{|\mathbf{r}_i - \mathbf{r}_j|} \tag{5.2.9}$$

For the London-van der Waals force between a particle and the filter medium, if the radius of curvature of the medium surface is large, then the surface can be considered as a sphere with radius of infinity. Equation (5.2.8) becomes

$$\mathbf{F}_{Lo} = -\frac{Ha_p}{6z^2}\mathbf{n} \tag{5.2.10}$$

where **n** is the unit normal vector of the medium surface and z is the separation distance between the particle and the medium.

All these three expressions (Equations 5.2.8, 5.2.9 and 5.2.10) are for the so-called unretarded London-van der Waals force and may give overestimations. To account for this overestimation, a correction factor α_{sp}, called the retardation factor, may be introduced into the expression. Approximate expressions for estimating α_{sp} has been developed by Payatakes (1973). It can also be found in the monograph by Tien (1989).

The expressions given above refer to the forces between two entities. To account for the presence of other particles, the contributions from all pairs of interactions should be included; or

$$(\mathbf{F}_{Lo})_i = -\sum_j (\mathbf{F}_{Lo})_{ij} \tag{5.2.11}$$

That is, the total London-van der Waals force experienced by the i-th particle is the sum of all interactions between the i-th particle and its neighboring particles (or medium) given by any one of the three expressions.[1] The number of pairs included vary with the size of the neighborhood taken. However, since the London-van der Waals force declines rapidly with the separation distance, it suffices to use an enclosure with a dimension of one order of magnitude greater than that of the particle.

It is also obvious that the London-van der Waals force becomes undefined as $z \to 0$. As all the forces considered in simulation need to be finite, the separation distance of two contacting bodies is often assumed to be 4 Å according to common practice (Hinds, 1999).

(ii) *Double layer force*. When a surface is placed in an aqueous solution, because of adsorption and/or ionization, the surface may acquire charges. The surface charge is balanced by countercharged ions present in the aqueous solution. Thus, a double layer of charge is established characterized by an electrical potential between the outer portion of the double layer and the bulk of the solution. When two surfaces are placed next to each other, the interaction of the two double layers gives rise to a force which may be either repulsive (if the surface charges are of the same sign) or attractive (if the surface charges are opposite in sign). For two spheres of radii $(a_p)_i$ and $(a_p)_j$ with the same surface potential, the repulsive force due to Coulombic repulsion is

$$(\mathbf{F}_{DL})_{ij} = \frac{64 n_o k T \gamma^2}{\kappa} \frac{2\pi (a_p)_i (a_p)_j}{(a_p)_i + (a_p)_j} \exp(-\kappa z_{ij}) \frac{\mathbf{r}_i - \mathbf{r}_j}{|\mathbf{r}_i - \mathbf{r}_j|} \tag{5.2.12}$$

[1] The inclusion of the particle–medium interaction is only necessary for those particles in close proximity of the filter medium.

where z_{ij} is the distance of the i-j pair, n_o is the electrolyte concentration (in number of ions/m^3). γ is the surface charge constant given by following equation

$$\gamma = \tan\mathrm{h}(z^* \psi/kT) \tag{5.2.13}$$

with z^* being the charge per indifferent electrolyte ion, ψ, the surface potential, k, the Boltzmann constant and T, the absolute temperature. κ in Equation (5.2.12) is the Debye-Huckel reciprocal double layer thickness defined as

$$\kappa = \frac{4\pi e^2}{\hat{\varepsilon} kT} \sum_j z_j^2 m_j \tag{5.2.14}$$

where e is the charge of an electron, m_j is the concentration of the j-th ion species present in the solution with valence z_j.

The double layer force between a particle and the medium may be approximated by that of a sphere and a plate. The expression by Hogg *et al.* (1966) with $|\psi| < 60\,\mathrm{mV}$ and $\kappa a_p > 10$ is

$$\mathbf{F}_{\mathrm{DL}} = \frac{\hat{\varepsilon} a_p (\psi_p^2 + \psi_c^2) \kappa e^{-\kappa z}}{2[1 - \exp(-2\kappa z)]} \left[2\frac{\psi_p \psi_c}{\psi_p^2 + \psi_c^2} - \exp(-\kappa z) \right] \mathbf{n} \tag{5.2.15}$$

with $\hat{\varepsilon}$ being the dielectric constant of the liquid, ψ_p and ψ_c are the surface potentials of the particle and the plate (i.e. medium), respectively. z is the distance between the particle and the medium.

For an assembly of particles, the double layer force acting on the j-th particle is the sum of the forces between the i-th particle and all the other ones. This can be calculated in a manner similar to Equation (5.2.11). The inclusion of the force between the i-th particle and the medium becomes necessary if the i-th particle is not far away from the medium.

(3) *Brownian diffusion force*

For a particle placed in a fluid medium, the Brownian diffusion force it experiences results from the bombardment of fluid molecules on the particle. It is a stochastic quantity and cannot be described deterministically. In integrating the trajectory equation [i.e. Equation (5.1.2)], the contributions to the velocity and displacement increments due to the Brownian diffusion force $\mathbf{R}_v$ and $\mathbf{R}_z$ are also stochastic.[2] $\mathbf{R}_v$ and $\mathbf{R}_z$ can be shown to be two random deviates which are bivaviate Gaussian

[2] Detailed discussion can be seen from Tien (1989).

distributed (Chandrasekhar, 1943). The components of $\mathbf{R}_v$ and $\mathbf{R}_z$ can be calculated as follows:

$$\begin{bmatrix} R_{vi} \\ R_{zi} \end{bmatrix} = \begin{bmatrix} \sigma_{vi} & 0 \\ \frac{\sigma_{vz_i}}{\sigma_{v_i}} & \left(\frac{\sigma_{vz_i}^2 - \sigma_{vz_i}^2}{\sigma_{z_i}^2}\right)^{1/2} \end{bmatrix} \begin{bmatrix} n_i \\ m_i \end{bmatrix} \tag{5.2.16}$$

where n_i and m_i are two normally distributed numbers. n_i (and m_i) can be determined from the following equation

$$N_i = \frac{1}{\sqrt{2\pi}} \int_{-\infty}^{n_i} e^{-\zeta^2/2}\,\mathrm{d}\zeta \tag{5.2.17}$$

In other words, a random number N_i ($0 < N_i < 1$) is generated first. Based on the generated N_i, a corresponding n_i can be determined from the above expression. The same procedure can be used to obtain m_i.

The values of σ_{vi}, σ_{zi} and σ_{vzi} can be found from the following equations:

$$\sigma_{v_i}^2 = \frac{\overline{q}}{\beta}\left(1 - e^{-2\beta\Delta t}\right) \tag{5.2.18a}$$

$$\sigma_{z_i}^2 = \frac{\overline{q}}{\beta^3}\left(2\beta\Delta t - 3 + 4e^{-\beta\Delta t} - e^{-2\beta\Delta t}\right) \tag{5.2.18b}$$

$$\sigma_{vz_i} = \frac{\overline{q}}{\beta^2}\left(1 - e^{-\beta\Delta t}\right)^2 \tag{5.2.18c}$$

and

$$\overline{q} = \frac{\beta kT}{m_p} \tag{5.2.19a}$$

where m_p is the mass of the particle, or $(4/3)\pi a_p^3 \rho_s$ and β is the friction coefficient per unit mass. At low Reynolds numbers, β is

$$\beta = \frac{6\pi\mu a_p}{m_p} \tag{5.2.19b}$$

To simplify the inclusion of the Brownian diffusion force in determining particle trajectories and the stability criterion, an approximation of the Browning diffusion force can be made as follows. Along a given direction of an orthogonal coordinate system, the particle energy associated with the Brownian diffusion motion is $(1/2)kT$. The characteristic velocity component along the direction can be assumed to be $\pm\sqrt{kT/m_p}$ with the velocity being either positive or negative with equal probability. If the friction coefficient may be assumed to be $6\pi\mu a_p$, the Brownian diffusion force becomes

$$\mathbf{F}_{DM} = \pm 6\pi\mu a_p \sqrt{\frac{3}{2}\frac{kT}{m_p}}\,\mathbf{n} \qquad (5.2.20)^{3,4}$$

where **n** is a unit vector of arbitrary direction.

(4) *Compressive force*

Cake particles (namely, those particles constituting a filter cake), in addition to the various forces mentioned above, are also subject to the compressive stress of the cake phase p_s. The compressive stress of the cake phase can be calculated from the pressure drop across the cake (see Section 3.2). If the fluid pressure at the downstream side of the medium is zero and the relationship between p_s and p_ℓ is that of Type 1, with filter cake divided into N sections, the compression stress at the j-th section (numbered from the medium) is

$$(p_s)_j = \sum_{i=j}^{N} (\Delta p_\ell)_i \qquad (5.2.21)$$

where $(\Delta p_\ell)_j$ is the pressure drop across the j-th section. $(\Delta p_\ell)_j$ can be found from Equation (5.1.1). The calculation is straightforward if the filtration rate is specified. Otherwise, a trial and error procedure may be applied.

5.3 STABILITY CRITERIA

In simulating cake formation, growth and compaction, information about the movement of particles participating in the process is required. This information can be obtained from the solution of the particle trajectory equation [i.e. Equation (5.1.2)] and from the knowledge of the outcome when a given particle makes contact, during its motion, with the filter medium or a deposited particle. The condition under which the contact leads to immobility of the particle involved is referred to as the stability criterion. Three such criteria which have been used in previous studies are described below.

(1) *Criterion involving two particles.* A schematic representation of this case is shown in Fig. 5.3a. The lower particle is assumed to be immobile (e.g. situated at the

---

[3] Fu and Dempsey (1998), based on dimensional argument, assumed that the diffusion force is $kT/2a_p$. Based on this estimation, they concluded that the Brownian diffusion force is unimportant for particles with diameter greater than 30 nm. However, if Equation (5.2.21) is used, the Brownian diffusion force under these conditions may be two or three orders of magnitude greater than their estimated value and inclusion of the Brownian diffusion force in simulation studies could become necessary.

[4] I am grateful to R. Rajagopalan for suggesting the procedure for estimating $\mathbf{F}_{DM}$.

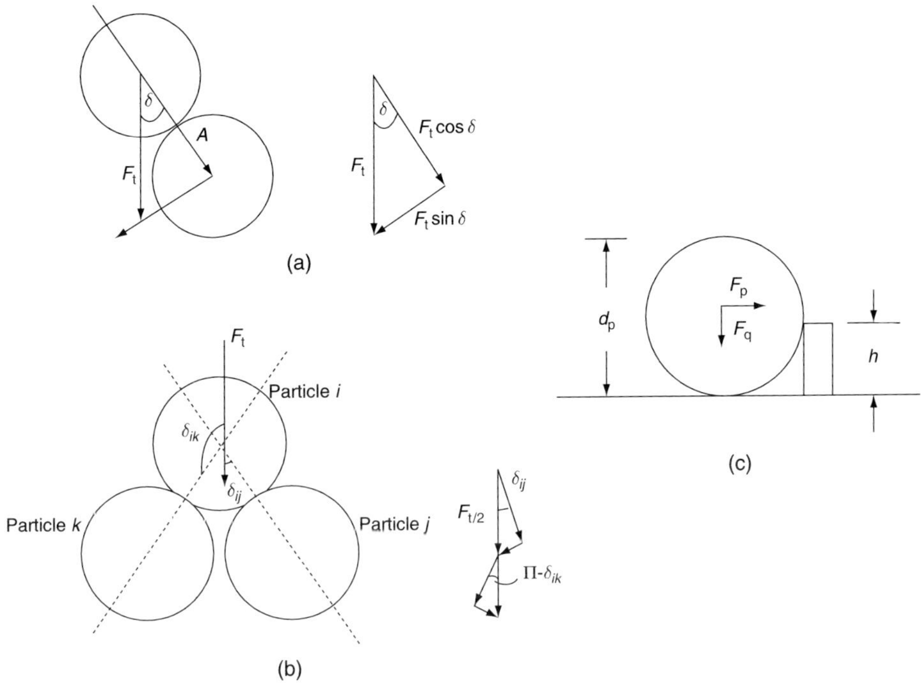

Figure 5.3 Stability condition based on (a) force (b) torque balances and (c) force and torque.

top layer of the cake). The upper particle, in its movement, makes contact with the lower one at point A. If δ is the angle between the line connecting the two contacting particles and the direction of the total force acting on the upper particle, F_t, $F_t \sin \delta$ is the force acting on the upper particle along the tangential direction (at point A) to the lower particle and $F_t \cos \theta$ is the force component along the normal direction. The condition that the two particles will remain in contact is

$$|F_t \sin \delta| \leq |(F_t \cos \delta)| f \tag{5.3.1}$$

where f is the friction coefficient.

In the event that Equation (5.3.1) is not obeyed, the upper particle may be assumed to roll off the lower particle and continue its movement until another contact is made. Equation (5.3.1) can then be applied to determine the outcome.

(2) *Criterion involving more than two particles*. It is conceivable that a moving particle may come into contact with more than one deposited particle. The case of a moving particle (the i-th particle), in its motion, making contact with two deposited particles (the j-th and k-th particle) is depicted in Fig. 5.3b. The stability criterion of Equation (5.3.1) is applied if one replaces F_t by the expression of $\overline{\alpha} F_t$ where $\overline{\alpha}$

is the fraction of F_t sustained by either the j-th or k-th article. If one follows the assumption used by Hoflinger *et al.* (1994) that the forces sustained by the j-th and the k-th particles are the same, the stability criterion becomes, for the contact between the i-th and the j-th particles

$$\left|(F_t/2)\sin\delta_{ij}\right| \leq \left|(F_t/2)\cos\delta_{ij}\right| f$$

or

$$\left|F_t \sin\delta_{ij}\right| \leq \left|F_t \cos\delta_{ij}\right| f \tag{5.3.2a}$$

Similarly, for the contact between the i-th and the k-th particles

$$|F_t \sin\delta_{ik}| \leq |F_t \cos\delta_{ik}| f \tag{5.3.2b}$$

where δ_{ij} and δ_{ik} are the angles formed between the center-to-center lines of the i-j pair of particles and F_t and that between the center-to-center lines of the i-k pair of particles of F_t.

For three-dimensional simulations, contact with more than two deposited particles is possible. If all the particles involved share equally to sustain the load of the contacting article, the expression of Equation (5.3.2a) and (5.3.2b) can be easily generalized and applied.

(3) *Criterion with cross-flow*. In conventional cake filtration, the direction of the main flow of the suspension coincides with that of the filtration flow. As an approximation, one may assume that all the suspended particles convectively transported to the medium are retained as filter cake. In contrast, in cross-flow filtration, the suspension flows along a direction normal to the flow of the filtrate. Retention of particles at membrane surface requires the consideration of the torques acting on the particle.

To account for the deposition of particles in cross-flow filtration, Stamatakis and Tien (1993) proposed a criterion based on moment balance. Referring to Fig. 5.3c which depicts the interaction between a cake surface and a particle transported to that surface over the length scale comparable to the particle size, the cake/suspension interface is not smooth and its roughness may be described in terms of the protrusions of various heights. When a particle makes contact with the interface against a protrusion of height h, the condition of attachment is given as

$$F_q \sqrt{\left(\frac{d_p}{2}\right)^2 - \left[\left(\frac{d_p}{2}\right)^2 - h\right]^2} \geq F_p \left(\frac{d_p}{2} - h\right) \tag{5.3.3}$$

where F_p and F_q are the forces acting on the particle along the tangential and normal directions and d_p is the particle diameter. The above expression is applicable for $h < \frac{d_p}{2}$. For $h > d_p/2$, deposition is assured.[5]

5.4 RESULTS OF SIMULATION STUDIES

A number of simulation studies on cake formation, growth and compaction have been made in the past. An account of these studies and their results is presented below.

5.4.1 Study by Hoflinger *et al.* (1994)

Hoflinger *et al.* presented one of the earlier investigations on the simulation of dust cake formation and compaction in gas filtration. A schematic diagram depicting the principle used is shown in Fig. 5.4. As shown in this figure, the simulation of Hoflinger *et al.* was two dimensional and the physical domain within which cake formation takes place is a rectangle represented by a number of horizontal segments of height of d_p (or "partial layers" in Hoflinger's words). A particle is considered to be present within a given segment if its center is within that segment.

Particles participating in cake formation began their movement toward the medium (i.e. the bottom lines of Fig. 5.4) from the entry line. Simulation of particle motion was made with the following assumptions:

(1) Particle movement toward the medium occurs sequentially (i.e. one particle at a time).
(2) The starting positions of particles at the entry line were randomly distributed.
(3) Particle trajectories were rectilinear.
(4) Once a particle makes contact with the medium, the particle is assumed to be deposited on the medium and its motion ceases.
(5) In the event that a particle makes contact with a previously deposited particle, the outcome of the contact is determined by an appropriate criterion.

The stability criterion used was similar to that of Equation (5.3.1) with some modifications. The forces considered include the compressive force (f_1) and particle–particle cohesive force (f_2) with the former acting along the vertical direction and the latter along the direction of the line segment connecting the centers of the two particles. The quantities present in Equation (5.3.1) are:

$$F_t \cos \delta = f_1 \cos \overline{\delta} + f_2$$
$$F_t \sin \delta = f_1 \sin \overline{\delta}$$

[5] A more detailed discussion on particle adhesion in cross-flow filtration is given in Chapter 8.

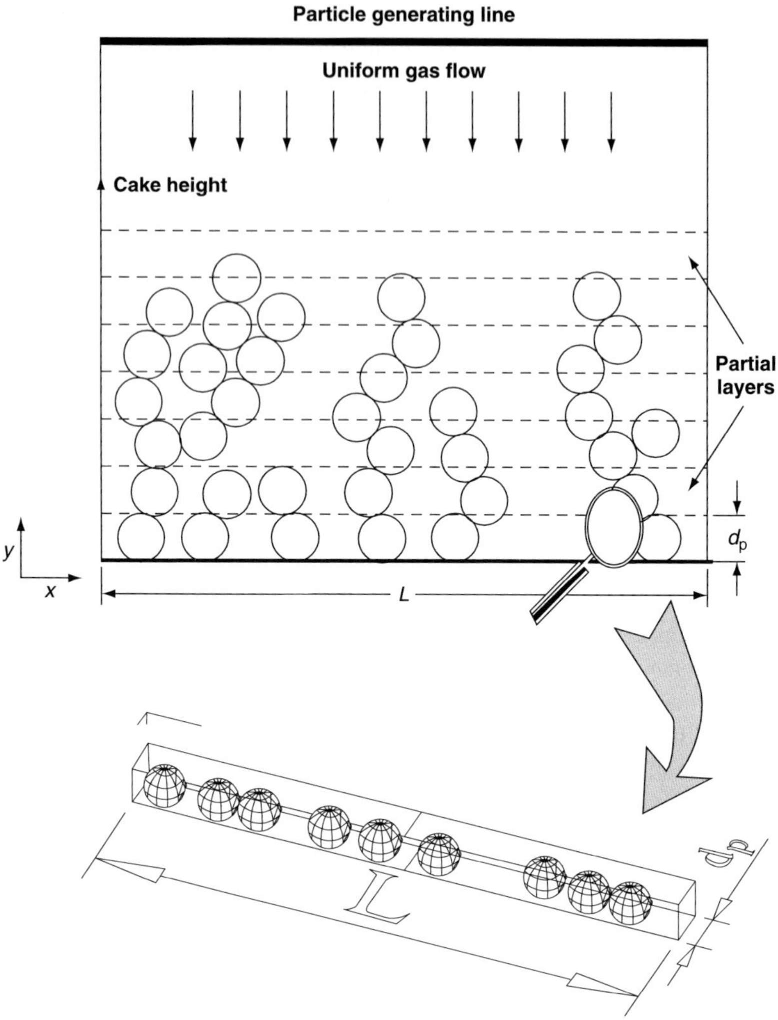

Figure 5.4 Schematic of two-dimensional simulation of cake formation. (Courtesy of Professor J.-W. Jung).

The stability criterion becomes

$$f_1 \sin \overline{\delta} \leq \left[f_1 \cos \overline{\delta} + f_2 \right] f$$

or

$$f_1 \leq \frac{f_2 \cdot f}{\sin \overline{\delta} - \left(\cos \overline{\delta} \right) f} \tag{5.4.1.1}$$

where $\overline{\delta}$ is the angle formed between the vertical line and the line connecting the two particle centers.

Both f_2 and f are assumed to be known quantities. The compressive force f_1 can be calculated from the pressure drops across the various segments. f_1 of the i-th segment is given as

$$(f_1)_i = \sum_{j=i}^{M} (\Delta p_\ell)_j \, (A)/n_i \tag{5.4.1.2}$$

where A is the cross-sectional area of a segment which can be taken as $L \cdot d_\mathrm{p}$ (see Fig. 5.4), $(\Delta p)_j$ is the pressure drop across the j-th segment, n_i is the number of particles present in the i-th segment, and M is the total number of segments. The pressure drop across a given segment can be determined from the Carman-Kozeny equation [i.e. Equation (5.4.1.3)] if the filtration velocity and the porosity of the segment are known.

In the event that Equation (5.4.1.1) is not obeyed, the particle is then assumed to roll off the particle with which it made contact and reassumed its rectilinear motion until it makes contact with either the medium or another previously deposited particle. This process is repeated until the particle comes to its rest. The change of the number of particles present in the horizontal segment necessitates the recalculation of the pressure drop, Δp_1, which, in turn, may affect the compressive stress to which the deposition particles are subject. A stability test is then applied to all the particles to determine whether Equation (5.4.1.1) is obeyed or not. Afterward, another new particle is introduced and the procedure repeats.

One set of simulation results obtained by Hoflinger *et al.* is shown in Fig. 5.5. The conditions of simulation were $d_p = 15\,\mu\mathrm{m}$, $\rho_\mathrm{s} = 2000\,\mathrm{kg\,m^{-3}}$ and $\mu = 1.85 \times 10^{-5}$ Pa s. The cake shown was formed with 2000 particles over a width of 0.8 mm. The friction coefficient, f, used was 0.364 and the cohesive force, f_2 (or $z_{\max}$ as shown in the figure), was 1×10^{-8} N. Also shown in the figure are the local cake porosity (denoted as ε_j), compressive stress (denoted $\Delta \mathrm{P}_{kj}$) and the strength of the cake layer (denoted as f_j) which is the product of the marginal cohesive force and the cake surface area divided by the number of particles in a given layer. The compressive stress was calculated based on Type 1 p_ℓ–p_s relationship (i.e. $\mathrm{d}p_\ell + \mathrm{d}p_\mathrm{s} = 0$) and the pressure drop according to Equation (5.4.1.3).

$$(q_\ell)_i = \frac{\varepsilon^3 d_\mathrm{p}^2}{k_1 (1-\varepsilon)^2} \frac{1}{\mu} \left[\frac{(\Delta p_\ell)_i}{d_\mathrm{p}} \right] \tag{5.4.1.3}$$

The procedure developed by Hoflinger *et al.* requires the knowledge of f_2 and f. While the friction coefficient f, in principle, can be determined experimentally, the necessary measurement is difficult to find. The cohesion force f_2 may include many types of interparticle forces and it may be incorrect to consider it to be a constant value under all circumstances.

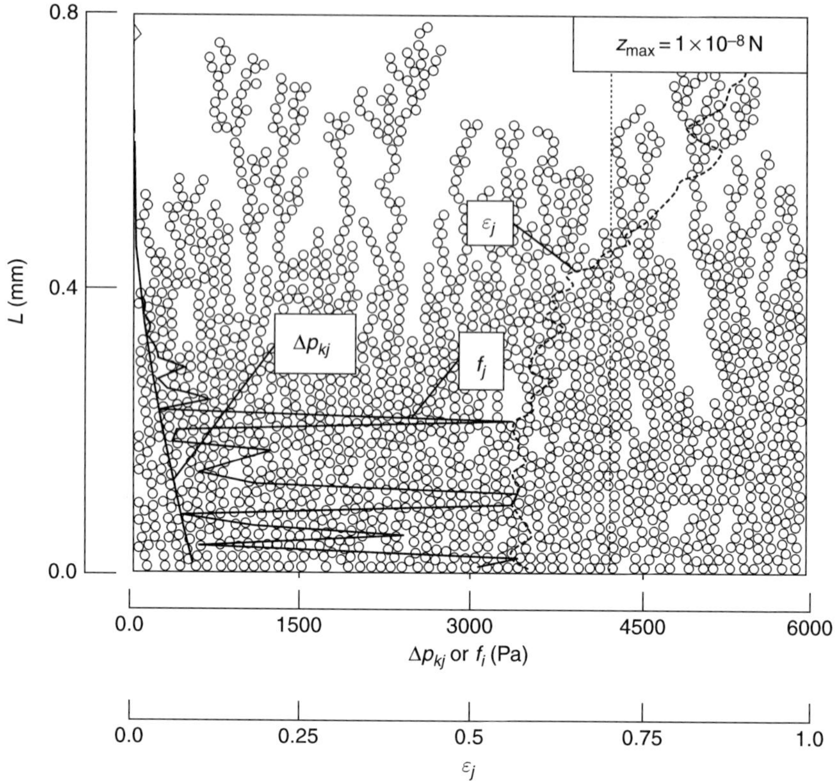

Figure 5.5 Simulation results of Hoflinger *et al.*: Effect of cohesive force on cake structure. (Courtesy of Professor W. Hoflinger).

5.4.2 Study by Jeon and Jung (2004)

A simulation study of cake formation which parallels closely the work of Hoflinger *et al.* was recently made by Jeon and Jung (2004). The basic principle used by Jeon and Jung was the same as that stated in the preceding section (see also Fig. 5.4). There are, however, differences. The value of k_1 of the flow rate–pressure drop relationship [i.e. Equation (5.4.1.3)] was assumed to be 180 by Jeon and Jung. Furthermore, the interparticle attractive force between two contacting particles of the same size was the London-van der Waals force given by Equation (5.2.9). With the compressive force acting along the direction of the suspension flow and the London-van der Waals force along the direction of the line connecting the *i*-th and *j*-th particles, the stability criterion given by Equation (5.3.1) becomes

$$f_1 \sin \overline{\delta} \leq \left[f_1 \cos \overline{\delta} + \mathbf{F}_{\mathrm{Lo}} \right] f \qquad (5.4.2.1)$$

where $\mathbf{F}_{\mathrm{Lo}}$ is the London-van der Waals force.

In applying Equations (5.4.2.1) as the criterion for stability, the magnitude of $\mathbf{F}_{\mathrm{Lo}}$ must be known. For its evaluation, Jeon and Jung assumed the separation distance between the two contacting particles to be 4 Å. In addition, the value of the friction coefficient is also required. Thus the two quantities which must be specified in Jeon and Jung's simulation are the separation distance and the friction coefficient. In contrast, the two parameters which require specification in the work of Hoflinger were the cohesive force f_2 and the friction coefficient f. Hoflinger *et al.*, however, did not offer any explanation about the nature of f_2 nor the methods for its estimation.

To relate the simulation results to cake filtration, if the two-dimensional domain of Fig. 5.4 is viewed as a box with a depth of d_{p}, the time of filtration and the number of particles introduced into the domain, N, are related by the expression

$$N = \frac{6L}{\pi d_{\mathrm{p}}^2} \varepsilon_{\mathrm{s_o}} \int_{\mathrm{o}}^{t} q_{\ell_{\mathrm{m}}} \,\mathrm{d}t \tag{5.4.2.2}$$

where $\varepsilon_{\mathrm{s_o}}$ is the particle volume fraction of the suspension and $q_{\ell_{\mathrm{m}}}$ is the instantaneous filtrate rate. For the case of constant rate filtration, Equation (5.4.2.2) becomes

$$N = \frac{6L\varepsilon_{\mathrm{s_o}} q_{\ell_{\mathrm{m}}} t}{\pi d_{\mathrm{p}}^2} \tag{5.4.2.3}$$

The mass of dry cake per unit medium surface w_{o} is

$$w_{\mathrm{o}} = \frac{N \pi d_{\mathrm{p}}^2 \rho_{\mathrm{s}}}{L} \tag{5.4.2.4}$$

and the porosity of the i-th segment ε_i is

$$\varepsilon_i = 1 - (\pi/6) n_i (d_{\mathrm{p}}/6) \tag{5.4.2.5}$$

where n_i is the number of particles present in the i-th segment. The total number of particles introduced is

$$N = \sum_{i=1}^{M} n_i \tag{5.4.2.6}$$

The conditions used by Jeon and Jung in their simulation are listed in Table 5.2. The results of the simulation shown in Figs 5.6–5.10 may be summarized as follows. Figures 5.6 and 5.7 show the effect of the London-van der Waals and the friction coefficient on cake structure. The cakes shown in these figures were formed with 2000 particles of diameter 15 μm over a medium of length 1 mm and depth 15 μm (or a cake areal mass density of 0.4712 kg m^{-2}). Figure 5.6 shows that more compact cake was formed if the Hamaker constant is smaller (or the attractive or adhesive force is less). The results of Fig 5.7 indicate that a large friction coefficient favors more porous cake

Table 5.2

Conditions used by Jeon and Jung in their simulation study

Hamaker constant (H)	6.0×10^{-20} J, 6.0×10^{-19} J
Friction coefficient	0.1, 0.3
Filtration velocity	0.2, 0.1, 0.05 m/s
Fluid viscosity	1.85×10^{-5} Pa s
Particles density	2000 kg/m^3
Separation distance between contacting particles	4 Å
Particle diameter	7.5, 15.0 μm
Entry length	1 mm

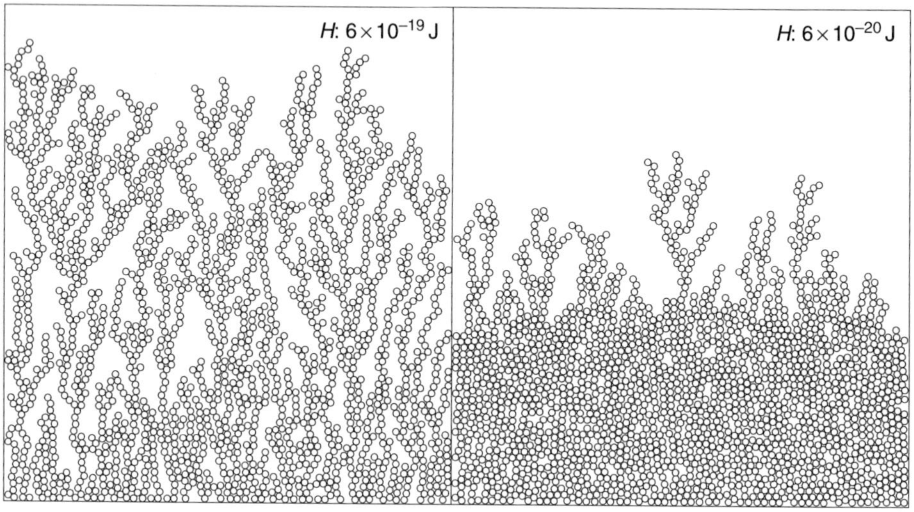

Figure 5.6 Simulation results of Jeon and Jung: Effect of the London-van der Waals force on cake structure; $d_p = 15\,\mu m$, $q_{\ell_m} = 0.1\,m/s$, $f = 0.1$. Number of particles 2000; Cake mass areal density 0.4712 kg/m^2. (Courtesy of Professor Y.-W. Jung).

structure. In other words, factors which retard adhesion between contacting particles lead to the formation of more dense particle assemblies. This is the same conclusion obtained from the results of Hoflinger *et al.* (1994).

The porosity of the cake formed is shown in Fig. 5.8 in which the porosities of various segments of a cake corresponding to two different mass areal densities are shown. As expected, significant porosity change occurs in the top segment while those of the bottom segments were found to be well compacted.

Other results obtained include the history of the pressure drop and the relationship between the cake porosity and the compressive stress. In Fig. 5.9, the results of the pressure drop required to maintain a fixed filtration rate vs. the cake mass areal density

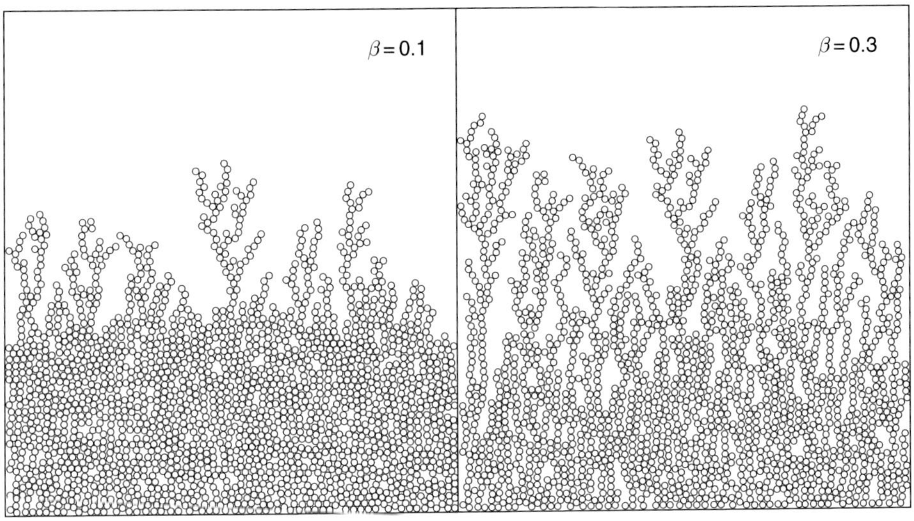

Figure 5.7 Simulation results of Jeon and Jung: Effect of the friction coefficient on cake structure, $H = 6.0 \times 10^{-20}$ J. Other conditions being the same as those of Fig. 5.6. (Courtesy of Professor Y.-W. Jung).

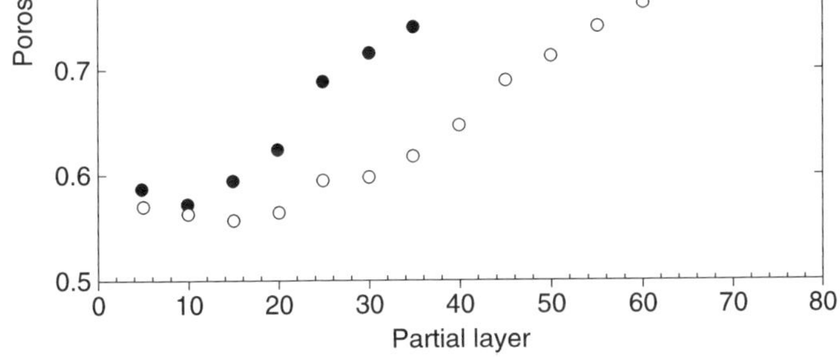

Figure 5.8 Simulation results of Jeon and Jung: Cake porosity vs. cake segment corresponding to two mass areal density values; $d_p = 15\,\mu m$, $q_{\ell_m} = 0.1$ m/s, $H = 6.0 \times 10^{-19}$ J, $f = 0.1$. Total number of particles, 2000 ($w_o = 0.4712\,kg/m^2$), 3000 ($w_o = 0.7069\,kg/m^2$). (Courtesy of Professor Y.-W. Jung).

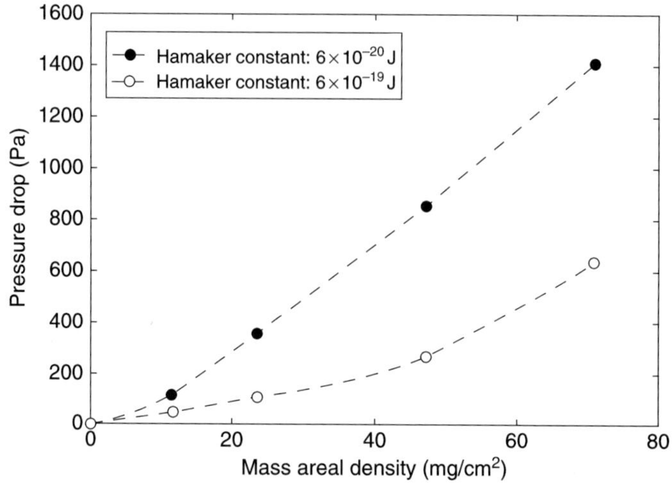

Figure 5.9 Simulation results of Jeon and Jung: Effect of the London-van der Waals force on pressure drop vs. w_o for constant rate. (Courtesy of Professor Y.-W. Jung).

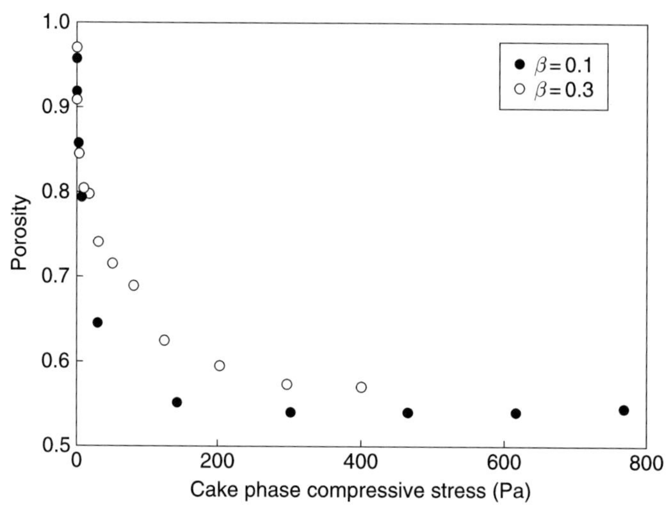

Figure 5.10 Simulation results of Jeon and Jung: Effect of the friction coefficient effect on the relationship of cake porosity vs. compression stress; β: friction coefficient. (Courtesy of Professor J.-W. Jung).

are shown. Notice that for constant rate filtration, the cake mass areal density is directly proportional to time. The relationship between the cake porosity and the compressive stress shown in Fig. 5.10 can be found from the knowledge of the number of particles present in each of the cake segments. The porosity was calculated according to

Equation (5.4.2.5) and the pressure drops across the segments from Equation (5.4.1.3). With $(\Delta p_\ell)_i$ known, $(p_s)_i$ can then be calculated from Equation (5.2.21). The results, however, cover only a rather limited range of the compressive stress. For the case of $H = 6 \times 10^{-19}$ J and $f = 0.1$, the power-law relationship of ε vs. p_s does not seem to be applicable (see Fig. 5.13 of Jeon and Jung (2004)).

5.4.3 Study by Fu and Dempsey (1998)

The study by Fu and Dempsey was an investigation on the effect of particle size and charge on cake structure in ultrafiltration. It differs from the two studies described above in several aspects including:

(i) The simulation is three dimensional.
(ii) The interaction forces considered include both the London-van der Waals force and the double layer force. Contributions from all particle pairs (with separation distance less than 30 nm) were considered.
(iii) No compressive stress is present in the cake.
(iv) Simulations were made through collective rearrangement instead of sequential addition used by Hoflinger *et al.* and Jeon and Jung.

The procedure used may be described as follows: A three dimensional domain – for example, a box of dimensions $x_1 \times y_1 \times z_1$, where x_1, y_1 and z_1 are multiples of particle diameter with z along the negative direction of filtrate flow – is first specified. A number of particles are placed randomly into the box subject to certain constraints. Rearrangement of the particles are then made through the integration of the particle trajectory equation. This rearrangement continues until the structure of the cake formed becomes constant. Based on the final position of the particle present in the core of the box, the cake porosity (or solidosity), permeability and specific cake resistance can then be calculated. Some of the details of these steps are described below.

(1) *Specification of the physical domain and placement of particles*
The cross section of the box has a dimension of $x_1 \times y_1$ with $x_1 = y_1$ and the bottom of the box coincides with the filter medium. Possible particle positions (x_i, y_i, z_i) are generated by a random number generator. Placement of particles into the box are made according to these randomly generated positions subject to the condition that particles cannot overlap with each other or with the box surfaces. There must be a separation distance of 30 nm or greater between adjacent particles.
For the results to be presented later, $x_1 = y_1 = 10\, d_p$, $z_1 = 50\, d_p$ and $d_p = 30$ nm. Based on the initial particle positions generated, 417 particles were placed in the box. If one defines the core of the box to be that part of the box with a distance of $2\, d_p$ from the side of the box, the core represents $(6 \times 6)/(10 \times 10)$ or 36% of the box volume with 149 particles present within the core.

(2) *Rearrangement of the particles*

Rearrangement can be made according to the particle trajectory equation. With the drag force given by the Stokes law [Equation (5.2.4)] and neglecting the Brownian diffusion force and the particle velocity on the ground that $|u_p| \ll |u|$, and considering particle interactions due to the London-van der Waals attraction force and the double layer repulsive force [Equations (5.2.9), (5.2.11) and (5.2.12)], Equation (5.1.2) may be written as

$$\frac{\pi}{6}(d_p)^3 \rho_s \frac{d^2\mathbf{r}_i}{dt^2} = (\mathbf{F}_D)_i + (\mathbf{F}_{Lo})_i + (\mathbf{F}_{DL})_i \qquad (5.4.3.1)$$

with

$$(\mathbf{F}_D)_i = -3\pi\mu(d_p)_i |\mathbf{u}| \mathbf{n}_i \qquad (5.4.3.2)$$

$$(\mathbf{F}_{Lo})_i = \sum_j \frac{H}{24 z_{ij}^2}(d_p)\frac{\mathbf{r}_i - \mathbf{r}_j}{|\mathbf{r}_i - \mathbf{r}_j|} \qquad (5.4.3.3)$$

$$(\mathbf{F}_{DL})_i = \sum_j \frac{32\pi n_o kT\gamma^2}{\kappa}(d_p)\frac{\mathbf{r}_i - \mathbf{r}_j}{|\mathbf{r}_i - \mathbf{r}_j|} \exp[\kappa z_{ij}] \qquad (5.4.3.4)$$

with κ and γ given by Equations (5.2.13) and (5.2.14). The summation includes all particles within a distance of 30 nm from the i-th particle.

Particle rearrangement is monitored by following particle trajectories. First, the particle position vector $\mathbf{r}_i$ is normalized by its diameter or

$$\mathbf{r}_i^* = \frac{(\mathbf{r}_i)}{d_p} \qquad (5.4.3.5)$$

and the dimensionless separation distance δ_{ij}^* is

$$\delta_{ij}^* = |(\mathbf{r}_i^* - \mathbf{r}_j^*)| - 1 \qquad (5.4.3.6)$$

Equation (5.4.3.1) may be rewritten as

$$\frac{d^2\mathbf{r}_i^*}{dt^2} = (\mathbf{f}_t)_i = (\mathbf{f}_D)_i + (\mathbf{f}_{Lo})_i + (\mathbf{f}_{DL})_i \qquad (5.4.3.7)$$

where

$$(\mathbf{f}_D)_i = \frac{-18\mu|u|}{\rho d_p^3}\mathbf{n}_3 \qquad (5.4.3.8)$$

$$(\mathbf{f}_{L_o})_i = \sum_j \frac{-H}{4\pi d_p^4 \rho z_{ij}^{*2}} \frac{\mathbf{r}_i - \mathbf{r}_j}{|(\mathbf{r}_i - \mathbf{r}_j)|} \qquad (5.4.3.9)$$

$$(\mathbf{f}_{DL})_i = \sum_j \frac{192 n_o kT\gamma^2}{K\rho d_p^3} \exp[-(\kappa d_p) z_{ij}^*] \frac{\mathbf{r}_i - \mathbf{r}_j}{|(\mathbf{r}_i - \mathbf{r}_j)|} \qquad (5.4.3.10)$$

Particle trajectories can then be obtained from the integration of Equation (5.4.3.7). Assuming that the positions of the i-th particle are known at $t = t - \Delta t$ and t, its position at $t + \Delta t$ is simply

$$\mathbf{r}_i^*(t+\Delta t) = -r_i^*(t-\Delta t) + 2r_i^*\ (t) + (\Delta t)^2\ (f_t)_i \tag{5.4.3.11}$$

Rearrangement is conducted particle by particle. At certain time intervals in the course of the rearrangement, the structure of the particles was examined by calculating the average particle height (i.e. the average z-coordinate of particle position) or the temporal porosity of the bottom layer of particles. The smallest time interval used in the work of Fu and Dempsey was 0.25×10^{-9} s.

(3) *Calculation of cake porosity, permeability and specific cake resistance*
Once the cake structure reaches a constant state, rearrangement calculation ceases and the positions of the particles are recorded. The distribution of the final particle positions may be written as

$$z = G(n_a) \tag{5.4.3.12}$$

where n_a is the number of particles with height equal to or less than z. If these particles are uniformly distributed, $G(n_a)/n_a$ is constant or Equation (5.4.3.12) is a linear function of n_a.

For a collection of particles of height z and cross-sectional area S, if the total number of particles is N, by definition, the porosity of the collection of particles is

$$\varepsilon = 1 - \frac{N\pi d_p^3}{6Sz} = 1 - \frac{\pi d_p^3}{6S(dG/dn_a)_{n_a=0}} \tag{5.4.3.13}$$

Accordingly, from the final particle height distribution data, ε can be found from the above expression. Once ε is known, the permeability k can be calculated as

$$k = \frac{\varepsilon^3 d_p^2}{k_1(1-\varepsilon)^2} \tag{5.4.3.14}$$

The specific cake resistance α is defined as $[k(1-\varepsilon)\rho_s]^{-1}$. Accordingly, α is

$$\alpha = \frac{k_1(1-\varepsilon)}{\varepsilon^3 \rho_s d_p^3} \tag{5.4.3.15}$$

The conditions used by Fu and Dempsey in their simulation are given in Table 5.3. To minimize the errors due to the presence of side surfaces, only the information of the core of the domain (i.e. the part of the domain at least $2d_p$ away from the surface) was used. Among the 417 particles present in the domain, 149 particles were in the core initially. This is shown in Fig. 5.11.

Table 5.3

Conditions used by Fu and Dempsey in their simulation study

Hamaker constant	4.12×10^{-20} J
Filtration velocity	6.55×10^{-5} m/s
Particle diameter	30, 60, 90 nm
Particle density	3900 kg/m^3
Particle potential	0, 10, 20, 30, 40, 50 mV
Number of particles considered in rearrangment	417
Physical domain of simulation	$10 \times 10 \times 50$ (unit, one particle diameter)

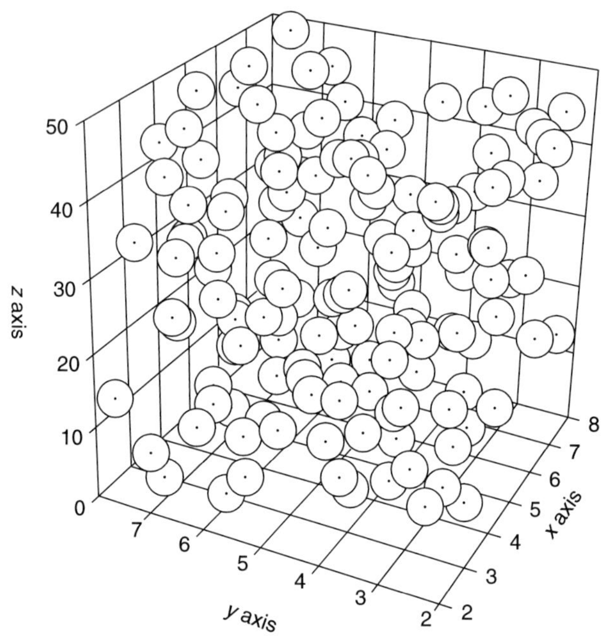

Figure 5.11 Initial position of particles in the central core used by Fu and Dempsey in their simulation study. (Fu and Dempsey, 1998. Reprinted by permission of Elsevier.)

The degree of rearrangement can be seen by tracking particle trajectories. In Fig. 5.12, trajectories of particles with three different surface potential (10, 20, 40 mV) are shown. At distances away from the medium, the trajectories of particles under different repulsive forces were similar. Close to the medium (bottom surface), the effect of the repulsive force became apparent. The 10 mV particles became immobilized easily and lateral movement along the medium surface was slight. The 20 mV particles displayed horizontal displacement before reaching its final position while the migration of the 40 mV particles was significant, even to the point that it moved against the filtrate flow at times.

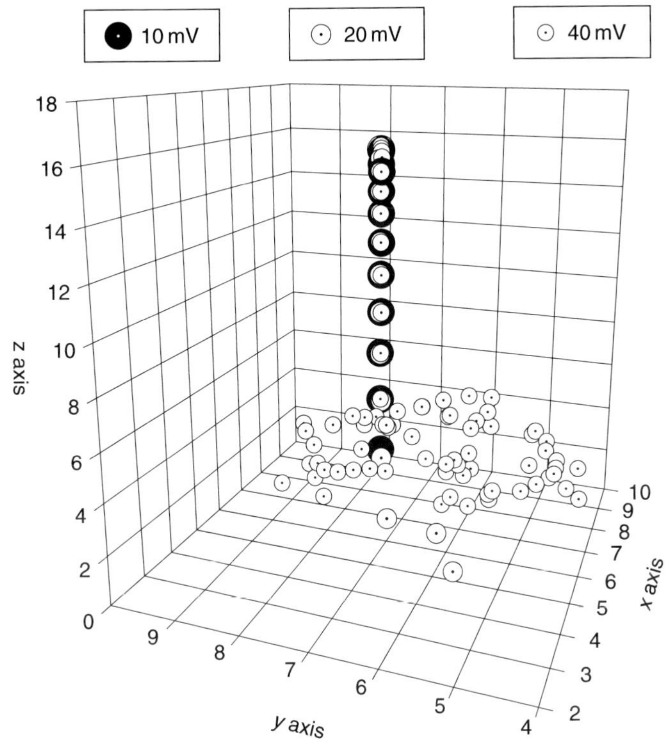

Figure 5.12 Simulation results of Fu and Dempsey: Trajectories of particle with surface potentials of 10, 20 or 4 mV. (Fu and Dempsey, 1998. Reprinted by permission of Elsevier.)

As stated before, the average particle height was used to indicate the establishment of the steady state. Figure 5.13 gives the average particle height as a function of time. There was a rapid decrease in particle height initially. For particles with no charge (0 mV) or low potential (10, 20 mV), they reached a constant average height rather quickly. For 40 and 50 mV particles, the average height oscillated and the number of particles in the core decreases with time. The 30 mV particles gives the lowest average particle height at the steady state.

The particle height distribution results at the end of the rearrangement (i.e. steady state) are shown in Fig. 5.14. The unit used in describing the particle height is the particle diameter. The main features shown in this figure include: (1) the number of particles present in the core at the end of the rearrangement is less than the initial value (i.e. 149), and (2) $G(n_a)$ can be approximated as a linear function of n_a up to z equal to the average particle height. Thus, by fitting the distribution data (up to z equal to the average height) with a linear expression, $dG(n_a)/dn_a$ of the linear part of the distribution curve was obtained and used to calculate ε, k and α according to Equations (5.4.2.13), (5.4.3.14) and (5.4.3.15), respectively. The calculated results of ε and α as functions of the particles surface potential are shown in Figs 5.15 and 5.16. These figures indicate

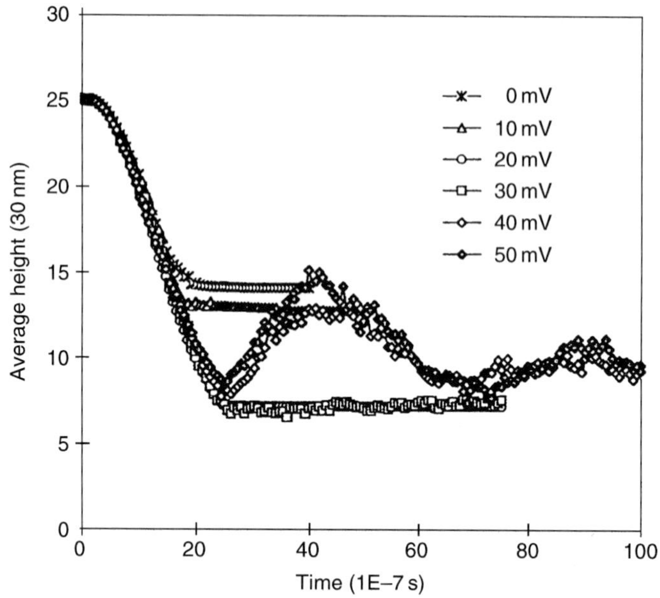

Figure 5.13 Simulation results of Fu and Dempsey: Progress of cake compaction as shown by the change of particle height vs. time for 30 nm Particles with different surface potential. (Fu and Dempsey, 1998. Reprinted by permission of Elsevier.)

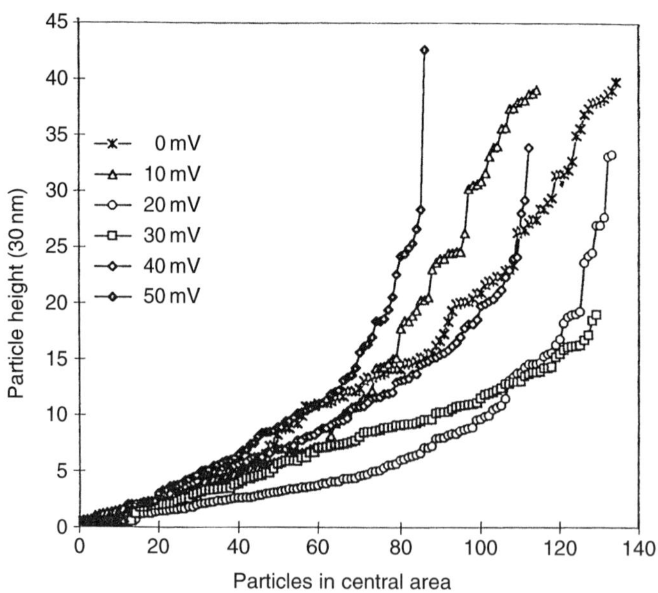

Figure 5.14 Simulation results of Fu and Dempsey: Particle height distribution at the end of rearrangement; 30 nm particles with different surface potential. (Fu and Dempsey, 1998. Reprinted by permission of Elsevier.)

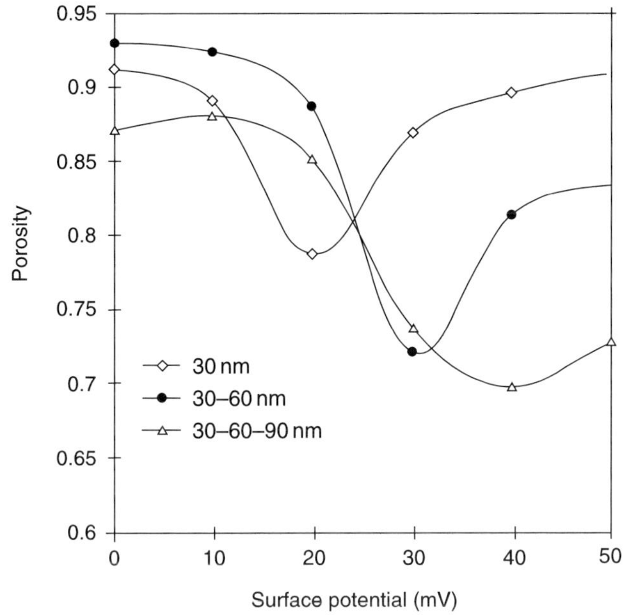

Figure 5.15 Simulation results of Fu and Dempsey: cake porosity vs. surface potential, 30 nm particles. (Fu and Dempsey, 1998. Reprinted by permission of Elsevier.)

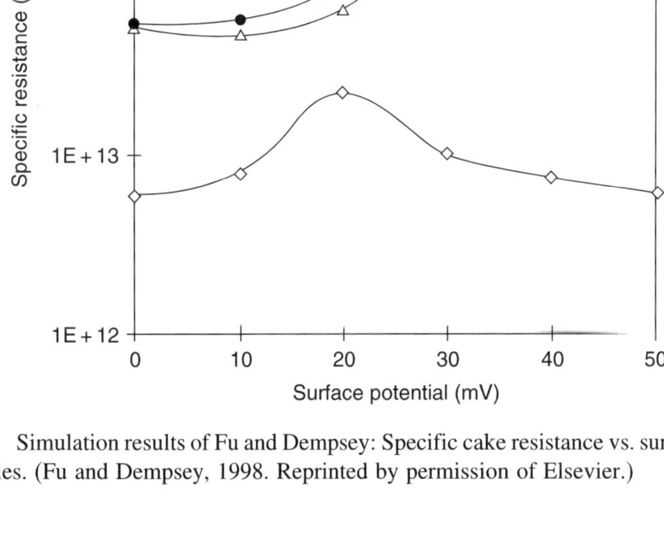

Figure 5.16 Simulation results of Fu and Dempsey: Specific cake resistance vs. surface potential; 30 nm Particles. (Fu and Dempsey, 1998. Reprinted by permission of Elsevier.)

that the most compact cake was formed with 20 mV particles. The 20 mV particle cake, therefore, had the lowest porosity but highest values of α.

Some comments regarding the work of Fu and Dempsey may be in order. Their simulation was carried out in the absence of any compressive stress. The ε and α results they obtained may therefore be considered as those at the zero compressive stress state: namely, $(1-\varepsilon_s^o)$ and α^o according to the power-law expression of the constitutive equations. As shown in Fig. 5.15, ε is greater than 0.8 except in the immediate neighborhood of $\psi_p = 20$ mp and the corresponding specific cake resistance is of the order of 10^{-13}–10^{-12} m kg^{-1}. These values are of the same order of magnitude of those of several filter cakes ($CaCO_3$, Kaolin, TiO_2 and Kromasil) as reported by Teoh recently (2003).

Another problem of the study of Fu and Dempsey was the omission of the Brownian diffusion force in their trajectory equation [i.e. Equations (5.4.3.1) and (5.4.3.7)]. This omission was based on the argument that the Brownian diffusion force, for the particles they studied, were relatively unimportant as compared with the surface interaction force (see Fig. 1 of Fu and Dempsey). This comparison, however, was not appropriate since the Brownian diffusion force is a stochastic quantity (both in magnitude and direction) and that it cannot be compared with any deterministic quantity of specified direction. The correct way of including the Brownian diffusion force is given in Section 5.2. Inclusion of the Brownian diffusion, of course, would significantly increase the computational effect.

The question regarding the possible effect of the initial particle placement also needs to be addressed. Intuitively, one may argue that as the steady state is reached only after substantial rearrangement, the effect of initial particle positions may not be significant. More exactly, this question can be resolved if replicate runs are to be made with the respective initial positions corresponding to different sets of random numbers generated. This, of course, will further increase the required computation effort.

The results and discussions regarding the previous studies given above demonstrate rather clearly the potential of the simulation approach in studying cake formation and cake structure and examining the effect of variables such as the nature and magnitude of the surface interaction. It is also clear that the studies so far completed are preliminary and none of the results have been tested against experiments. Much more effort is therefore required before the potential of the simulation approach can be fully exploited.

REFERENCES

Brinkman, H.C., *App. Sci. Res.* **A1**, 81 (1947).
Carman, P.C., *Trans. Inst. Chem. Engrs.* (London), **15**, 150 (1937).
Chandrasekhar, S., *Rev. Mod. Phys.*, **15**, 1 (1943).
Davis, C.N., *Proc. Phys. Soc.*, **57**, 529 (1945).
Davis, C.N., *Proc. Inst. Mech. Eng.* **B1**, 185 (1952).
Davis, C.N., *Air Filtration*, Academic Press, London (1973).
Fu, L.F., "Charge Effect in Ultrafiltration", Ph.D. Thesis (Environmental Engineering), The Pennsylvania State University (1996).
Fu, L.F. and Dempsey, B.A., *J. Membrane Sci.*, **149**, 221 (1998).

Happel, J., *AIChE J.*, **4**, 197 (1958).
Happel, J. and Brenner, H., *Low Reynolds Hydrodynamics*, Kluwer, Dordrecht (1991).
Hinch, E.J., *J. Fluid Mech.*, **83**, 695 (1977).
Hinds, W.C., *Aerosol Technology*, John Wiley & Sons, New York (1999).
Hoflinger, W., Stocklmayer, Ch., and Hachl, A., *Filtration and Separation*, p. 807, (December 1994).
Hogg, R., Healy, T.W. and Fuerstenan, D.W., *Trans. Faraday Soc.*, **62**, 1938 (1966).
Howells, I.D., *J. Fluid Mech.*, **64**, 449 (1974).
Jackson, G.W. and James, D.F., *Can. J. Chem. Eng.*, **64**, 364 (1986).
Jeon, K.-J and Jung, Y.-M., *Powder Tech.*, **141**, 1 (2004).
Kim, A.S. and Stolzenbach, K.D., *J. Colloid Interface Sci.*, **253**, 315 (2002).
Kim, S. and Russel, W.B., *J. Fluid Mech.*, **154**, 269 (1985).
Kozeny, J., *Bez. Wien Akad*, **136A**, 271 (1927).
Lu,W.-M. and Hwang, K.-J., *Sep Technol.*, **3**, 122 (1993).
Lu, W.-M., and Hwang, K.-J., *AIChE J.*, **41**, 1443 (1995).
Neals, G.H. and Nader, W.K., *AIChE J.*, **20**, 530 (1974).
Payatakes, A.C., "A New Model for Granular Media: Application to Filtration through Packed Beds", Ph.D. Dissertation (Chemical Engineering), Syracuse University, Syracuse, New York (1973).
Reed, C.C. and Anderson, J.L., *AIChE J.*, **26**, 816 (1980).
Schmidt, E., *Powder Tech.*, **86**, 113 (1996).
Stamatakis, K. and Tien, C., *AIChE J.*, **39**, 1292 (1993).
Teoh, S.-K., "Studies in Filter Cake Characterization and Modelling", Ph.D. Thesis (Chemical Engineering), National University of Singapore, Singapore (2003).
Tien, C., *Granular Filtration of Aerosols and Hydrosols*, Butterworths, Stoneham, MA (1989).

PART II
Experiments and Measurements

Experiments and measurements are essential to the study and application of filtration technology. The results obtained from filtration experiments and measurements provide the information necessary for the design and scale-up of filtration systems. The results are also required for analyzing filtration process as shown in the chapters of Part I. Equally, if not more important, experimental and measurement results assist investigators to gain insight into the process and make it possible to improve existing filtration models and/or formulate new ones.

The first chapter of this part (Chapter 6) gives a presentation of cake filtration experiments. Included in the presentation are the experimental apparatus and procedures used for determining filtration performance, and the development of methods for probing cake structure and its evolution. The effects of various variables on filtration performance observed are also described.

Solidosity and permeability are the characteristic properties in the formation and growth of filter cakes. In Chapter 7, the determinations of these properties and the equipments and procedures used are discussed including the C-P principle suggested by Ruth, the development of the C-P cell over the past five decades; the problems associated with C-P measurement and their possible solutions. The procedures which may be applied to obtain these quantities from filtration experiments carried out under different conditions are also described. As solid/liquid separation is often based on the relative motion between suspended particles and suspending liquid, different expressions have been used to describe the relative motion. However, there exists equivalence among these expressions. The existence of the equivalence was further demonstrated through the establishment of a general constitutive relationship applicable to different solid/liquid separations.

– 6 –

CAKE FILTRATION EXPERIMENTS

Notation

A_I	amplitude of a propagating wave at the reference point I (V)
A_o	amplitude of a propagating wave after traveling a distance d from point I (V)
b_1	coefficient in Equation (6.4.3.2) ($m\,s^{-1}\,Pa^{-1}$)
b_2	coefficient in Equation (6.4.3.2) ($m\,s^{-3/2}\,Pa^{-2}$)
d	distance (m)
F	formation factor defined by Equation (6.3.2) (-)
f	wave frequency (Hz)
f'	defined by Equation (6.4.2.4) (-)
k	cake permeability (m^2)
L	cake thickness (m)
$\overline{m}$	wet to dry cake mass ratio (-)
p_ℓ	filtrate pressure (Pa)
p_s	cake compressive stress (Pa)
p_{sm}	value of p_s at cake/medium interface (Pa)
R_m	medium resistance (m^{-2})
$R_m(o)$	initial medium resistance (m^{-2})
s	mass fraction (-)
t	time (s)
V	cumulative filtrate volume (m^3/m^2)
$\overline{x}$	coefficient in Equation (6.4.3.1a) $(m^3/m^2)s^{-1}$
$\overline{y}$	coefficient in Equation (6.4.3.1a) $(m^3/m^2)t^{-3/2}$

Greek letters

α	attenuation coefficient (-) or specific cake resistance ($m\,kg^{-1}$)
α_{av}	average specific cake resistance defined by Equation (6.4.2.3) ($m\,kg^{-1}$)
$\overline{\beta}$	exponent of Equation (6.3.3) (-)
Δp_c	pressure drop across filter cake (Pa)
$\overline{\overline{\varepsilon}}_s$	average cake solidosity (-)

μ filtrate viscosity (Pa s)
ρ filtrate density ($kg\,m^{-3}$)
ρ_s particle density ($kg\,m^{-3}$)

A general description of experimental studies of cake filtration conducted in the past is given in this chapter. The topics presented include the apparatus used, the various methods for the determination of filtration performance, the techniques developed for probing the internal properties of filter cakes and some of the results obtained. The presentation is not intended to provide an exhaustive literature review but to give an overview of cake filtration experimental research, its historical development and present status.

6.1 APPARATUS

Experimental collection of basic cake filtration data using simple apparatus has been a practice for nearly a century. Almay and Lewis (1912) conducted filtration experiments of chromium hydrate sludge using a laboratory press under pressure (up to $100\,lb/in.^2$ or 6.89×10^5 Pa) and established an empirical relationship between the filtration rate, the operating pressure and the cumulative filtrate volume. The so-called SRF (specific resistance to filtration) test commonly used for assessing sludge dewaterability is essentially a filtration experiment conducted with a Buchner funnel under vacuum (Christensen and Dick, 1985a; Vesilind, 1988).

Contemporary design and development of cake filtration apparatus can be traced to the work of Grace (1953). Grace fabricated a test cell for conducting cake filtration experiment under relatively high operating pressure ($500\,lb_f/in.^2$ or 3.645×10^6 Pa). The cell bodies were made of Lucite acrylic resin and stainless steel of various depth in order to keep the cell volume available for test suspension to a minimum, thus minimizing the effect of particle sedimentation. A specially designed blow case was used to deliver test suspensions under specified operating pressure and a stress-gauge balance was used for continuous recording of the filtrate collected. A disassembled view of the test cell is given in Fig. 6.2 and the overall filtration apparatus is shown in Fig. 6.1.

Considerable efforts have been made since the work of Grace to further improve the capability and precision of cake filtration apparatus by a number of investigators. Through the integration of electronics, computer control and technology, a mechatronic cake filtration apparatus was developed by Tarleton and Hancock (1997) and Tarleton (2004). A schematic diagram of the mechatronic apparatus of Tarleton is shown in Fig. 6.3. The basic unit consisted of a Nutsche filter and a feed tank with computer-controlled electro-pneumatic valves. For cake thickness and cake solidosity measurements, diametric pairs of electrodes were placed into the cell at 1 mm intervals above the medium. By placing these electrodes approximately 2 mm from the cell wall, the intrusion effect of the electrode on cake formation was kept to a minimum.

The pressure required to conduct filtration experiments was provided by compressed air and an electronic pressure regulator over a range of 50–600 kPa. The control algorithm

Figure 6.1 Photograph of the test cell used by Grace. (Grace, 1953. Reprinted by permission of The American Institute of Chemical Engineers.)

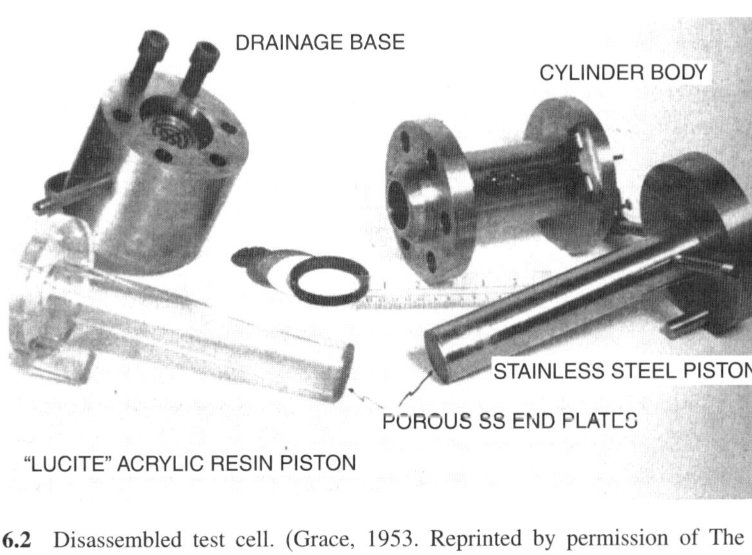

Figure 6.2 Disassembled test cell. (Grace, 1953. Reprinted by permission of The American Institute of Chemical Engineers.)

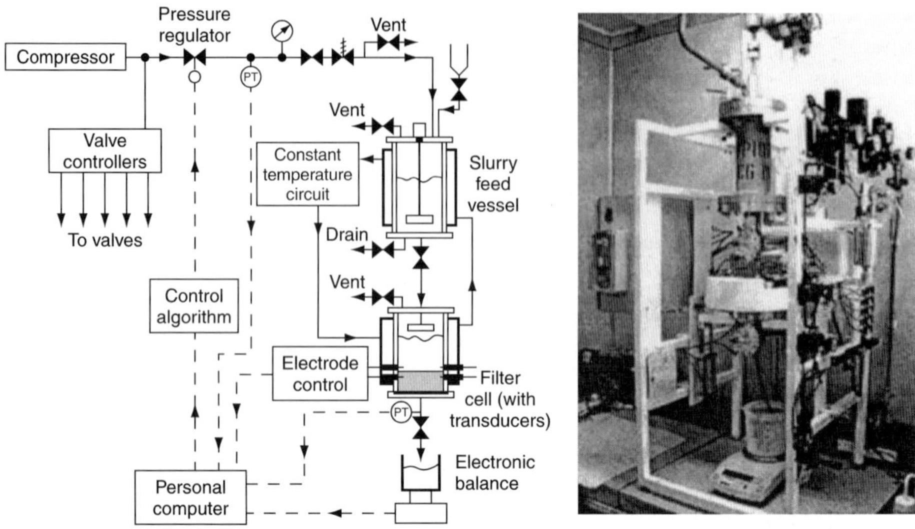

Figure 6.3 Schematic diagram and photograph of the mechatronic filtration apparatus of Tarleton. (Tarleton, 2004. Reprinted by permission of The American Filtration and Separation Society.)

developed allowed filtration to be carried out at constant as well as variable pressure including the case of increasing the operating pressure stepwise with each specified pressure lasting a given period of time.

The liquid pressure profile throughout the cake was determined by placing a number of micro-pressure transducers above the filter medium. These transducers were placed at a distance of 0.5, 0.8, 1.0, 1.3, 1.8, 2.3, 3.3, 5.3, 9.3 and 15.3 mm above the medium. Among the 10 transducers, 7 were positioned within 3.3 mm above the medium. Transducers were attached to custom-designed holders and micropore tubes. To obtain accurate liquid pressure measurement, a liquid bridge was created between the tip of a transducer and the cake by injecting water from a separate water reservoir. The transducers were placed around the cell periphery and at a distance of 2 mm away from the cell wall. A diagram demonstrating the details of the transducer construction and their arrangement is shown in Fig. 6.4.

Another example of the recently developed cake filtration apparatus is the multifunction test cell of the National University of Singapore (Tan *et al.*, 1998; Teoh *et al.*, 2001). This apparatus may function as a C-P cell as well as a variable-volume filtration chamber thereby enabling a direct comparison and corroboration of data obtained from filtration experiments and C-P measurements.

The multifunction test cell was modified from a commercially available precision universal testing machine, a Shimadzu Authograph AGS-10 kNG which is typically used for measuring mechanical properties of materials up to a maximum loading of 10 kN (or 2000 kPa). As shown in Fig. 6.5, the test cell consists of an upper piston, a

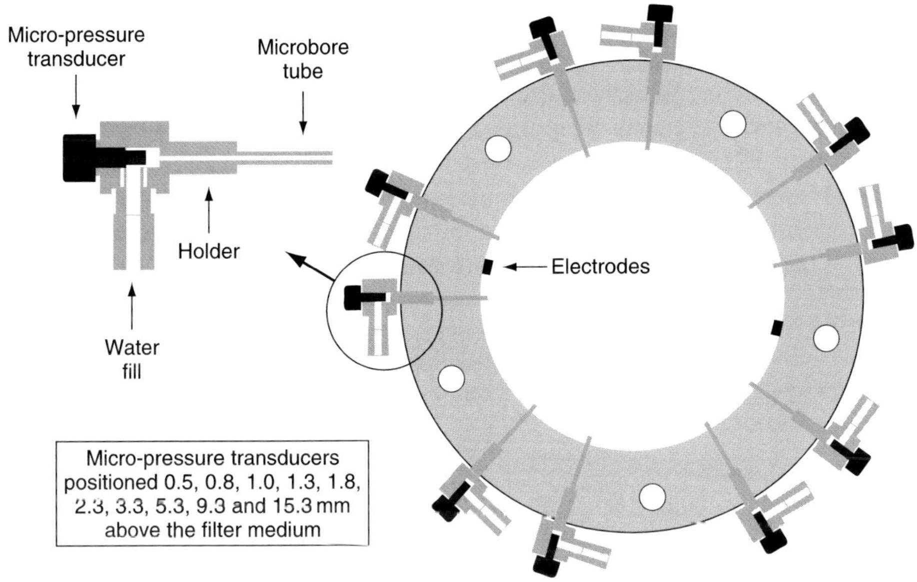

Figure 6.4 Micro-pressure transducer arrangement of the mechatronic apparatus of Tarleton. (Tarleton, 2004. Reprinted by permission of The American Filtration and Separation Society.)

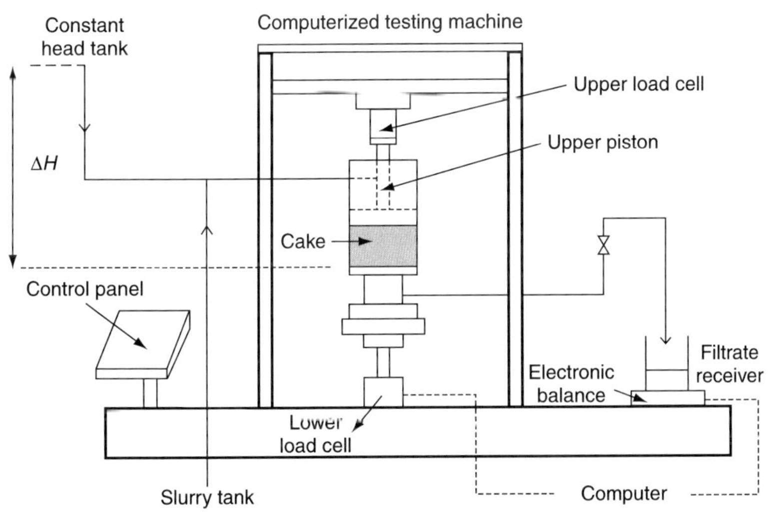

Figure 6.5 Schematic diagram of the multifunction test cell of Tan *et al.* (Teoh, Tan, He and Tien, 2001. Reprinted by permission of the Filtration Society.)

complementing lower piston and a test cell made of stainless steel. The vertical travel and applied load of the upper piston are controlled by a local frame device. The lower piston rests on a local cell and pressure transducers are placed at different locations along the cell surface. With these features, the transmitted load and the friction force during cake compression may be measured.

When the multifunction test cell is used for C-P measurements (see Fig. 6.6), the upper and lower pistons enclose a preformed filter cake of solid particles to be tested. The upper piston incorporates a liquid inlet through which liquid can be delivered to the cake. The bottom of the upper piston is provided with a distribution region consisting

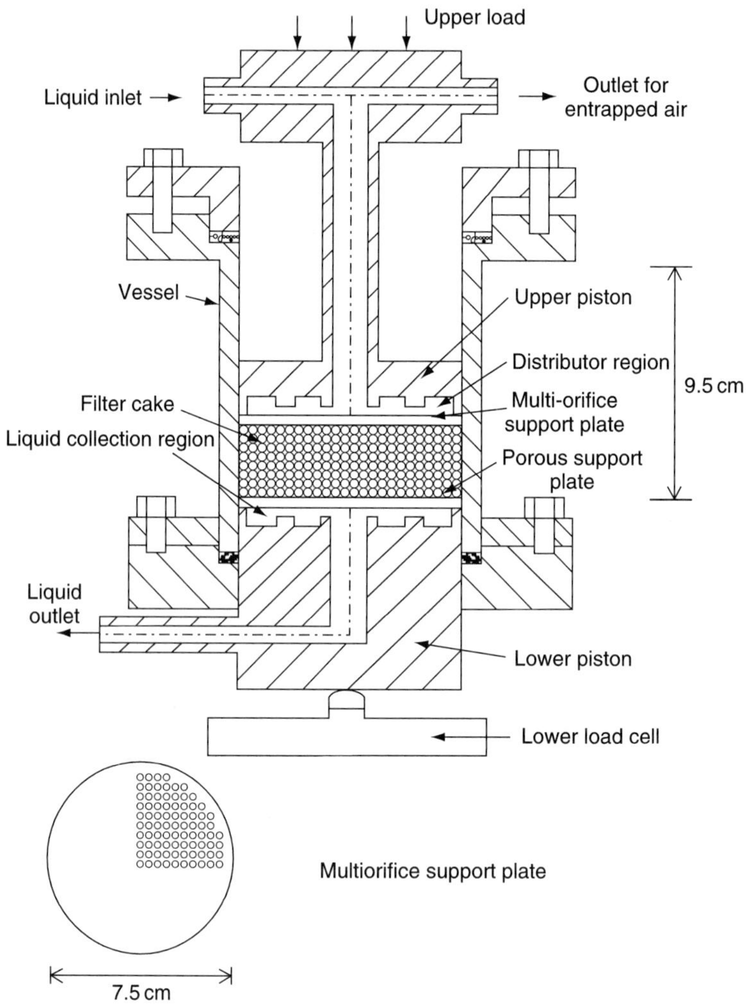

Figure 6.6 Multifunction test cell used as a filtration cell. (Teoh, Tan, He and Tien, 2001. Reprinted by permission of the Filtration Society.)

of a series of grooves. A separate opening serves as an outlet for trapped air. The bottom of the upper piston is fitted with an interchangeable multiorifice support plate. The distributor region is left open to allow an even distribution of incoming liquid over the entire cross section of the test cell. The lower piston contains a liquid outlet and a multiorifice support plate below which locates a liquid collection region.

The multifunction test cell may also be configured for use in filtration experiments (see Fig. 6.7). In this case, the bottom of the upper piston is fitted with an interchangeable insert plate with a sharp orifice opening. The insert plate was designed with a cylindrical portion surrounding the central orifice such that when it is fitted into place, the distribution region of the upper piston becomes closed off, thus allowing the

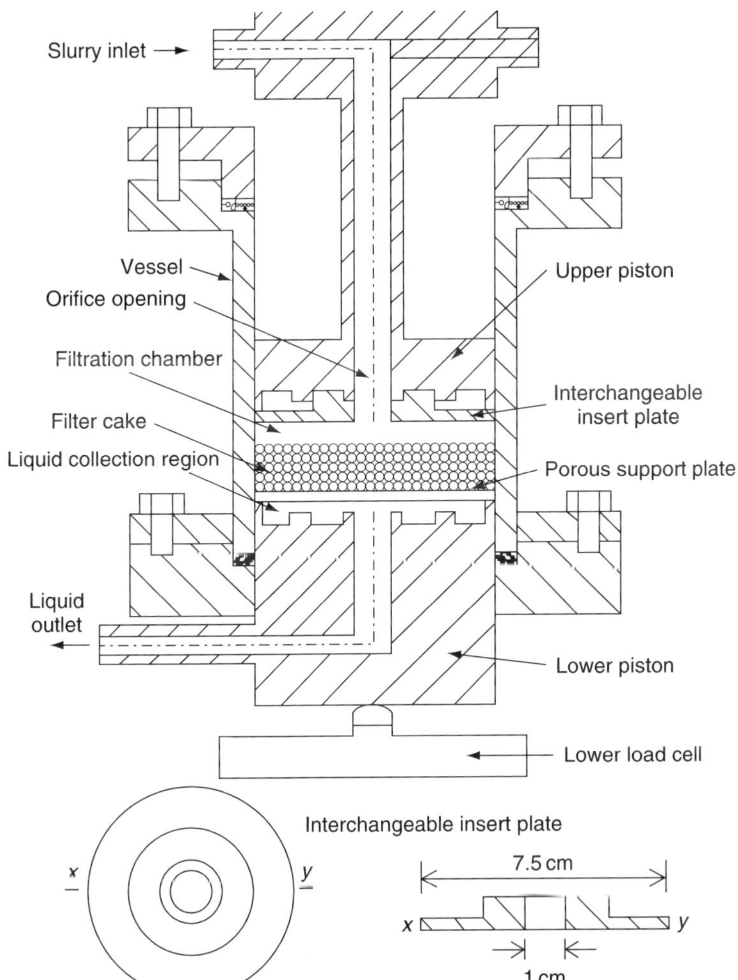

Figure 6.7 Multifunction test cell used as a C-P cell. (Teoh, Tan, He and Tien, 2001. Reprinted by permission of the Filtration Society.)

entry of the test suspension into the filtrate chamber solely through the central orifice opening.

In a typical filtration experiment, a circular filter medium is placed on the support plate of the lower piston. The position of the upper piston is fixed; thereby enclosing a filtration chamber of known volume between the two pistons. Test suspension is introduced into the chamber at a fixed constant pressure by means of compressed N_2 gas. During filtration, the filtrate collected is recorded on line. At the same time, particles present in the test suspension are retained by the filter media to form a cake. The cake thickness increases with time. Finally, the cake may reach the orifice opening in the upper piston. This instant is detected by a sharp decrease in filtration and signifies the cake thickness reaches a value corresponding to the position of the upper piston. This point will be discussed in the following section.

6.2 DETERMINATION OF FILTRATION PERFORMANCE

As a solid/liquid separation process, the performance of cake filtration may be given by the rate of filtration or the filtrate volume collected (V) and the cake thickness (L) as functions of time (V vs. t and L vs. t). As a measure of the quality of the cake formed, the wet to dry cake mass ratio $\overline{m}$ is often used. Methods of determining these variables are described below.

6.2.1 Cumulative filtrate volume

This is a comparatively simple measurement. The filtrate obtained during an experiment can be collected and weighed with precision and recorded continuously. Control of ambient conditions (temperature and humidity) may be necessary especially for cases when filtration rate is low and error due to filtrate evaporation should be minimized.

6.2.2 Cake thickness

Measurement of cake thickness, by comparison, is a relatively difficult task. Almay and Lewis (1912) assumed that the cake thickness is proportional to the filtrate volume collected on the assumption that the liquid entrained within a cake was insignificant. Visual observation of cake growth is often difficult even with transparent test cells since demarcation of the cake and suspension phase is not always clear.

Over the years, a number of methods have been developed for the determination of cake thickness during the course of filtration. Description of some of these methods are presented below.

(a) *Method of Murase et al.*

Murase *et al.* (1987) developed a method of infering cake thickness through a sudden change in filtration rate due to a reduction of filtration area. The method requires the placement of a disk (an insert plate) with a sharp central hole within an experimental cell at a specified height (or distance from the filter medium) as shown in Fig. 6.6. During an experiment, test suspension enters into the cell through the inlet at the top of the cell, passes through the orifice opening of the disk and moves toward the medium. However, as the cake thickness increases and approaches the disk, the filtration area suddenly reduces. The instant at which this sharp decrease occurs may be taken as the time when the cake thickness is the same as the distance between the disk and the medium. By carrying out experiments with the disk placed at different heights (or cake thickness setting), the results of the cake thickness as a function of time can be obtained.

The hypothesis used by Murase *et al.* can be seen from the conventional cake filtration theory. From Equation (2.1.20), one has

$$\frac{dV}{dt} = \frac{p_0}{\mu\left[[\alpha_{av}]_{\Delta p_c}\frac{V\rho s}{1-\overline{m}s} + R_m\right]} \tag{6.2.2.1}$$

If the quantities $R_m, \overline{m}$ and $[\alpha_{av}]_{\Delta p_c}$ are relatively constant, $(dV/dt)^{-1}$ can be expected to be a linear function of V. However, if there is a sudden decrease in filtration area, a decrease in dV/dt should result.

Experimental results obtained by Murase *et al.* confirmed their hypothesis. In Fig. 6.8, the data they obtained (constant pressure filtration of Korean Kaolin Suspension; mass fraction = 0.393 with a cake thickness setting of 1 cm) were presented in the form of the reciprocal of the filtration rate, dt/dV against the cumulative volume, V. The first part of the data gives a linear relationship between dt/dV and V. Sharp departure from linearity was observed at certain values of V (or time) and the transition point corresponds to the time when the outer thickness was 1 cm.

The method of Murase *et al.* was incorporated into the fabrication of the multi-function test cell described previously. Teoh (2003) developed a somewhat different procedure for determining the time when cake thickness reached the cake thickness settling value. The procedure consists of plotting the data in the form of t/V vs. V (see Fig. 6.9). The data can be represented by two line segments, and the transition time (in terms of the corresponding cumulative filtration volume value) is given by the interacting points of the two line segments.

The linearity of the t/V vs. V plot is based on the conventional cake theory that the flow of the suspension is one dimensional and uniform. The procedure of

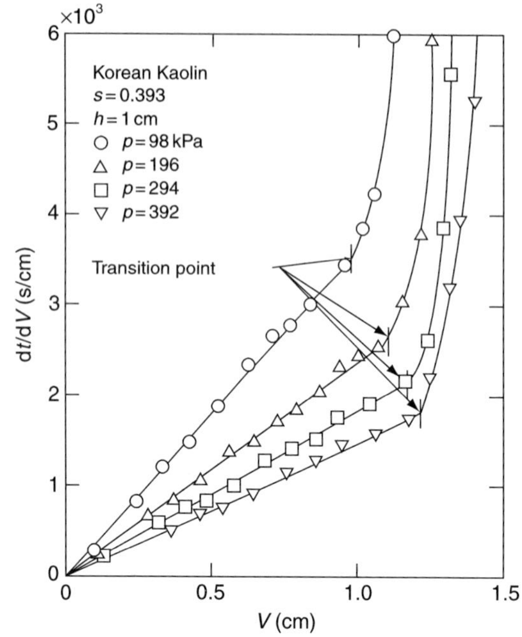

Figure 6.8 Relationship between $(dV/dt)^{-1}$ and V observed experimentally with the placement of the inserted plate. (Murase, Eritani, Cho, Nakanomori and Shirato, 1987. Reprinted by permission of The Society of Chemical Engineering, Japan.)

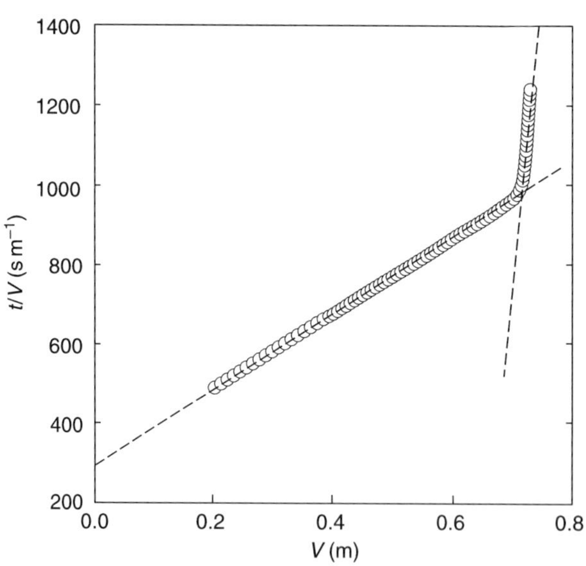

Figure 6.9 Plot of t/V vs. V for the determination of cake thickness reached at a given time. (Teoh, 2003. Reprinted by permission of Dr S.-K. Teoh.)

Murase *et al.* assumes that this condition is met until the cake thickness reaches the position of the insert plate (or disk). For the multifunction test cell case, test suspension enters into the cell through the inlet and passes through the central orifice of the insert. If there is sufficient space between the inert plate and the cake/suspension interface, the assumption of uniform one-dimensional flow may be valid. Otherwise, the effect due to the placement of the insert plate may be significant, causing uncertainties of the procedure of Murase *et al.*

In a recent study, Teoh conducted constant-pressure experiments with different cake thickness settings. The results obtained for the cases of 2% $CaCO_3$ suspension at $P_o = 100$, 500 and 800 kPa are shown in Fig. 6.10. It is apparent that the cake

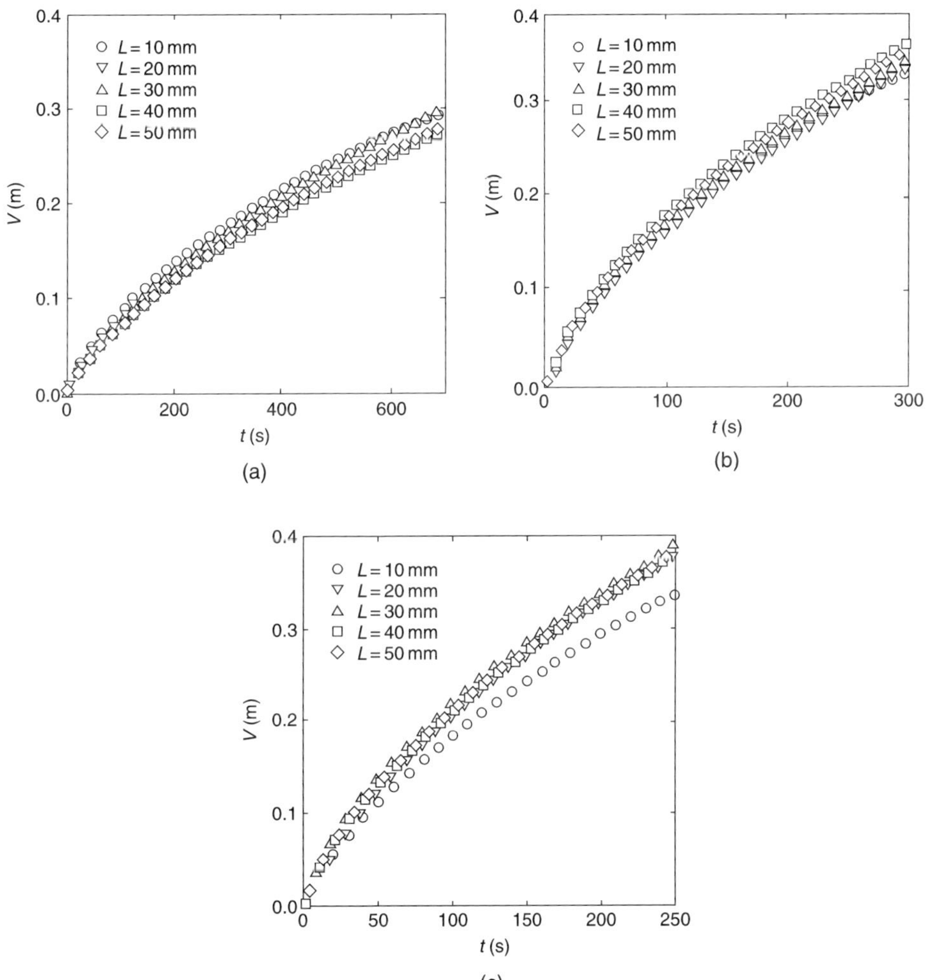

Figure 6.10 Possible effect due to the placement of inserted plate (Teoh, 2003): (a) $p_o = 100$ kPa; (b) $p_o = 500$ kPa; (c) $p_o = 800$ kPa. (Teoh, 2003. Reprinted by permission of Dr S.-K. Teoh.)

thickness setting does have some effect on the results of V vs. t. For example, the cumulative filtration volume, V, with a thickness setting of 10 mm was the highest at $p_o = 100\,\text{kPa}$, but the lowest for the two other cases. There also appears to be considerable merging of data and this fact becomes more pronounced if the data are plotted in the form of t/V vs. V. It is reasonable to assume that the data of L vs. t obtained by using the procedure of Murase *et al.* may not have a high degree of accuracy. But as shown in later discussions, this deficiency of low accuracy may be shared by other methods as well.

(b) *Method based on pressure measurements*

During cake filtration, the test cell is composed of two regions: the cake formed above the filter medium and the test suspension present outside the cake. For practical purpose, the liquid pressure in the suspension phase may be considered uniform and equal to the operating pressure. On the other hand, there is significant pressure drop across the cake and the pressure drop increases as the cake thickness increases. This phenomena was first noted by murase *et al.* (1987) and a method based on this phenomena for determining cake thickness history was developed in recent years by Theliander and coworkers (Fathi-Najafi and Theliander, 1995; Sedin *et al.*, 2003; Johansson *et al.*, 2004). The experimental apparatus used by these investigators is shown schematically in Fig. 6.11. Seven probes were placed

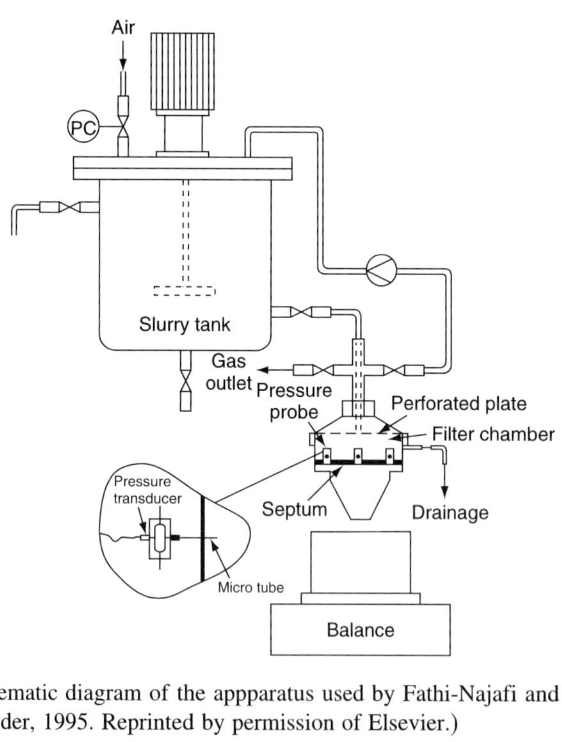

Figure 6.11 Schematic diagram of the appparatus used by Fathi-Najafi and Theliander. (Fathi-Najafi and Theliander, 1995. Reprinted by permission of Elsevier.)

inside the filter chamber and each probe consisted of a microtube and a pressure transducer. These probes were placed at a distance of 3.9, 6.8, 9.7, 13.0, 16.0, 19.0 and 22.0 mm away from the filter medium and numbered numerically according to the ascending order of their distances from the medium (i.e. p_1 is the probe located closest to the medium, or 3.9 mm). Figures 6.12a–c gives the pressure histories obtained in the work of Fathi-Najafi and Theliander (1995). As shown in Fig. 6.12a, the readings of these probes displayed an orderly transition from the

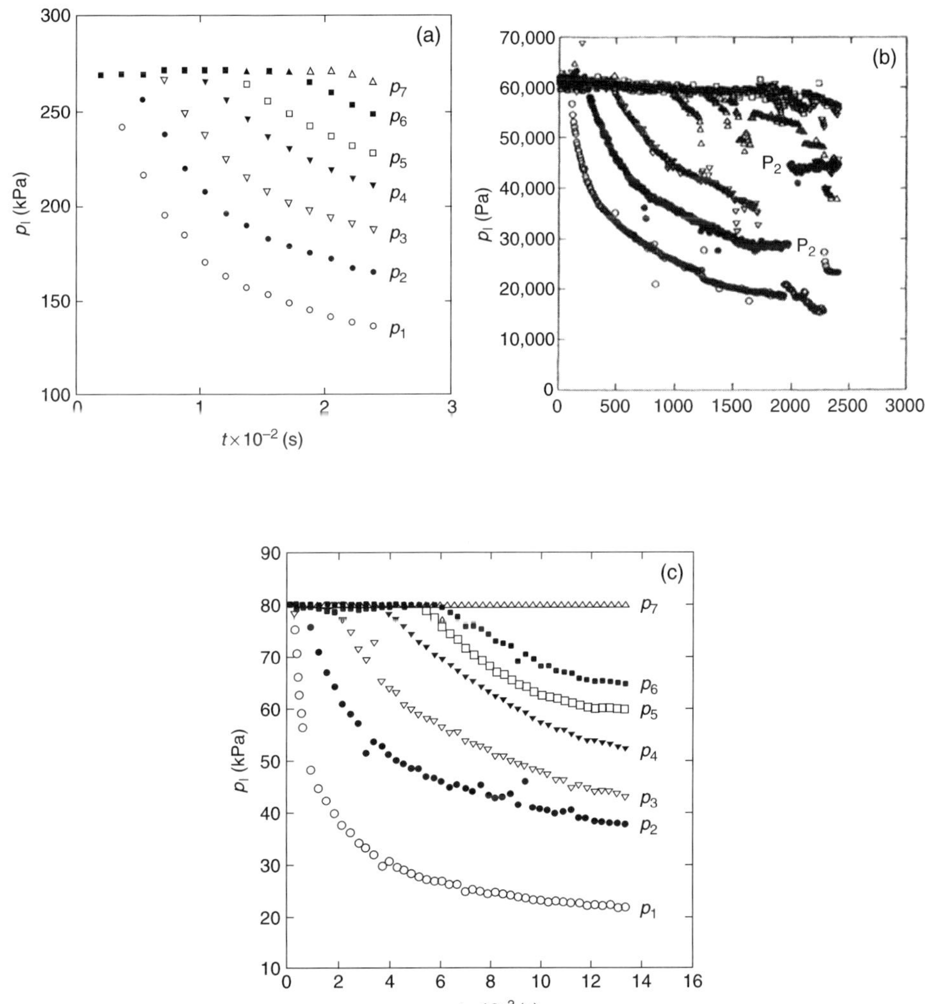

Figure 6.12 Pressure profile across filter cake obtained by Fathi-Najafi and Theliander for cake thickness measurement: (a) successful test; (b) occurrence of channels; (c) uneven cake function. (Fathi-Najafi and Theliander, 1995. Reprinted by permission of Elsevier.)

value of the operating pressure (270 kPa) and the transition gave the time when the cake thickness reached the particular probe location.

The method also has the advantage of detecting the anamolies of cake formation as shown in Fig. 6.12b,c. In Fig. 6.12b, the readings of the pressure probes were erratic and overlapping, probably due to channel formation within the cake. Figure 6.12c gives the results of a case of possible uneven cake formation. It can be seen that the response of the sixth probe was earlier than expected. The readings of the seventh probe indicated that cake thickness never reached the value of 22 mm (the location of the probe) although visual observations by the investigators indicated otherwise.

Application of the pressure measurement method is not limited to constant-pressure filtration experiment. Earlier, Murase *et al.* showed the possibility of applying the method to experiments conducted with the operating pressure increasing stepwise. One set of the results is shown in Fig. 6.13. The results shown in Fig. 6.13 were those of filtration of Korean Kaolin Suspension with operating pressure increasing incrementally. The results were plotted in the form of the pressure at a given location against the cumulative filtrate volume. For a given problem, the transition of the pressure reading from the operating pressure to a lower value signifies the cake thickness reaching the location of the probe. Based on results shown in Fig. 6.13 and the results of V vs. t, the cake thickness history of L vs. t can be readily obtained.

Similar to the method of Murase *et al.* described before, the method based on pressure measurement depends on the determination of the transition of experimental readings. Since subjective judgement was often used in determining the transition,

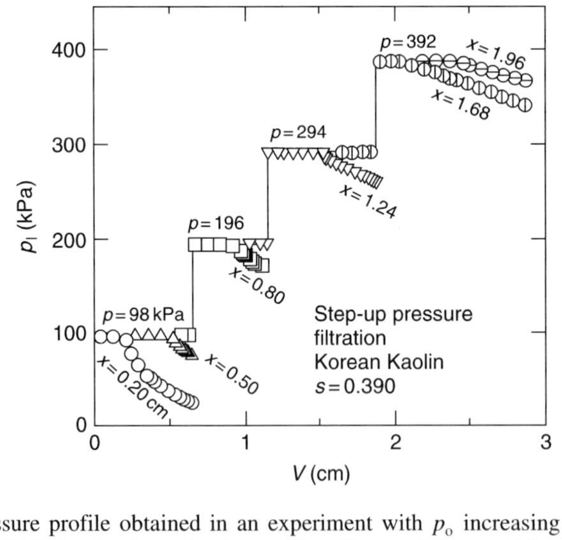

Figure 6.13 Pressure profile obtained in an experiment with p_0 increasing stepwise and used by Murase *et al.* (1987) for the determination of filter history. (Murase, Eritani, Cho and Shirato, 1989. Reprinted by permission of The Society of Chemical Engineering, Japan.)

the results obtained may not have a high degree of accuracy. The intrusive effect of the pressure probes is another problem. These are the inherent disadvantages of both the method of Murase *et al.* and that based on pressure measurement.

(c) *Non-intrusive methods of cake thickness measurement*

In addition to the two methods described above, a number of non-intrusive methods have been developed for online cake thickness measurement in recent years. Generally speaking, these methods measure cake thickness by passing a signal (acoustical or optical) through or onto the cake to be measured and monitoring the change in amplitude or intensity of the signal. As a rule, they require the use of reasonably elaborate equipment and calibration relationships between the measurement results, and cake thickness must be established first. In certain cases, significant efforts of signal processing may also be necessary. In the following, two of these methods which have shown to be of practical use are briefly described.

Takahashi *et al.* (1991) developed a method of determining cake thickness based on ultrasonic measurements. The method employed the use of specially designed ultrasonic polymer concave transducer (see Fig. 6.14), the experimental set-up shown in Fig. 6.15. The transmitting transducer was located inside the filtration cell and the receiving transducer was mounted on a precision stage assembly outside the cell. The amplitude of the ultrasonic wave emitted by the transmitting transducer and that of the receiving transducer was measured and recorded as functions of time.

The attenuation coefficient of a medium, α, is defined as

$$\frac{\alpha}{f^2} = \frac{20}{df^2} \log(A_I/A_0) \qquad (6.2.2.2)$$

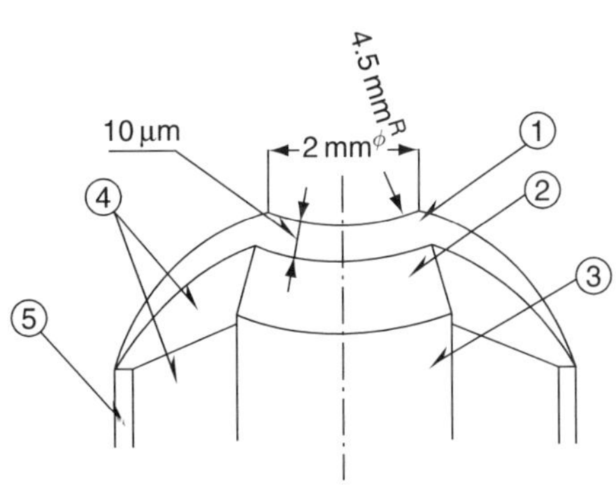

Figure 6.14 Polymeric concave transducer used by Takahashi *et al.* (1991) for cake thickness measurement: (1) Al Electrode, (2) polymer film (VDF-TREE), (3) metal rod, (4) polymer film, (5) metal cylinder. (Takahashi, Kobayashi, Yokota and Koyama, 1991. Reprinted by permission of The Society of Chemical Engineers, Japan.)

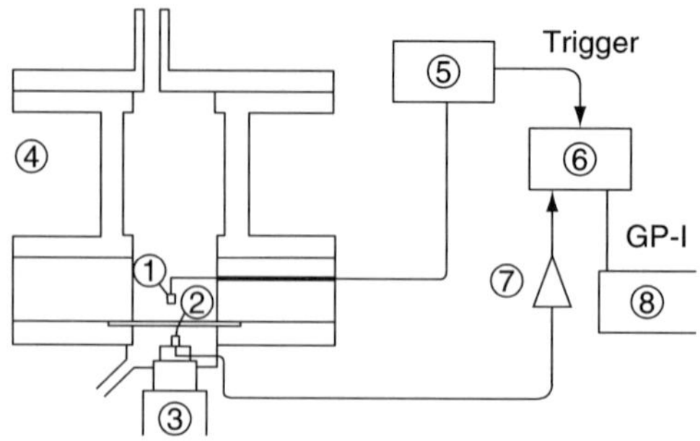

Figure 6.15 Block Diagram of the online measurement of cake thickness used by Takahashi *et al.* (1991): (1) transducing tranceducer, (2) receiving tranceducer, (3) precision stage assembly, (4) batch cell, (5) pulse function generator, (6) digital oscilloscope, (7) amplifier, (8) personal computer. (Takahashi, Kobayashi, Yokota and Koyama, 1991. Reprinted by permission of The Society of Chemical Engineers, Japan.)

where f is the wave frequency, A_I is the amplitude of the propagating wave at reference point I and A_o is the amplitude of the wave after propagating a distance of d.

If the attenuation coefficient of the cake is significantly different from that of the suspension, the cake thickness L can be expressed as

$$L = \frac{20 \log(A_\mathrm{I}/A_0)}{(\alpha/f^2)_{\mathrm{sus}} - (\alpha/f^2)_{\mathrm{cake}}} \tag{6.2.2.3}$$

where A_I and A_o are now the amplitudes of the transmitting and receiving waves, respectively. (α/f^2) can be determined experimentally. The subscripts "sus" and "cake" denote suspension and cake, respectively.

More recently, an optical method of determining cake thickness was developed by Hamachi and Mietton-Peuchot (2001). This method made use of a He–Ne beam and an optical captor and was based on increasing light absorption of a cake due to its increase in thickness. A schematic diagram of the optical apparatus used is shown in Fig. 6.16. With the use of a series of lenses, microscopic apertures and a diaphragm, a laser beam was focused tangentially onto a point of filter medium. The image of the focal point was received onto a photomultiplier and the intensity measured. Through proper signal processing, the cake thickness as a function of time was determined.

It is apparent that the non-intrusive methods have distinct advantages over the methods of Murase *et al.* and that based on pressure measurements. Nevertheless, the non-intrusive methods do have their problems. First, the methods described above make measurements at a particular point of the cake. The results obtained may be considered valid if cake growth is uniform over the entire cake surface. As discussed previously (see Section 6.2.2b and Fig. 6.11), this condition is not always satisfied.

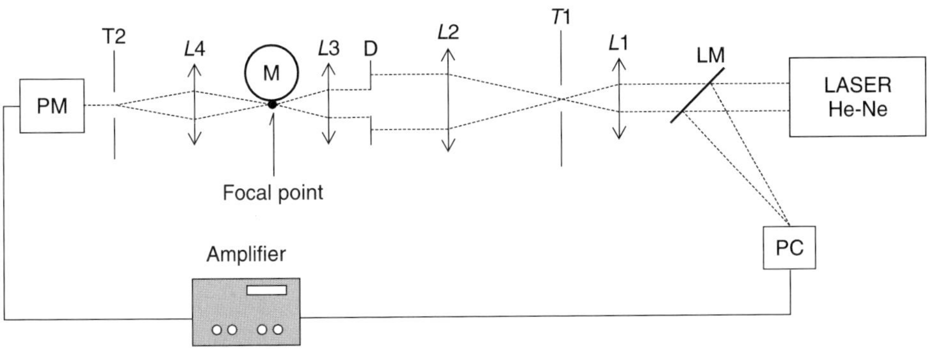

Figure 6.16 Schematic diagram of the laser optical apparatus used by Hamachi and Mietton-Peuchot (2001) for cake thickness measurement. *L*1, *L*2, *L*3, *L*4: convex on biconvex lenses; T1, T2: microscopic holes; D: diaphragm; PM: photomultiplier; PC: photocell; M: membrane (filter medium); LM: glass strip. (Hamachi and Mietton-Peuchot, 2001. Reprinted by permission of The Institution of Chemical Engineers.)

The non-intrusive methods inevitably require the knowledge of the relationship between cake thickness and the measuring signal. And it is often time-consuming to obtain the calibration results. In the case of the ultrasonic method of Takahashi *et al.*, the method was applied only to cakes with insignificant compressible effect. To a lesser degree, this was also true for the method of Hamachi and Mietton-Peuchot. The cross-flow filtration experiments used to test the optical method was carried out under relatively low pressure.

6.3 MEASUREMENT OF INTERNAL CAKE PROPERTIES

For more detailed examination of cake formation and growth, information such as the filtrate pressure (and therefore cake compressive stress) and cake solidosity (or porosity) profiles and their evolutions are required. The various methods which can be used for their determination are described below.

Pressure profile measurements was discussed to some extent previously (see Section 6.2.2). Generally speaking, accurate filtrate pressure measurements can be made by placing pressure transducers inside test cell. The problem encountered earlier, namely, the intrusive effect of the probes, can be minimized through proper probe design and placements as shown in the work of Johansson *et al.* (2004) and Tarleton (2004).

For methods of determining cake solidosity, following Meeten (1993), these methods may be divided into two categories: the destructive and the non-destructive methods. For destructive methods, filtration experiments is terminated at a particular time and the cake formed is removed and examined. For example, the wet to dry cake mass ratio $\overline{m}$ can be determined by weighing the cake first, dewatering the cake by drying and weighing the dried cake.[1] The average cake solidosity $\bar{\bar{\varepsilon}}_s$ can then be determined from Equation (2.1.19) or

$$\bar{\bar{\varepsilon}} = \left[\frac{\rho_s}{\rho} (\overline{m} - 1) + 1 \right]^{-1} \tag{6.3.1}$$

More detailed information such as the solidosity profile can be obtained by dissecting the cake formed and determining the solidosities of each segment by weighing. Smiles and Rosenthal (1968) developed such a dissection method with a segment height of 5 mm. A more refined procedure was reported by Meeten (1993). Meeten's method enabled the determination of cake solidosity over a thickness of 0.5 mm, thus providing a much better spatial resolution of the cake solidosity than the earlier work. The dissection method also allows complete examination of the cake formed (for example, the possible variation of cake particle size distribution spatially). The disadvantage of the destructive method is obvious; the information obtained from one experiment is restricted to that at a particular time. For the purpose of obtaining the evolution of the solidosity profile, the results of a number of similar filtration experiments terminated and different times are required.

A large number of non-destructive methods for probing cake structure have been developed in recent years. By the principles and/or the probing signals used, they can be further divided as follows.

(a) *Methods based on electrical conductivity measurements*
Several investigators have explored the possible determination of cake porosity (or solidosity) based on electrical conductivity measurements (Baird and Perry, 1967; Shirato *et al.*, 1971; Wakeman, 1981; Chase and Willis, 1992; Tarleton, 2004). The basic problem here is to establish a relationship between the cake conductivity and solidosity. Shirato *et al.* (1971) presented an empirical expressions relating the porosity with the so-called formation factor F defined as

$$F = \frac{\text{Electrical resistance of a saturated porous medium, } R}{\text{Electrical resistance of the saturating liquid, } R_o} \tag{6.3.2}$$

to be

$$\varepsilon = F^{\overline{\beta}} \tag{6.3.3}$$

[1] Strictly speaking, $\overline{m}$ (and $\bar{\bar{\varepsilon}}_s^0$) is a function of time. For a given experiment, $\overline{m}$ is nearly constant as shown in Fig. 6.19.

where $\overline{\beta}$ is an empirical constant. However, for certain cases, Equation (6.3.2) may not be valid and different expressions should be used. That the relationship between ε and F is system specific is a major disadvantage of the methods based on electrical conductivity measurements.

(b) *Methods based on NMR imaging*
Nuclear Magnetic Resonance (NMR), which is best known for its diagnostic applications and to a lesser degree in molecular structural analyses, can be readily applied for determining filter cake structure (Horsfield *et al.*, 1989; La Heij *et al.*, 1996). An absolute accuracy of 1% was claimed by La Heij *et al.* for their porosity profile measurement with ε ranging from 0.84 to 0.97 (constant pressure filtration of flocculated sludge with $p_o = 50 \sim 100\,\text{kPa}$). A thorough discussion of using NRM imaging in solid/liquid separation studies (including cake filtration) was recently given by Hall *et al.* (2004).

(c) *γ- and x-ray attenuation*
A number of investigators have developed methods based on radiation attenuation for measuring the composition of solid/liquid systems including filter cakes (Been and Sills, 1981; Bierk *et al.*, 1988; Bergstrom, 1992; Sedin *et al.*, 2003). Sedin *et al.* obtained solidosity results for four kinds of filter cakes with an experimental error estimated at 0.005–0.04. The measurements were made using 241 Am as the radiation sources and a NaI scintillation counter.

Computer-assisted tomography (CAT) is a more elaborate way of applying x-ray attenuation to the study of porous media. Both Shen *et al.* (1994) and Tiller *et al.* (1995) conducted CATSCAN studies of filtration/sedimentation. The CATSCAN results were largely limited to the identification of the position and movement of interface boundaries (suspension/clear liquid; cake/suspension). The full potential of the method – its capability of providing detailed information of cake structure such as pore size and shape, cake particle size distribution, etc – however, was not explored.

It is evident that these non-destructive methods have enormous advantages over the destructive methods. But similar to the non-intrusive methods for cake thickness measurements discussed in Section 6.2.2, these methods are complex and their applications require considerable efforts in calibrations and/or in the development of the necessary algorithms for signal processing. The cost and efforts required may not be proportional to the needs of users. Selection of appropriate methods can only be made on a case-by-case basis.

6.4 CAKE FILTRATION RESULTS

As may be expected, the results of the rather large number of experimental studies reported in the literature are not always consistent. A brief description and discussion of some of these results are given below.

6.4.1 Cake filtration performance

Filtration performance which can be defined by the histories of the cake thickness, the filtrate volume recovered (for the specified operating pressures) or the operating pressure required (for specified filtration rate) depends upon a number of variables. For convenience, one may divide these variables into two categories: operating and system variables. The operating variables include the operating pressure (or the filtration rate) and the medium used. The system variables encompass properties of the suspension to be treated, such as particle size and size distribution, the particle surface charges, the pH of the suspending liquid and the electrolyte/surfactant present. The effect due to these variables may be summarized as follows:

(a) In a recent study, Teoh (2003) obtained constant-pressure filtration results of a number of systems. The experiments were conducted using the multifunction test cell described in Section 6.1 and the method previously developed by Murase *et al.* (1987) was applied for cake thickness measurements. The cake thickness results (L vs. t) of two systems (2% by weight $CaCO_3$ and 5% by weight Kaolin suspension) obtained under three operating pressures are shown in Fig. 6.17. Figure 6.18 gives the results of V vs. t for the same two systems. The expected effect of the operating pressure, increasing V and L with the increase of p_o, is clearly displayed.

A measure of the quality of the cake obtained can be seen from the value of the wet to dry cake mass ratio $\overline{m}$, which is given by Equation (2.1.19) or

$$\overline{m} = 1 + \frac{\rho(1-\overline{\overline{\varepsilon}}_s)}{\rho_s \overline{\overline{\varepsilon}}_s} \tag{6.4.1.1}$$

where ρ and ρ_s are densities of the filtrate and particle, respectively. $\overline{\overline{\varepsilon}}_s$ is the average (spatial) cake solidosity defined by Equation (2.1.8). By mass balance, $\overline{\overline{\varepsilon}}_s$ can be found to be

$$\overline{\overline{\varepsilon}}_s = \frac{\rho(V+L)s}{\rho_s L(1-s) + \rho L s} \tag{6.4.1.2}$$

One can therefore determine the value of $\overline{m}$ as a function of time from the V vs. t and L vs. t data. As an example, Fig. 6.19 shows the results of $\overline{m}$ vs. t of $CaCO_3$ filter cake with $p_o = 800\,\text{kPa}$ based on the data presented in Figs. 6.17 and 6.18. As seen from this figure, $\overline{m}$ was nearly constant except for the very beginning of filtration. The assumption that $\overline{m}$ can be treated as a constant in the conventional cake filtration theory is largely justified.

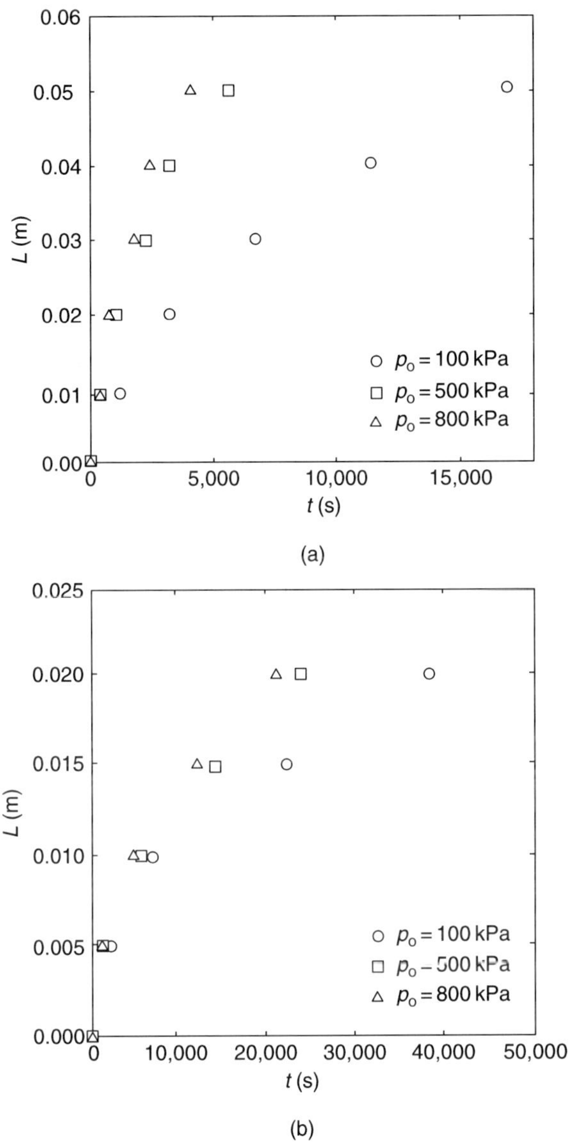

Figure 6.17 Cake thickness vs. time; Constant pressure filtration: (a) 2% $CaCO_3$ suspension; (b) 5% Kaolin suspension. (Teoh, 2003. Reprinted by permission of Dr S.-K. Teoh.)

(b) The effect of the medium resistance on filtration performance is shown in Fig. 6.20. This figure gives the results of V vs. t collected by Teoh (2003) for the filtration of 2% $CaCO_3$ suspensions under $p_o = 100\,\text{kPa}$. The filter media were composed of a number of Whatman No. 1 filter papers. The results demonstrate clearly that

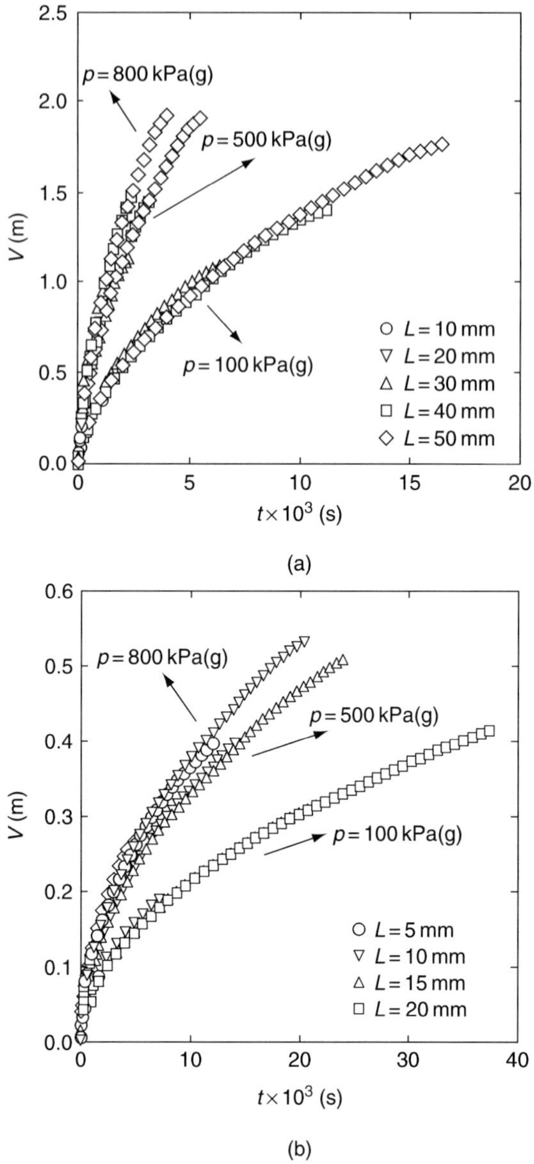

Figure 6.18 Cumulative filtrate volume vs. time, constant pressure filtration at various operating pressure and with different cake thickness Setting: (a) 2% $CaCO_3$ suspension; (b) 5% Kaolin suspension. (Teoh, 2003. Reprinted by permission of Dr S.-K. Teoh.)

with the increase of medium resistance (increasing number of filter papers), the rate of filtration and the cumulative filtrate volume decrease. More important, these data on medium resistance are useful in the discussion of the parabolic behavior of constant pressure filtration to be given in Section 6.3.2.

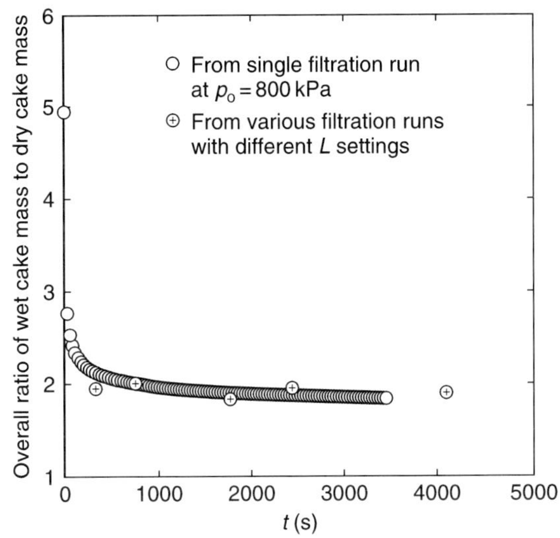

Figure 6.19 Wet to dry cake mass ratio vs. time; constant pressure filtration of 2% $CaCO_3$; $p_o = 800\,\text{kPa}$. (Teoh, 2003. Reprinted by permission of Dr S.-K. Teoh.)

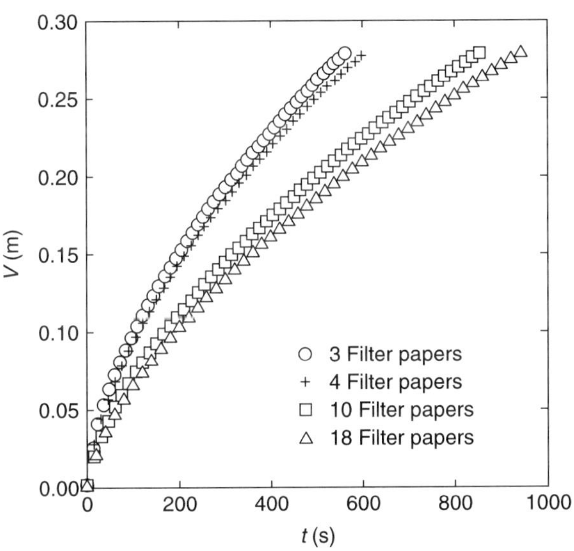

Figure 6.20 Effect of Medium Resistance; Cumulative filtrate volume vs. time, constant pressure filtration of 2% $CaCO_3$, $p_o = 100\,\text{kPa}$. (Teoh, 2003. Reprinted by permission of Dr S.-K. Teoh.)

(c) The effect of particle size can be seen from the fact that the quantity k/a_p^2 (k: cake permeability, a_p: particle radius) is a function of ε_s (see Fig. 5.1) or that the specific cake resistance is inversely proportional to a_p^2. The effect of suspension particle concentration has been examined by a number of investigators (for example,

Christensen and Dick, 1985b; Meeten, 2000; Johansson, 2003). The results of Christensen and Dick showed that the average specific cake resistance decreased with the increase of particle concentration until $s = 0.005$. For $s > 0.005$, particle concentration was found to have no effects. Meeten found that cakes of polyethylene particles were less permeable (or show greater specific cake resistance) if formed from suspension of less particle concentration. The opposite behavior was observed for sephadex particles. The more recent work of Johansson was in agreement with Meeten's findings of polyethylene cakes. However, the differences in all these cases were not significant.

Taking into account that these differences could be attributed to experimental errors and uncertainties in data interpretation, and the lack of any theoretical argument for particle concentration effect, the possible effect of suspension particle concentration probably should be discarded at least for the time being.

(d) The nature and magnitude of the inter-particle interaction forces are controlled, to a large degree, by the ionic strength and pH of the suspending liquid. The effect on cake filtration performance can be seen from the data reported by Larue *et al.* (2003), which are shown in Fig. 6.21. The results shown in this figure are those of constant pressure filtration of latex (styrene-acrylic [SA] and vinyl-versatate [VV]) particle suspensions electrically coagulated (with iron electrodes) or dosed with chemicals ($FeSO_4$ or $FeCl_3$) at constant pressure with specified pH values and ionic strength. The results are plotted in the form of $(t - t_o)/(V - V_o)$ vs. $V - V_o$ to account for the fact that it took a finite time (t_o) for the operating pressure to reach the specified value. The effect of the particle–particle interaction is clearly displayed.

6.4.2 Parabolic behavior of constant-pressure filtration

According to the conventional cake filtration theory of Chapter 2, for constant-pressure filtration, the relationship between the cumulative filtrate volume and time is given by Equations (2.2.2) and (2.2.3), or

$$\mu s \rho \overline{(1 - \overline{m}s)^{-1}[\alpha_{av}]_{p_{s_m}}} \frac{V^2}{2} + \mu R_m V = p_o t \tag{6.4.2.1}$$

with

$$\overline{(1 - \overline{m}s)^{-1}[\alpha_{av}]_{p_{s_m}}} = \frac{2}{V^2} \int_0^V (1 - \overline{m}s)[\alpha_{av}]_{p_{s_m}} V \, dV \tag{6.4.2.2}$$

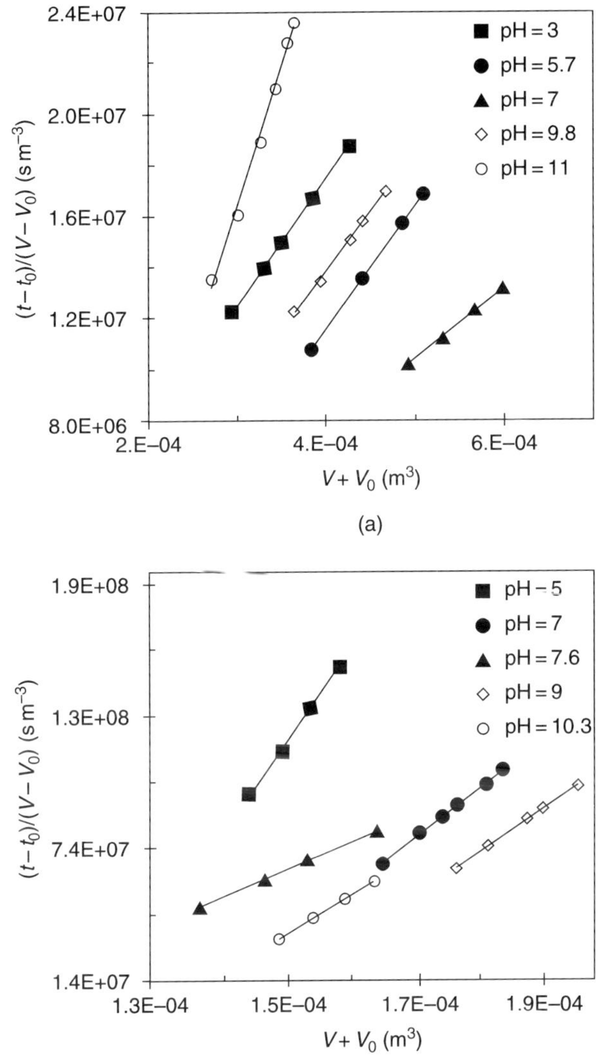

Figure 6.21 Effect of solution pH on constant-pressure filtration (a) EC-treated SA suspensions (b) EC-treated VV particles. (Larue, Vorobiev, Vu and Durand, 2003. Reprinted by permission of Elsevier.)

The definition of $[a_{av}]_{p_{s_m}}$ is given by Equation (2.1.13)

$$[\alpha_{av}]_{p_{s_m}} = \frac{\Delta p_c / p_{s_m}}{\left[\int_0^{p_{s_m}} \left(\frac{1}{\alpha} \right) (-f') \mathrm{d}p_s \right] \Big/ p_{s_m}} \tag{6.4.2.3}$$

where α is local specific cake resistance $[= 1/(k\rho_s\varepsilon_s)]$. Δp_c is the pressure drop across the cake. p_{s_m} is the compressive stress p_s corresponding to $p_\ell = p_0 - \Delta p_c$. The relationship between p_ℓ and p_s is given by f' defined by Equation (2.1.11) or

$$\frac{\partial p_\ell}{\partial p_s} = f' \tag{6.4.2.4}$$

f', in general, is a function of ε_s.

The quantity $\overline{(1-\overline{m}s)[\alpha_{av}]_{p_{s_m}}}$ is a function of V (or time). However, if the cake resistance is dominant, one may replace Δp_c by p_o. (This is not the same as neglecting the medium resistance, or letting $R_m = 0$ as stated before.) With this assumption, p_{s_m} becomes constant and so does $[\alpha_{av}]_{p_{s_m}}$. Since $\overline{m}$ is largely constant as shown in Fig. 6.18, the quantity $\overline{(1-\overline{m}s)[\alpha_{av}]_{p_{s_m}}}$ becomes

$$\overline{(1-\overline{m}s)^{-1}[\alpha_{av}]_{p_{s_m}}} = (1-\overline{m}\,s)^{-1}[\alpha_{av}]_{p_{s_m}} \tag{6.4.2.5}$$

with p_{s_m} being the value of p_s corresponding to $p_\ell = 0$. Equation (6.4.2.1) now becomes

$$\mu s\rho(1-\overline{m}s)^{-1}[\alpha_{av}]_{p_{s_m}}\frac{V^2}{2} + \mu R_m V = p_0 t \tag{6.4.2.6}$$

and $[\alpha_{av}]_{p_{s_m}}$ becomes

$$[\alpha_{av}]_{p_{s_m}} = \frac{\Delta p_c/p_{s_m}}{\int_0^{p_{s_m}}\left(\frac{1}{\alpha}\right)(-f')\mathrm{d}p_s/p_{s_m}} \tag{6.4.2.7}$$

If $f' = -1$, $p_\ell + p_s = p_0$ and $p_{s_m} = p_0$, $[\alpha_{av}]_{p_{s_m}}$ becomes

$$[\alpha_{av}]_{p_{s_m}} = \frac{p_0}{\int_0^{p_0}\left(\frac{1}{\alpha}\right)\mathrm{d}p_s} \tag{6.4.2.8}$$

Equation (6.4.2.6) gives t as a second order polynomial of V: or there is a linear relationship between t/V and V. The expression is commonly known as the parabolic law of constant pressure filtration.

The extent to which filtration data can be described by Equation (6.4.2.6) can be seen by replotting the results of Fig. 6.20 in the form of t/V and V as shown in Fig. 6.22. It is clear that linearity between t/V and V is observed in the latter part of data for all cases. On the other hand, deviation from linearity is more pronounced for the case with greater medium resistance. Both Ruth *et al.* (1933) and Tiller *et al.* (1981) previously argued that deviation of the parabolic behavior (or the linearity between t/V and V) was due to interior medium clogging. This argument does not appear likely in light of the data of Fig. 6.22. If one assumes that the fine particle penetration into filter medium is a phenomenon similar to what is known as deep bed filtration, additional resistance due

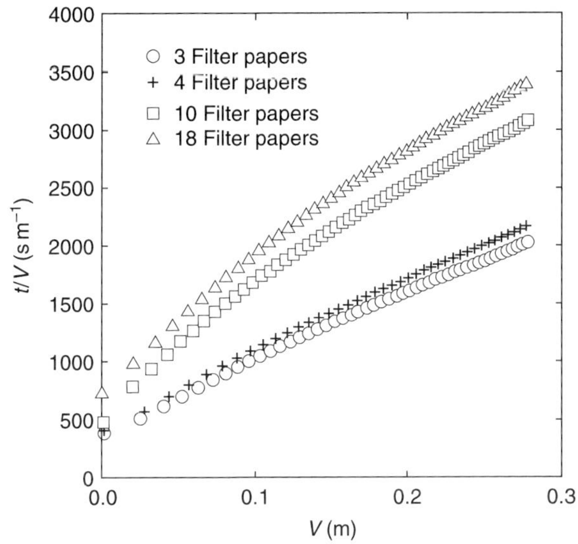

Figure 6.22 Replot of data of Fig. 6.20 in the form of t/V vs. V.

to particle clogging of a medium should depend upon the amount of particles retained by the medium. Consequently, the fractional increase of medium resistance would be higher if the inherent medium resistance is less. The results of Fig. 6.20 do not support this hypothesis.

Willis *et al.* (1983) contended that failure to obey the parabolic law of filtration invalidates the use of the conventional cake filtration theory and a more rigorous approach is needed to describe cake filtration. What they failed to realize was that the conventional theory is not empirical, as pointed out before. The more "rigorous approach" they used actually had a serious error; they assumed that the compressive stress across a filter cake was uniform [i.e. their Equation (44) of Willis *et al.*, 1983]. Equally important, the experimental evidence they presented to substantiate their argument seems not to have sufficient accuracy. The conclusion that the average cake solidosity is constant was also not validated by the results of Fig. 6.19.

Christensen and Dick (1985a,b) presented evidence to demonstrate that a number of factors may contribute to the observed non-parabolic behavior. These investigators carried out extensive constant-pressure experiments (batch) using several types of suspensions with and without conditioning and nine different filter media. While the overwhelming majority of the results were found to follow the parabolic behavior, a number of cases displayed distinct non-parabolic behavior. Non-parabolic data were associated with the presence of significant sedimentation prior to filtration measurement (caused by time lapse of applying pressure once the test cell was filled with suspension), sedimentation concomitant with filtration (as seen by results by varying the time between the addition of conditioning chemicals and the start of an experiment) and suspension-media interaction. In many instances, the extent of non-parabolic behavior may be reduced or eliminated by applying pressure immediately following the introduction of the test suspension into the filtration cell, thickening the suspension to minimize the significance of sedimentation and using a different type of filter medium.

Another factor which may cause the non-parabolic behavior resides in the mechanics of conducting filtration experiments. For constant-pressure filtration experiments, a finite time interval is required for the applied pressure to reach its specified value, giving rise to the so-called "nonsynchronization of time-filtrate volume data". The results of one such example given by Christensen and Dick (1985b) are shown in Fig. 6.23a,b. The results shown in Fig. 6.23a were those of filtration of 25 ml of a 6% anaerobically digested sludge. It is clear that the data do not obey the parabolic law. However, after correcting the volume data for the 2 ml filtrate already present (i.e. the filtrate already present in the cell prior to the start of time-recording), the linear behavior between t/V and V was observed (see Fig.6.23b).

Based on the available experimental data, one may conclude that Equation (6.4.2.5), in spite of the many simplifications used for its derivation, in most cases, gives a reasonable description of constant-pressure filtration. One must, however, keep in mind of its limitations. It does not describe the behavior of the initial period of filtration with sufficient accuracy and fails in cases where sedimentation is significant.

6.4.3 Behavior of the initial period of filtration

Koenders and Wakeman, in several publications (1996, 1997a,b), analyzed cake formation and the effect of particle interaction on the process. According to their results, the initial period of filtration may be described as

$$V = \overline{x}t + \overline{y}t^{3/2} \tag{6.4.3.1a}$$

and

$$\overline{x} = \frac{p_0}{R_m}(0) \tag{6.4.3.1b}$$

where $R_m(0)$ is the initial medium resistance and R_m is assumed to be a function of time, dependent upon the cake solidosity value at the medium surface.

Similar but not identical expressions of $\overline{y}$ were given in their publications. In terms of its dependence of p_o, $\overline{y}$ can be written as

$$\overline{y} = b_1 p_0 + b_2 p_0^2 \tag{6.4.3.2}$$

where b_1 and b_2 are dependent upon the nature and magnitude of particle interactions.

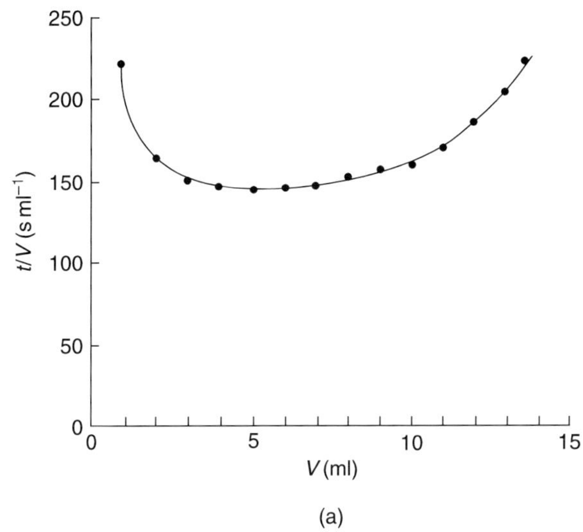

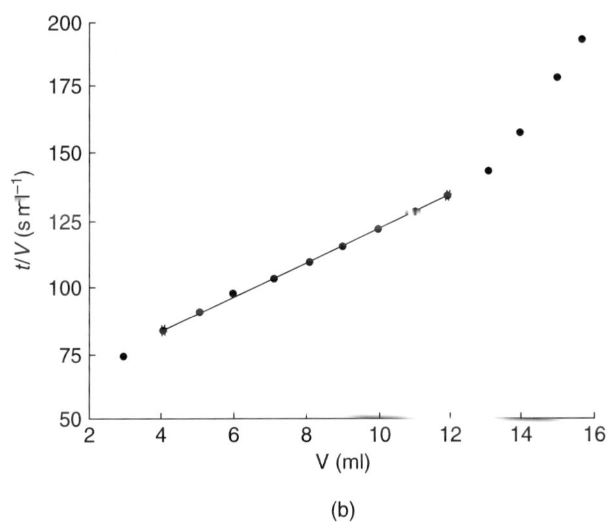

Figure 6.23 Effect of non-Synchronization of time-filtrate volume data: (a) nonsynchronous time-filtrate volume data; (b) corrected nonsynchronous time-filtrate volume data. Filtrate volume being total volume in ml. (Christensen and Dick, 1985b. Reprinted by permission of The American Society of Civil Engineers.)

Equation (6.4.3.1a) may be rearranged to give

$$\frac{V}{t} = \bar{x} + \bar{y}\, t^{1/2} \tag{6.4.3.3}$$

In other words, the relationship between (V/t) and $t^{1/2}$ should be linear. Koenders and Wakeman confirmed this with their own data.

More recently, both Meeten (2000) and Teoh (2003) applied their data to test the validity of Equation (6.4.3.3). Meeten's data were those of constant pressure filtration of polyethylene particles (non-deformable) and Sephadex particles (deformable). He defined the initial period as the period extending from $t = 0$ to the time when the instantaneous filtration velocity decreased to half of the initial value. Meeten's results showed that Equation (6.4.3.3) was obeyed for the case of polyethylene suspensions but not Sephadex suspensions. The results are shown in Fig. 6.24. Teoh, on the other

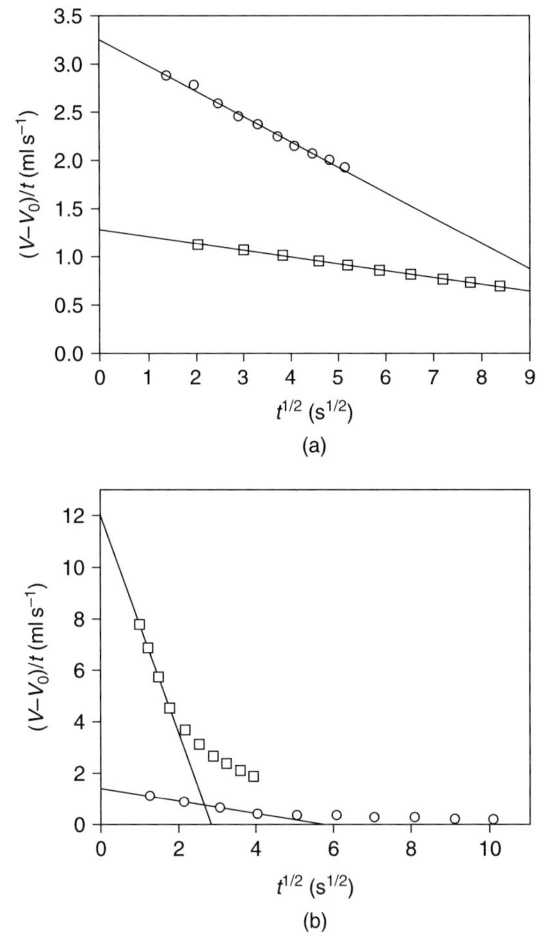

Figure 6.24 Experimental validation of the linear relationship between V/t and $t^{1/2}$: (a) constant pressure filtration of suspension of polyethylene particles ($\varepsilon_{s_o} = 0.8$) at $p_o = 2.88$ kPa (squares) and 10.3 kPa (circular); (b) constant pressure filtration of Sephadex suspensions of 109^{-1} (squares) and 42.6 g^{-1} (circles) at $p_o = 31.37$ kPa. (Meeten, 2000. Reprinted by permission of Elsevier.)

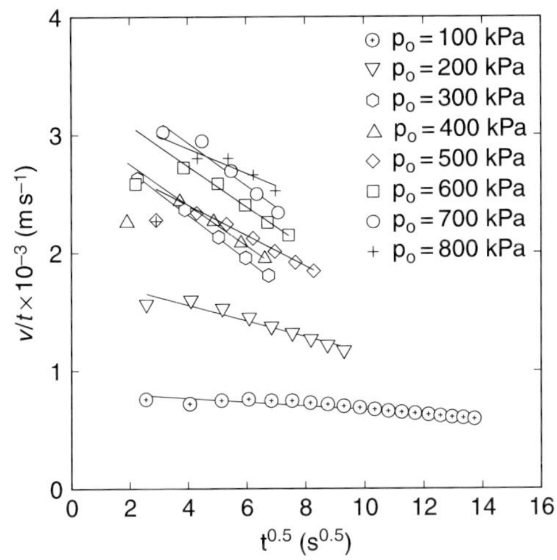

Figure 6.25 Experimental validation of the linear relationship between V/t and $t^{1/2}$, constant pressure filtration of 2% $CaCO_3$ suspensions. The last digit of the run no. gives the operating pressure in bar. (Teoh, 2003. Reprinted by permission of Dr S.-K. Teoh.)

hand, found that her data obeyed Equation (6.4.3.3) approximately but not Equation (6.4.3.2) (see Figs. 6.25 and 6.26).

6.4.4 Filter medium resistance

The medium resistance, R_m, in principle, can be determined from filtration data. If Equation (6.4.2.6) is assumed to be valid, R_m can be found from the intercept of the linear plot of t/V vs. V. Alternatively, R_m can be determined from the initial value of the filtration rate, $(dV/dt)_{t\to 0}$. From Equation (2.1.20) as $t \to 0$, one has

$$R_m = \frac{p_0}{\mu (dV/dt)_{t\to 0}} \tag{6.4.4.1}$$

The value of $(dV/dt)_{t\to 0}$ can be estimated according to the following procedure. From the data of V vs. t, values of (dV/dt) at various times can be obtained. Extrapolation of (dV/dt) vs. t then yields the value of $(dV/dt)_{t\to 0}$. The medium resistance can be found from Equation (6.4.4.1).

A comparison of the R_m values obtained from the two methods was made by Teoh (2003). Teoh's results (for the filtration of $CaCO_3$ and Kaolin suspensions) are reproduced in Table 6.1. Also included in this table are the intrinsic medium resistance determined by passing deionized water through the medium at various pressures.

The results shown in Table 6.1 can be summarized as follows: The values of R_m determined from the two methods are rather close – within 10% for the case of $CaCO_3$ and within a factor of two for Kaolin suspension. For both systems, R_m is shown to

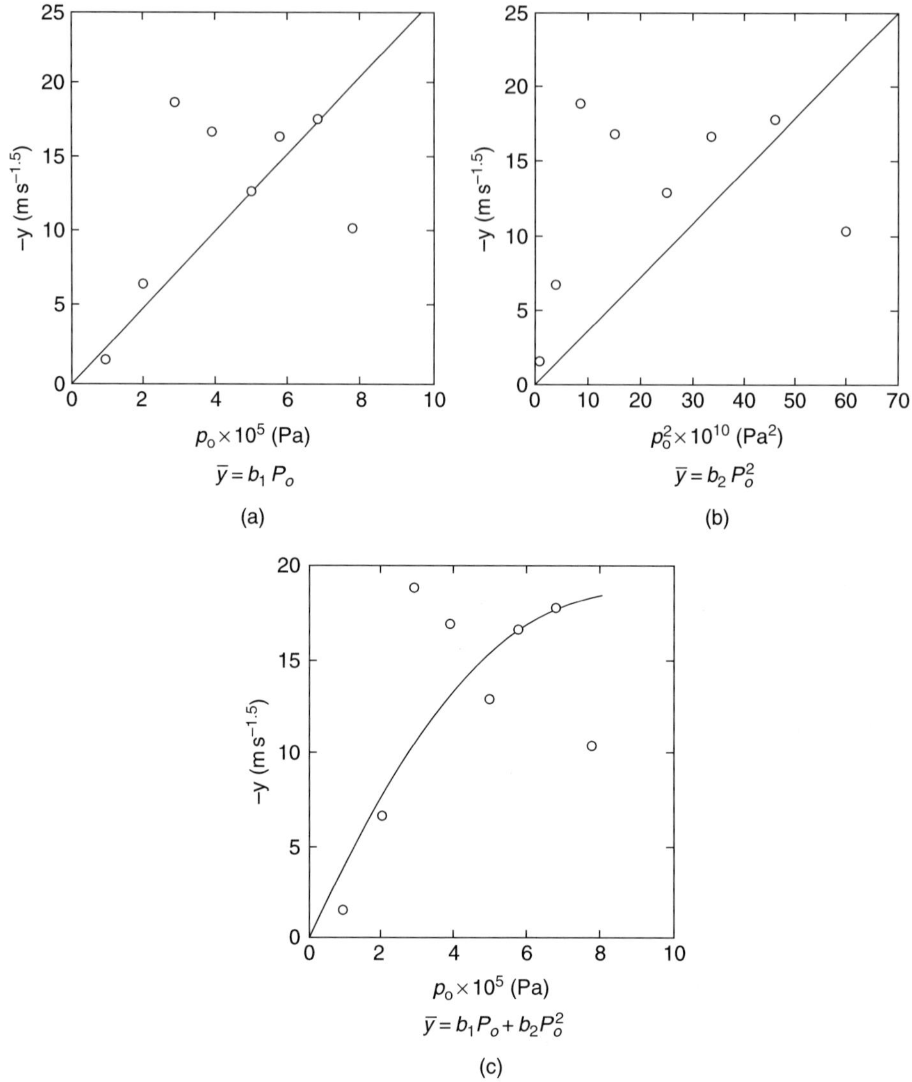

Figure 6.26 Relationship between $\bar{y}$ in p_o; Comparison of Equation (6.4.3.2) with experimental data shown in Fig. 6.25: (a) $b_2 = 0$; (b) $a_1 = 0$, (c) both b_1 and b_2 non-vanishing. (Teoh, 2003. Reprinted by permission of Dr S.-K. Teoh.)

increase with the increase of p_o. Furthermore, the results differ significantly (by one order of magnitude) from the intrinsic value of R_m, which exhibit only slight dependence on p_o. These findings are consistent with those obtained by Meeten (2000) but differ from the earlier work of Lew and Tiller (1983). The results of Lew and Tiller did not show any simple relationship between R_m and p_o.

Table 6.1

Medium resistance values determined from different methods (Teoh, 2003)

p_o (Pa)	Intrinsic* medium resistance (m^{-1})	Medium resistance, R_m (m^{-1})			
		$CaCO_3$ suspension		Kaolin suspension	
		a	b	a	b
1×10^5	1.00×10^{10}	1.07×10^{11}	1.15×10^{11}	5.21×10^{11}	2.15×10^{11}
2×10^5	1.14×10^{10}	1.25×10^{11}	1.36×10^{11}	8.09×10^{11}	4.52×10^{11}
3×10^5	1.11×10^{10}	1.23×10^{11}	1.37×10^{11}	10.1×10^{11}	7.37×10^{11}
4×10^5	1.07×10^{10}	1.61×10^{11}	1.86×10^{11}	11.2×10^{11}	11.6×10^{11}
5×10^5	1.31×10^{10}	1.98×10^{11}	2.05×10^{11}	10.1×10^{11}	5.75×10^{11}
6×10^5	1.50×10^{10}	2.10×10^{11}	2.17×10^{11}	8.91×10^{11}	5.51×10^{11}
7×10^5	1.53×10^{10}	2.15×10^{11}	2.34×10^{11}	10.0×10^{11}	8.19×10^{11}
8×10^5	1.54×10^{10}	2.26×10^{11}	2.36×10^{11}	9.19×10^{11}	5.72×10^{11}

*Determined by using deionized water.
a – Determined based on $(dV/dt)_{t \to 0}$.
b – Determined from t/V vs. V plot.

The difference between R_m determined from filtration data and its intrinsic value is commonly attributed to medium clogging (for example see Tiller *et al.*, 1981; Lew and Tiller, 1983; Lee, 1997). Medium clogging may occur in two different ways: interior clogging and surface clogging. For the former, if the suspended particles are not of uniform size, fine particles of sufficiently small sizes may penetrate into the medium and become deposited throughout the medium, thus causing interior clogging. On the other hand, deposited particles at the cake/medium interface may block or partially block various pore entrances of the medium. In either case, the effective medium resistance is increased.

Interior clogging was considered by Lee (1997) as the major cause for the increase in R_m. This may indeed be correct if fine particles (those with sizes considerably less than the pore openings of the medium) present in the suspension are significant. From the size distributions of the powders used by Meeten and Teoh, this condition was not met. On the other hand, the extent of surface clogging of a medium is determined largely by the compaction of the deposited particle, which, in turn, is dependent upon the local compressive stress of the cake phase. For constant pressure filtration, the compressive stress at the cake/medium interface increase with time and approaches the value corresponding to $p_\ell = p_o$. The experimentally determined value of R_m, as an approximation, can therefore be considered as a function of p_o if surface clogging is the dominant clogging mechanism. The results given in Table 6.1 are consistent with this speculation.

If one considers that medium clogging is mainly of the surface clogging type and the extent of clogging is determined by the compressive stress at the cake/medium interface,

Table 6.2

Calculated filtration performance results using different R_m values

Conditions assumed

$$p_o = 800\,\text{kPa}$$
$$[\alpha_{av}] = 5 \times 10^{10}\,\text{m kg}^{-1}$$
$$s = 0.02$$
$$\rho = 10^3\,\text{kg m}^{-3}$$
$$\mu = 10^{-3}\,\text{Pa s}$$

Time (s)	V (m^3/m^2)		
	$R_m = 10^{10}$	10^{11}	$2 \times 10^{11}\,m^{-1}$
100	0.3901	0.3123	0.2472
200	0.5558	0.4775	0.4000
400	0.7901	0.7063	0.6248
1000	1.253	1.1689	1.0806

using a constant R_m in predicting filtration performance is, at best, an approximation. The error introduced by this approximation can be seen from calculations based on Equation (6.4.2.5). Using $[\rho s/2(1-\overline{m}s)] = 10\,\text{kg m}^{-3}$, $\mu = 10^{-3}\,\text{Pa s}$, $\alpha_{av} = 5 \times 10^{12}\,\text{m kg}^{-1}$ and $R_m = 10^{10}\,\text{m}^{-1}$ (intrinsic medium resistance) 10^{11}, $2 \times 10^{11}\text{m}^{-1}$ (typical values for the case of $CaCO_3$), the results obtained (V vs. t) are shown in Table 6.2. It is clear that for filtration calculation, the intrinsic value R_m should not be used. On the other hand, R_m determined from filtration data can be used for calculations as long as the applied pressure is of the same order of magnitude of the pressure with which the experimental data were obtained.

6.4.5 Comparison with theoretical predictions

As a measure of validation of the analysis presented in Chapter 3, predictions of filtrate performance based on the solutions of Equations (3.4.9)–(3.4.13c) and (3.4.15a)–(3.4.15b) with some of the experimental results of Teoh's are presented. The experimental data used include the cumulative filtrate volume vs. time of four kinds of suspensions ($CaCO_3$, Kaolin, TiO_2 and Kromasil) and cake thickness histories of two cases ($CaCO_3$ and Kaolin). The information required for predictions for a specified set of operating conditions include the values of the parameters of the constitutive relationships of the systems and the values of the medium resistance and the relationship between p_ℓ and p_s (i.e. expressions of f'). The parameter values used are those given in Table 6.3. The medium resistance values are those listed in Table 6.1. As R_m can be determined by two different methods, for the case of $CaCO_3$ suspensions, their averages were used since the values obtained from the two methods are very similar. For Kaolin suspensions, the

Table 6.3

Parameter values used to obtain prediction results shown in Figs 6.27–6.30

Quantity	$CaCO_3$	Kaolin	TiO_2	Kromasil
ε_s^0 (-)	0.20	0.34	0.31	0.79
$k^o(m^2)$	4.8915×10^{-14}	1.9885×10^{15}	1.4817×10^{-15}	1.8411×10^{-13}
α^o $(m\,kg^{-1})$	3.85×10^{10}	5.47×10^{11}	5.63×10^{11}	1.29×10^{10}
β (-)	0.13	0.17	0.17	0.03
η (-)	0.57	1.02	0.68	0.35
n (-)	0.44	0.85	0.51	0.32
p_a (Pa)	4.4×10^4	8.7×10^4	9.9×10^4	1.0×10^3

R_m values obtained showed significant scattering. It was decided that for the case of $p_o = 100\,kPa$, both of these two values (corresponding to methods *a* and *b*) were used. For $p_o = 500$ and $800\,kPa$, the two averages of R_m at $p = 500, 600, 700$ and $800\,kPa$ were used. By using two different values of R_m for the same prediction, the effect of R_m was also examined.

The medium resistance, R_m, values for the cases of Kromasil and TiO_2 were not determined by Teoh. Based on the fact that R_m depends, to a degree, on the extent of surface clogging due to cake compaction as argued previously, the R_m value used for predicting $CaCO_3$ suspension filtration were also used for the case of TiO_2 since the two systems are rather similar according to their C-P measurement results. Based on the same argument, predictions of the filtration of Kromasil suspension were made using the R_m values of the Kaolin case.

The comparison results shown in Figs 6.27–6.30 may be summarized as follows:

(a) *$CaCO_3$ suspensions*
Comparisons of the filtration results of V vs. t and L vs. t for $p_o = 100, 500$ and $800\,kPa$ are shown in Fig. 6.27a,b,c. For $p_o = 100\,kPa$, experimental data of V vs. t agree well with predictions using either Type 1 or Type 2 p_ℓ–p_s relationship, with Type 2 relationship yielding slightly better agreement. On the other hand, better agreement (especially initially) obtained using Type 1 relationship was found for the results of L vs. t. For $p_o = 500\,kPa$, the experimental results of L vs. t were shown to agree well with prediction using either Type 1 or Type 2 relationship. The results of L vs. t were lower than either predictions although they agree better with prediction using Type 1 p_ℓ–p_s relationship. The same trend was observed for the results obtained at $p_o = 800\,kPa$.

(b) *Kaolin suspensions*
The comparisons are shown in Fig. 6.28a–c. As stated before, two different values of R_m were used for each prediction. For the prediction of V vs. t, at $p_o = 100\,kPa$, the effect of using different R_m was slight and predictions based on Type 1 $p_\ell - p_s$

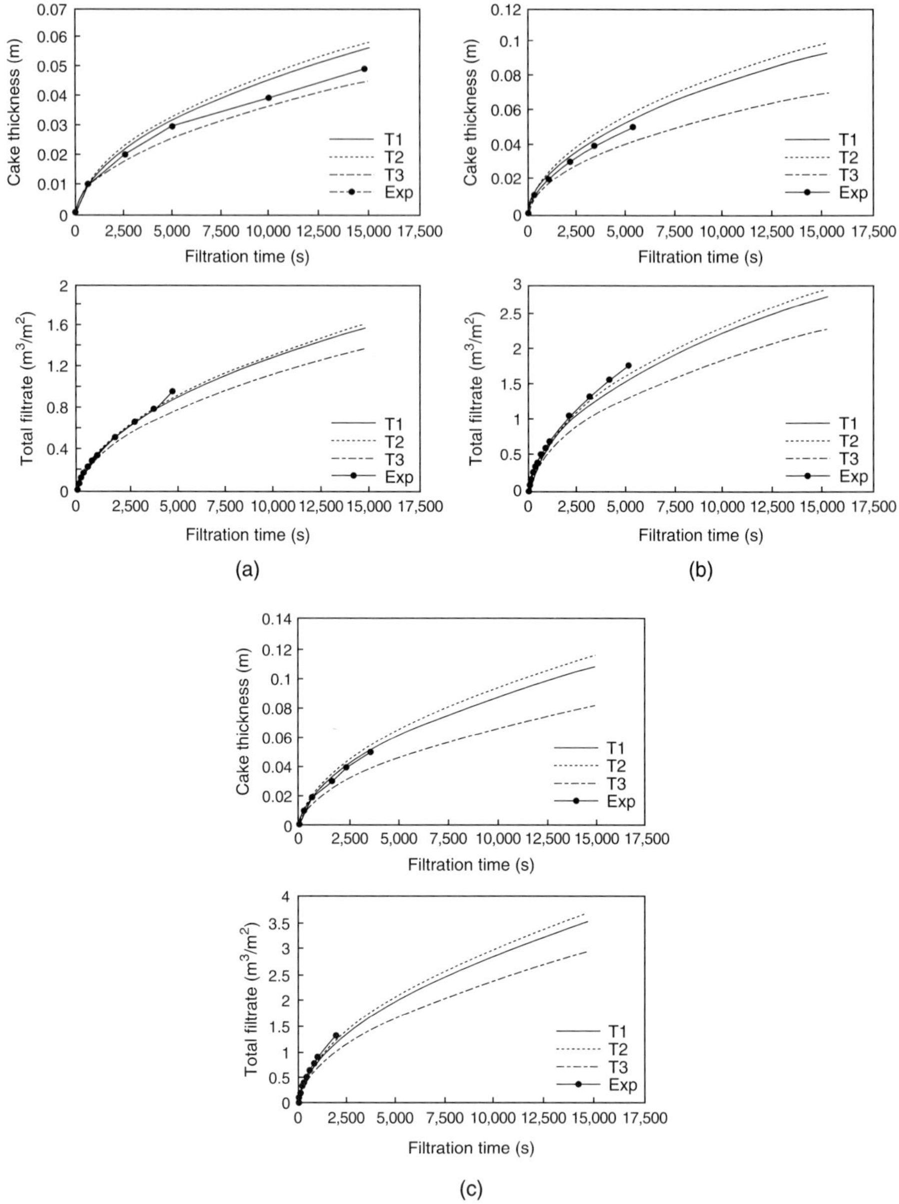

Figure 6.27 Comparison of predictions with experiments. Constant pressure filtration, 2% $CaCO_3$ suspensions: (a) $p_o = 100\,\text{kPa}$; (b) $p_o = 500\,\text{kPa}$; (c) $p_o = 800\,\text{kPa}$.

relationship agree well with experiments. The effect of using different R_m in predicting cake thickness history was also insignificant. However, the L vs. t data was found to agree with predictions using Type 3 p_ℓ–p_s relationship. For the results obtained at $p_o = 500\,\text{kPa}$, the L vs. t data agree well with predictions using Type 3

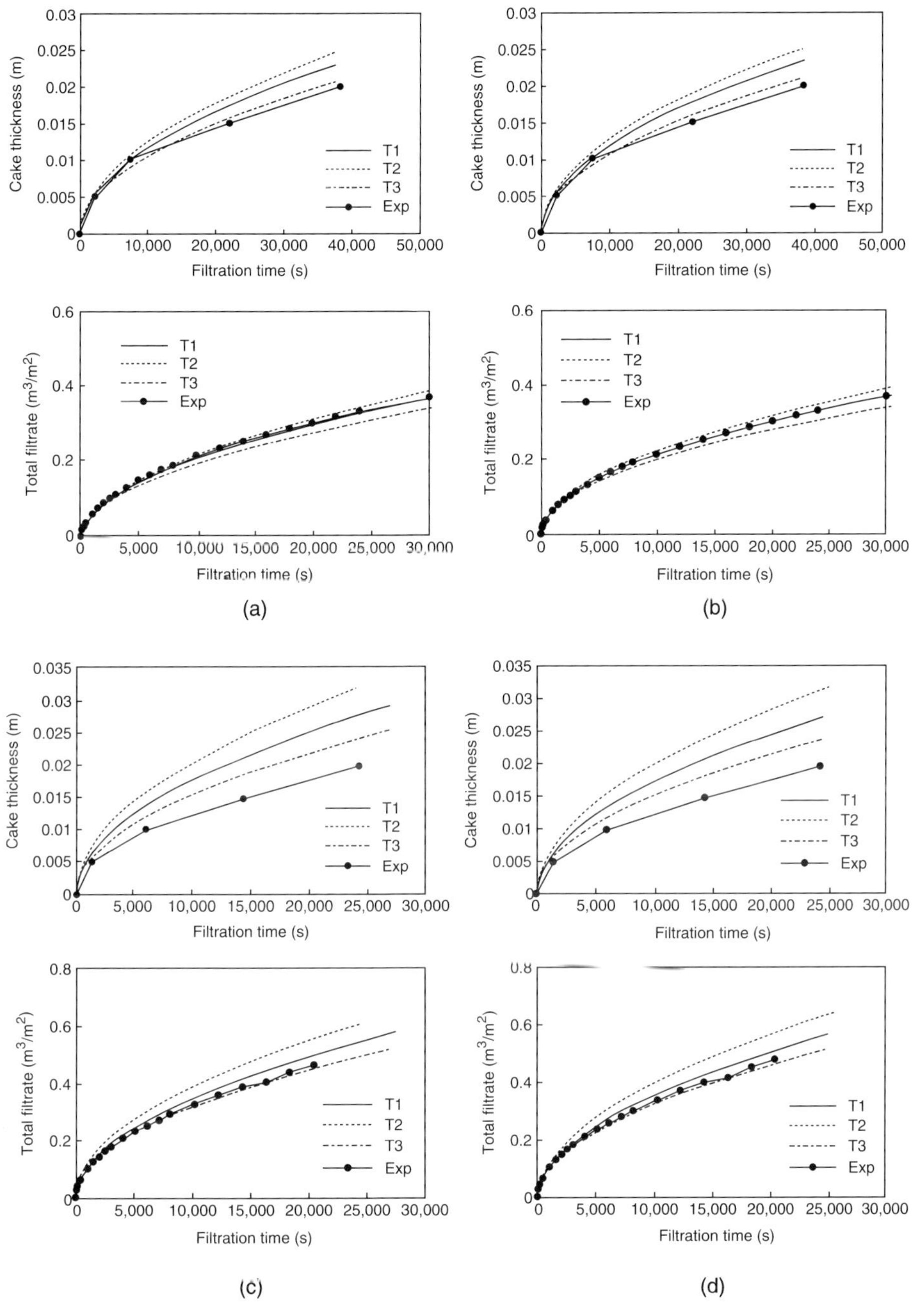

Figure 6.28 Comparisons of predictions with experiments. Constant pressure filtration, 5% Kaolin suspensions: (a) $p_o = 100\,\text{kPa}$, $R_m = 2.15 \times 10^{11}\,\text{m}^{-1}$, (b) $p_o = 100\,\text{kPa}$, $R_m = 5.21 \times 10^{11}\,\text{m}^{-1}$; (c) $p_o = 500\,\text{kPa}$, $R_m = 6.9 \times 10^{11}\,\text{m}^{-1}$; (d) $p_o = 500\,\text{kPa}$, $R_m = 9.5 \times 10^{11}\,\text{m}^{-1}$.

p_ℓ–p_s relationship. The effect of using different R_m was also slight. All predictions were found to overestimate the cake thickness although the predictions based on Type 3 p_ℓ–p_s relationship gave the least deviation.

(c) *TiO_2 suspensions*. The comparison results are given in Fig. 6.29. At $p_o = 100$ kPa, all predictions underestimate the V vs. t results although predictions based on Type 2 p_ℓ–p_s relationship gave the least differences. The results obtained at $p_o = 400$ kPa agreed best with predictions based on Type 1 p_ℓ–p_s relationship, and predictions using Type 2 gave the best comparisons with data obtained at $p_o = 800$ kPa.

(d) *Kromasil suspension*. As shown in Fig. 6.30, at $p_o = 400$ kPa, either Type 1 or Type 2 p_ℓ–p_s relationship gave results which agreed with experimental data of V vs. t. At higher pressures, all predictions underestimated although Type 2 relationship gave the least deviation.

The comparison results presented above indicate rather clearly the importance of the p_ℓ–p_s relationship. The p_ℓ–p_s relationship is also important when specific cake resistance values obtained from filtration experiments are compared with those from the C-P measurements (Tien *et al.*, 2001). This will be discussed in the following chapter.

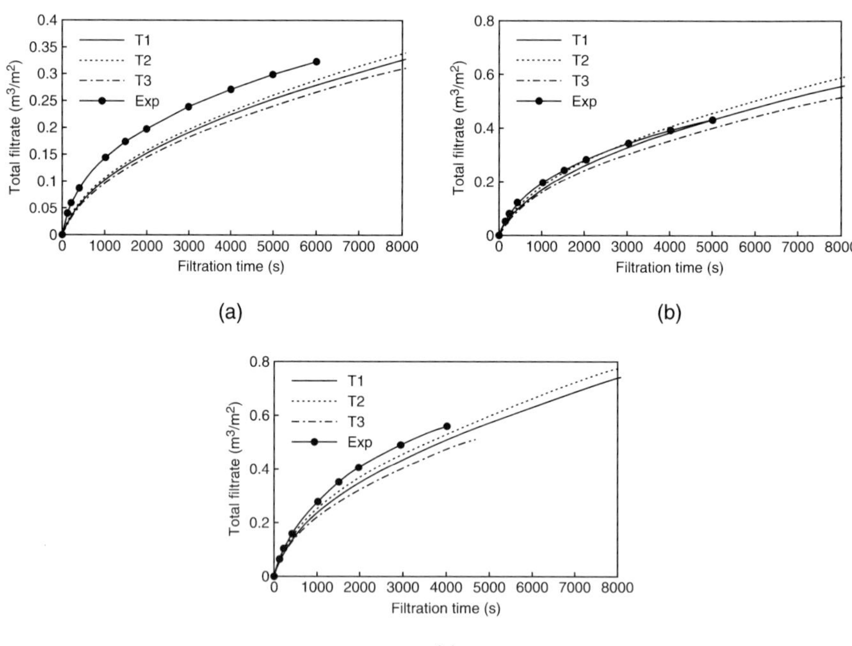

Figure 6.29 Comparisons of predictions with experiments. Constant pressure filtration, 2% TiO_2 suspensions: (a) $p_o = 100$ kPa, $R_m = 2.15 \times 10^{11}$ m^{-1}; (b) $p_o = 400$ kPa, $R_m = 6.9 \times 10^{11}$ m^{-1}; (c) $p_o = 800$ kPa, $R_m = 6.9 \times 10^{11}$ m^{-1}.

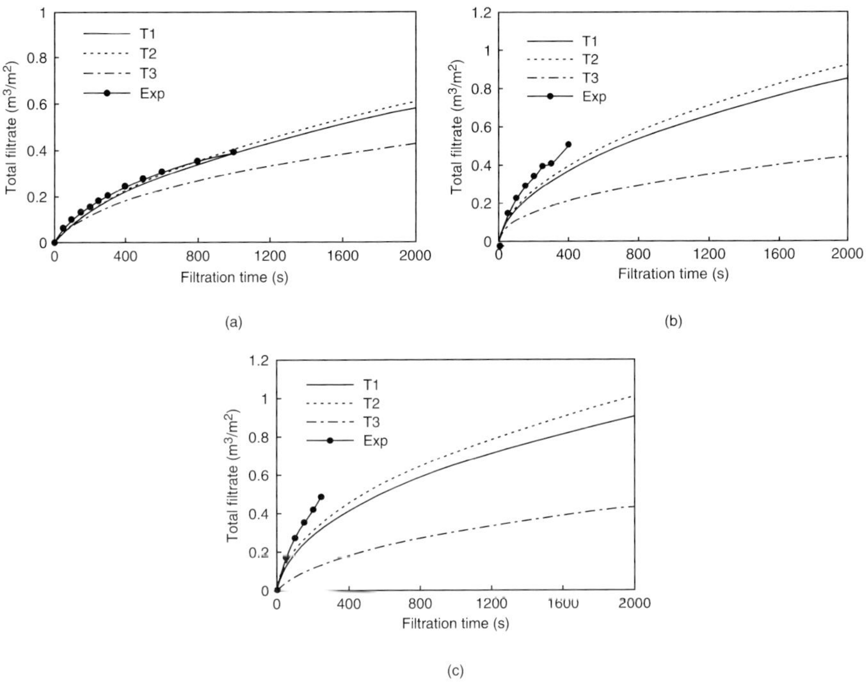

Figure 6.30 Comparisons of predictions with experiments. Constant pressure filtration, 2% Kromasil suspensions: (a) $p_o = 100\,\text{kPa}$; (b) $p_o = 400\,\text{kPa}$; (c) $p_o = 800\,\text{kPa}$.

REFERENCES

Almay, C. and Lewis, W.K., *Ind. Eng. Chem.*, **4**, 528 (1912).

Baird, R.L. and Perry, M.G., *Filtr. Sep.*, 471 (1967).

Been, K. and Sills, G.C., *Geotechnique*, **31**, 519 (1981).

Bergstrom, L., *Chem. Soc. Faraday Trans.*, **88**, 3201 (1992).

Bierck, B.R., Wells, S.A. and Dick, R.I., *J. WPCF*, 645 (1988).

Chase, G.G. and Willis, M.S., *Chem. Eng. Sci.*, **47**, 1373 (1992).

Christensen, G.L. and Dick, R.I., *J. Environ. Eng. Div.*, ASCE **111**, 243 (1985a).

Christensen, G.L. and Dick, R.I., *J. Environ. Eng. Div.*, ASCE **111**, 258 (1985b).

Fathi-Najafi, M. and Theliander, H., *Sep. Tech.*, **5**, 165 (1995).

Grace, H.P., *Chem. Eng. Prog.*, **49**, 303 (1953).

Hall, L., Heese, F., Acosta-Cabronero, J. and Robson, P., "MRI Quantitation of Filtration and Separation for Clinical Medicine and Industry", *World Filtration Congress 9*, New Orleans, Louisiana, USA (2004).

Hamachi, M. and Mietton-Peuchot, M., *Trans. I. Chem. E.*, **79**, 151 (2001).

Horsfield, M.A., Fordham., E.J., Hall, C. and Hall, L.D., *J. Magn. Reson.* **81**, 593 (1989).

Johansson, C., "Experimental Determination of Pressure and Solidosity Profiles in Cake Filtration", Thesis for Licentiate of Engineering, Chalmers University of Technology, Gothenburg, Sweden (2003).

Johansson, C., Sedin, P. and Theliander, H., "Determination of Local Filtration Properties", *World Filtration Congress 9*, New Orleans, Louisiana, USA (2004).

Koenders, M.A. and Wakeman, R.J., *Chem. Eng. Sci.*, **51**, 3897 (1996).
Koenders, M.A. and Wakeman, R.J., *Trans. I Chem E*, 75, Part A 309 (1997a).
Koenders, M.A. and Wakeman, R.J., *AIChE J.*, **43**, 946 (1997b).
La Heij, E.J., Kerkhof, P.A.J.M., Kopinga, K. and Pel, L., *AIChE J.*, **42**, 953 (1996).
Larue, O., Vorobiev, E., Vu, C. and Durand, B., *Sep. Pur. Tech.*, **31**, 177 (2003).
Lee, D.J., *AIChE J.*, **43**, 273 (1997).
Lew, W.F. and Tiller, F.M., *Sep. Sci. Tech.*, **18**, 1351 (1983).
Meeten, G.H., *Chem. Eng. Sci.*, **48**, 2391 (1993).
Meeten, G.H., *Chem. Eng. Sci.*, **55**, 1755 (2000).
Murase, T., Eritani, E., Cho, J.H., Nakanomori, S. and Shirato, M., *J. Chem. Eng. Japan*, **20**, 246 (1987).
Murase, T., Eritani, E., Cho, J.H. and Shirato, M., *J. Chem. Eng. Japan*, **22**, 373 (1989).
Ruth, B.F., Montillon, G.H. and Montanna, R.E., *Ind. Eng. Chem.* **25**, 153 (1933).
Sedin, P., Johansson, C. and Theliander, H., *Trans. I Chem E.*, **81**, Part A, 1393 (2003).
Shen, C., Russel, W.R. and Auzerais, F.M., *AIChE J.*, **40**, 1876 (1994).
Shirato, M., Aragaki, T., Ichimura, K. and Ootsuji, N., *J. Chem. Eng. Japan*, **4**, 60 (1971).
Smiles, D.E. and Rosenthal, M.J., *Aust. J. Soil Res.*, **6**, 237 (1968).
Takahashi, K., Kobayashi, Y., Yokota, T. and Koyama, K., *J. Chem. Eng. Japan*, **5**, 599 (1991).
Tan, R.B.H., He, D. and Tien, C., "Multifunction Test Cell for Filtration Studies", Singapore Patent No. 980482–7 (1998).
Tarleton, E.S., "Determination of Cake Structure During Stepped and Variable Pressure Filtration", *World Filtration Congress 9*, New Orleans, Louisiana, USA (2004).
Tarleton, E.S. and Hancock, D.L., *Trans. I. Chem. E.*, **75**, 298 (1997).
Teoh, S.K., "Studies in Filter Cake Characterization and Modeling", Ph.D. Thesis, National University of Singapore (2003).
Teoh, S.K., Tan, R.B.H., He, D. and Tien, C., *Trans. Filtration Soc.*, **1**, 81 (2001).
Tien, C., Teoh, S.K. and Tan, R.B.H., *Chem. Eng. Sci.*, **56**, 5361 (2001).
Tiller, F.M., Chow, R., Weber, W. and Davies, O., *Chem. Eng. Prog.*, **77**, 61 (1981).
Tiller, F.M., Hsyung, N.B. and Cong, D.Z., *AIChE J.*, 41, 1153 (1995).
Vesilind, P.A., *J. WPCF*, **60**, 215 (1988).
Wakeman, R.J., *Trans. I. Chem E.*, **59**, 260 (1981).
Willis, M.S., Collins, R.M. and Bridges, W.G., *Chem. Eng. Res. Des.*, **61**, 96 (1983).

– 7 –

CAKE PROPERTY MEASUREMENTS

Notation

A^E, A^P	quantities obtained experimentally and by prediction
a	exponent of Equation (7.4.1) or a quantity of Equation (7.4.3)
a_1, a_2	coefficients of Equations (7.4.4) or (7.4.5)
a_p	particle radius (m)
$\overline{a}$	coefficient of Equation (7.5.4)
B	coefficient of Equation (7.5.5) or the slope of the linear segment of t/V vs. V
b	a quantity of Equation (7.4.3)
$\overline{b}$	exponent of Equation (7.5.4)
c_w	particle cohesion stress (Pa)
D	diameter of C-P cell (m)
$D(e)$	filtration diffusivity ($m^2\,s^{-1}$)
$[D]_{\ell_{we}}$	filtration diffusivity defined by Equation (7.1.6) ($m^2\,s^{-1}$)
e	void ratio defined as $(1-\varepsilon_s)/\varepsilon_s$ (-)
e_o	suspension void ratio (-)
f	wall friction coefficient (-)
f'	defined by Equation (7.1.4) (-)
g	gravitational acceleration ($m\,s^{-2}$)
H	height of suspension (m)
H_o	initial value of H (m)
H_∞	ultimate value of H (m)
K	ratio between the vertical stress and the radial stress
k	cake permeability (m^2)
k^o	cake permeability at the zero stress state (m^2)
L	cake thickness (m)
M	total particle volumes per unit cross section area (m^3/m^3)
m	defined as $\int_o^x \varepsilon_x \, dx$ (m)
$\overline{m}$	wet to dry cake mass ratio (-)

N	total number of measurements (-)
n	exponent of Equation (7.3.1.2c)
P_A	applied compression load (Pa)
P_t	transmitted load (Pa)
P_Y	yield stress (Pa)
p_a	a quantity appearing in the constitutive relationship (Pa)
p_ℓ	filtrate pressure (Pa)
p_o	operating pressure (Pa)
$(p_o)_i$	operating pressure during the interval $t_{i-1} < t < t_s$
p_s	cake compressive stress (p_o)
p_{s_m}	value of p_s and the cake/medium interface (Pa)
$\overline{P}_A$	logrithmic average of P_A and P_t (Pa)
q_ℓ	liquid velocity $(\mathrm{m\,s^{-1}})$
$q_{\ell s}$	liquid/solid relative velocity $(\mathrm{m\,s^{-1}})$
q_s	particle velocity $(\mathrm{m\,s^{-1}})$
R	defined by Equation (7.1.8) (-)
R_m	medium resistance $(\mathrm{m^{-1}})$
r	a constant of Equation (7.3.5.2) (-)
$(r)_{\ell we}$	hindered settling factor defined by Equation (7.1.7) (-)
S	coefficient of Equation (7.3.5.1) $(\mathrm{m\,t^{-1/2}})$
s	mass fraction of particle (-)
t	time (-)
t_i	time corresponding to the end of the i-th period (s)
V	cumulative filtrate volume $(\mathrm{m^3/m^2})$
V_{i-1}	cumulative filtrate volume collected at the end of the period $t_{i-2} < t < t_{i-1}$ $(\mathrm{m^3/m^2})$
V_p	particle volume $(\mathrm{ft^3})$
v_o	initial settling velocity $(\mathrm{m\,s^{-1}})$
x	distance measurement from filter medium (m)
y	variable of Equation (7.4.1)
y_o	coefficient of Equation (7.4.1)
y_1, y_2	coefficients of Equation (7.4.2)

Greek letters

α	specific cake resistance $(\mathrm{m\,kg^{-1}})$
α^o	specific cake resistance at the zero-stress state $(\mathrm{m\,kg^{-1}})$
α_{av}	average specific cake resistance $(\mathrm{m\,kg^{-1}})$
β	exponent of Equation (7.3.1.2a) or (7.5.5) (-)
Δp_c	pressure drop across filter cake (Pa)
δ	exponent of Equation (7.3.1.2b) (-)
ε	cake porosity (-)

ε_s	cake solidosity (-)
ε_s^o	values of ε_s at zero-stress state (-)
ε_{s_o}	initial particle volume fraction (-)
$\varepsilon_{s_{p_o}}$	cake solidosity corresponding to a compressive stress of p_o (-)
$\overline{\varepsilon}_s$	average cake solodosity (-)
λ	equal to $6\pi\mu a_p$ (Pa s m) or similar variable defined as $m/t^{1/2}$ (m s$^{-1/2}$)
μ	filtration viscosity (Pa s)
ρ	filtration density (kg m^{-3})
ρ_s	particle density (kg m^{-3})
Φ	objective function defined by Equation (7.3.1.1)

Cake solidosity and cake permeability which describe and define cake structure and its resistance to filtrate flow are crucial information to the analysis and simulation of cake filtration. Accurate knowledge of these quantities are also essential to the design and scale-up of filtration systems. This chapter is devoted to a presentation and discussion of the methods which have been developed and applied for their determinations.

7.1 CAKE SOLIDOSITY, CAKE PERMEABILITY AND THEIR EQUIVALENTS

Beginning with the work of Donald and Hunneman (1933), cake filtration is often described as a problem akin to that of liquid flow through a growing and compacting porous medium characterized by its solidosity and permeability. Accordingly, the problem of cake filtration has often been analyzed using the concept of liquid permeation through porous media. Alternatively, the same problem has also been viewed as a diffusion process or considered analogous to particle settling. Although different parameters are used in expressing the fluid/particle relative velocity depending upon the approaches used, the existence of parameter equivalence is obvious as they are used to discuss the same phenomena (See discussion in Chapters 3 and 4). For the clarity of subsequent discussions, the equivalences of these model parameters are further reviewed and summarized in the following.

For analyses based on the permeability–solidosity concept, cake structure is described by ε_s, the cake solidosity which is a function of the cake compressive stress, p_s, or

$$\varepsilon_s = \varepsilon_s(p_s) \tag{7.1.1}$$

The relative fluid/solid velocity (liquid permeation) is defined by the generalized Darcy's law. For the one-dimensional case, it is given as [Equation (3.1.4)]

$$\frac{q_\ell}{1-\varepsilon_s} - \frac{q_s}{\varepsilon_s} = \frac{k}{(1-\varepsilon_s)\mu}\frac{\partial p_\ell}{\partial x} \tag{7.1.2}$$

In treating cake filtration as a diffusion problem, using the material coordinate instead of the spatial coordinate, filtrate flow through filter cake can be viewed as a diffusion problem with the void ratio e $[= (1-\varepsilon_s)/\varepsilon_s]$ as the dependent variable. The filtration diffusivity, $D(e)$, is given by Equation (4.1.12) or

$$D(e) = (f')\frac{k}{\mu(1+e)(\mathrm{d}e/\mathrm{d}p_s)} \tag{7.1.3a}$$

$$= (-f')\frac{k\varepsilon_s^3}{(\mu)(\mathrm{d}\varepsilon_s/\mathrm{d}p_s)} \tag{7.1.3b}$$

and

$$f' = \frac{\mathrm{d}p_\ell}{\mathrm{d}p_s} \tag{7.1.4}$$

and the fluid/solid relative velocity $q_{\ell s}$ is given by Equation (4.1.14) or

$$q_{\ell s} = q_\ell - \frac{1-\varepsilon_s}{\varepsilon_s} q_s = D(e)\frac{\partial \mathrm{e}}{\partial \mathrm{m}} \tag{7.1.5}$$

The filtration diffusivity is also referred to as the consolidation coefficient or expression coefficient. $D(e)$ is different from the filtration diffusivity used by Landman and White (1997). The filtration diffusivity of Landman and White $[D]_{\ell w}$ can be shown to be related to $D(e)$ by the expression

$$D(e) = \varepsilon_s^2[D]_{\ell w} \tag{7.1.6}$$

By comparing Equations (7.1.2) with (7.1.5), D(e) may be expressed in terms of k and ε_s. The equivalency relationships between these two sets of parameters are summarized in Table 7.1.

The so-called compressional rheological model of Landman *et al.* (1995), which considers cake filtration being analogous to particle settling, expresses the fluid/solid relative velocity as

$$\frac{q_\ell}{1-\varepsilon_s} - \frac{q_s}{\varepsilon_o} = -\frac{V_p}{\lambda}\frac{(1-\varepsilon_s)^2}{\varepsilon_s[r]_{\ell we}}\frac{\partial p_\ell}{\partial x} = -\frac{(1-\varepsilon_s)^2}{\varepsilon_s R}\frac{\partial p_\ell}{\partial x} \tag{7.1.7}$$

and

$$R = \frac{\lambda}{V_p}[r]_{\ell we} \tag{7.1.8}$$

where V_p is the particle volume and $\lambda = 6\pi\mu a_p$ with a_p being the particle radius. $[r]$ is the hindered settling factor and R, the hindered drag coefficient or the hindered

Table 7.1

Equivalence between ε_s–k and $D(e)$–e

	Permeability conception	Diffusion concept
Coordinate	Spatial coordinate, x	Material coordinate, m $m = \int_o^x \varepsilon_s \, dx$
Cake structure	Cake solidosity, $\varepsilon_s = \varepsilon_s(p_s)$	Cake void ratio, $e = e(p_s)$ $e = \frac{1-\varepsilon_s}{\varepsilon_s}$
Rate parameter	Cake permeability, $k = k(p_s)$	Filtration diffusivity, $D = D(e)$
Expression of $q_{\ell s}$	$q_{\ell s} = \frac{k}{\mu}\frac{\partial p_\ell}{\partial x}$	$q_{\ell s} = D(e)\frac{\partial e}{\partial m}$
Relationship between K and $D(e)$	$k = \frac{\mu(d\varepsilon_s/dp_s)}{\varepsilon_s^3(-f')}D(e)$	$D(e) = (-f')\frac{k\varepsilon_s^3}{\mu(d\varepsilon_s/dp_s)}$

settling function. The subscript ℓwe denotes that the definition was given by Landman *et al.* (1995). A slightly different definition of $[r]$ was given in the earlier publications (Landman *et al.*, 1991; Landman and Russell, 1993).

By equating Equations (7.1.2) and (7.1.7), one has

$$[r(\varepsilon_s)]_{\ell we} = \frac{\mu}{k}\frac{V_p}{\lambda}\frac{(1-\varepsilon_s)^2}{\varepsilon_s} \tag{7.1.9a}$$

or

$$R = \frac{\mu}{k}\frac{(1-\varepsilon_s)^2}{\varepsilon_s} \tag{7.1.9b}$$

Assuming that particles present in a suspension form a network displaying a yield stress, P_Y, which is a function of ε_s, furthermore, the yield stress manifests itself as the cake compressive stress, $P_Y(\varepsilon_s)$ is another fundamental quantity in cake characterization. It is clear that this yield stress function, $P_Y(\varepsilon_s)$, is just the inverse function of $\varepsilon_s(p_s)$ as stated previously (see Section 3.3), or

$$P_Y(\varepsilon_s) = \varepsilon_s^{-1}(p_s) \tag{7.1.10}$$

The equivalency between k–ε and $[r]_{\ell we}$–P_Y and that between $D(e)$–e and $[r]_{\ell we}$–P_Y are given in Tables 7.2 and 7.3.

Table 7.2

Equivalence between ε_s–k and P_Y–$[r]_{\ell we}$

	Permeability concept	Rheological model
Cake structure	Cake solidosity $\varepsilon_s = \varepsilon_s(p_s)$	Yield stress, $P_Y = P_Y(\varepsilon_s),\ P_Y = p_s$
Rate parameter	Cake permeability, $k_s = k(p_s)$	Hindered settling factor, $[r]_{\ell we} = r(\varepsilon_s)$
Expression of $q_{\ell s}$	$q_{\ell s} = \frac{k}{\mu}\frac{\partial p_\ell}{\partial x}$	$q_{\ell s} = \frac{V_p}{\lambda}\frac{(1-\varepsilon_s)^2}{\varepsilon_s[r]_{\ell we}}\frac{\partial p_\ell}{\partial x}$
Relationship between k and $[r]_{\ell we}$	$k = \frac{V_p}{\lambda}\frac{\mu}{[r]_{\ell we}}\frac{(1-\varepsilon_s)^2}{\varepsilon_s}$	$[r]_{\ell we} = \frac{V_p}{\lambda}\frac{\mu}{k}\frac{(1-\varepsilon_s)^2}{\varepsilon_s}$

Table 7.3

Equivalency between e–$D(e)$ and P_Y–$[r]_{\ell we}$

	Diffusion concept	Rheological model
Coordinate	Material coordinate, $m = \int_o^x \varepsilon_s \, dx$	Spatial coordinate, x
Cake structure	Cake void ratio $e = e(p_s)$ $e = \frac{1-\varepsilon_s}{\varepsilon_s}$	Yield stress $P_Y = P_Y(\varepsilon_s)$ $P_Y = p_s$
Rate parameter	Filtration diffusivity, $D(e)$	Hindered settling factor, $[r]_{\ell we}$
Expression of $q_{\ell s}$	$q_{\ell s} = D(e)\frac{\partial e}{\partial m}$	$q_{\ell s} = \frac{V_p}{\lambda}\frac{(1-\varepsilon_s)^2}{\varepsilon_s[r]_{\ell we}}\frac{\partial p_\ell}{\partial x}$
Relationship between $D(e)$ and $[r]_{\ell we}$	$D(e) = \frac{V_p}{\lambda}\frac{\varepsilon_s^2(1-\varepsilon_s)^2}{[r]_{\ell we}}(-f')\frac{dP_Y}{d\varepsilon_s}$	$[r]_{\ell we} = \frac{V_p}{\lambda}\frac{e^2}{(1+e)^2 D(e)}\left(\frac{f'}{de/dp_s}\right)$

7.2 COMPRESSION-PERMEABILITY (C-P) CELL AND C-P MEASUREMENTS

Ruth (1946) first proposed the use of the concept of local equilibrium for describing the state of filter cakes. Accordingly, one may obtain the constitutive relationships of ε_s vs. p_s and k vs. p_s by conducting measurements in which uniform (or nearly uniform) cakes of known thickness and solidosity are formed and their solidosities and permeabilities determined. Such measurements, known as the compression-permeability (C-P) measurements, can be made using a C-P cell, which is basically a cell of constant cross section with a movable piston through which a known compression load can be applied to the top of a confined volume of solid/liquid mixture to be tested. Grace (1953) designed and fabricated the first modern-day C-P cell capable of high-compression operation. Figure 7.1 gives an assembled view and a disassembled view of the cell he developed.

Grace conducted a systematic study of filter cake characterization by carrying out measurements of a large number of systems. Regarding the degree of homogeneity of the cake formed, Grace dissected the cake formed and concluded that uniform cake porosity was achieved as long as the cake thickness/cell diameter (L/D) ratio was less than 0.6. A typical set of data obtained by Grace is shown in Fig. 7.2.

Since the seminal work of Grace, improved versions of C-P cells have been developed in recent years (Haynes 1966; Shirato *et al.*, 1968; Tiller *et al.*, 1972; Tiller and Lu, 1972; Lu *et al.*, 1998; Tan *et al.*, 1998; Reichmann and Tomas, 1999). The multifunction cell developed by Tan *et al.* functions as both an experimental filter and a C-P cell. The press-shear cell developed by Reichmann and Tomas is a combination of a laboratory filter, C-P cell and rheometer which can be used for carrying out expression (filtration and consolidation) experiments, solidosity/permeability measurements and shear tests for the determinations of the wall friction angle and the stress ratio. A schematic diagram of the apparatus of Reichmann and Tomas is given in Fig. 7.3.

As the accuracy and usefulness of C-P measurements depend upon the formation of homogeneous cake, the presence of surface friction between the confining surface and

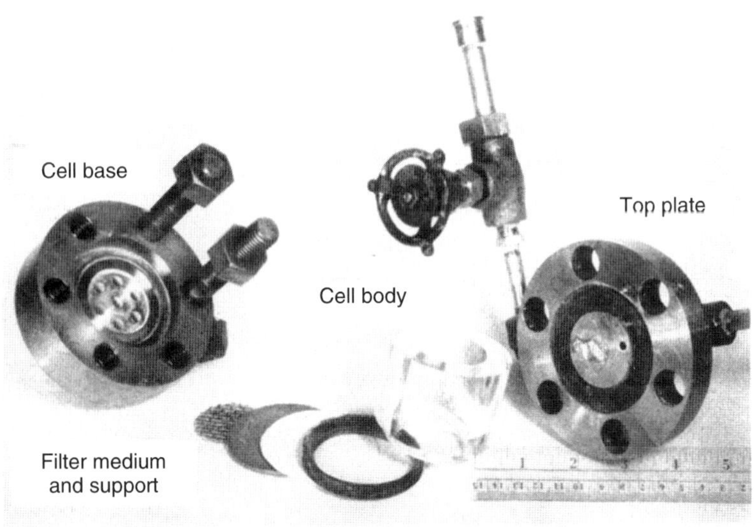

(a)

Figure 7.1 The first modern-day compression-permeability cell developed by Grace (1953): (a) disassembled view; (b) assembled view. (Grace, 1953. Reprinted by permission of The American Institute of Chemical Engineers.)

(b)

Figure 7.1 *Continued.*

the cake tested was a major concern. Unlike the cell developed by Grace, the newer versions of C-P cells were often equipped with floating pistons at both the cell top and bottom. For the cell developed by Haynes, Tiller *et al.* (1972) obtained results indicating that the effect of surface friction not only depends on the ratio L/D, but is a function of the wall material as well as the cell diameter. These results expressed as the percentage of the applied load transmitted from top to bottom are shown in Figs 7.4 and 7.5.

Based on the assumption that the compression load is distributed uniformly over the cross-sectional area of the cake, constant wall friction coefficient and stress ratio, the distribution of the compression load along the axial distance of the cell is found to be (Shirato *et al.*, 1968; Tiller *et al.*, 1972)

$$(p_s)_z = \left(P_A + \frac{c_W}{4fK}\right) \exp\left(-4fKz/D\right) - \frac{c_W}{fK} \tag{7.2.1}$$

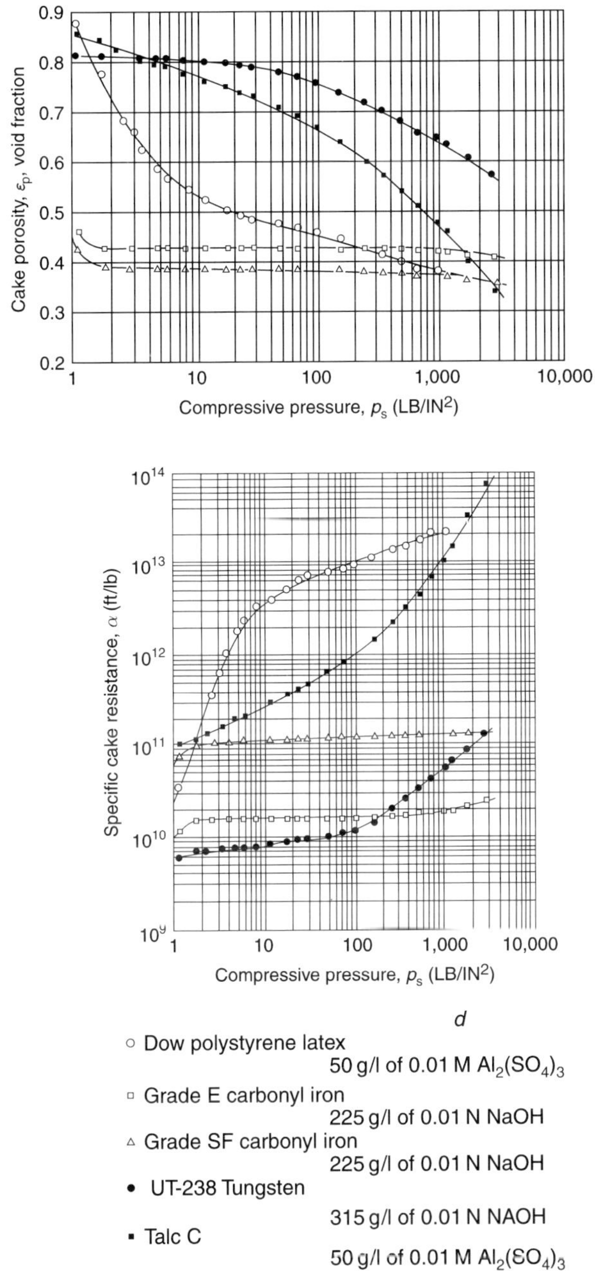

Figure 7.2 Constitutive relationships (ε vs. p_s; α vs. p_s) of a few systems studied by Grace. (Grace, 1953. Reprinted by permission of The American Institute of Chemical Engineers.)

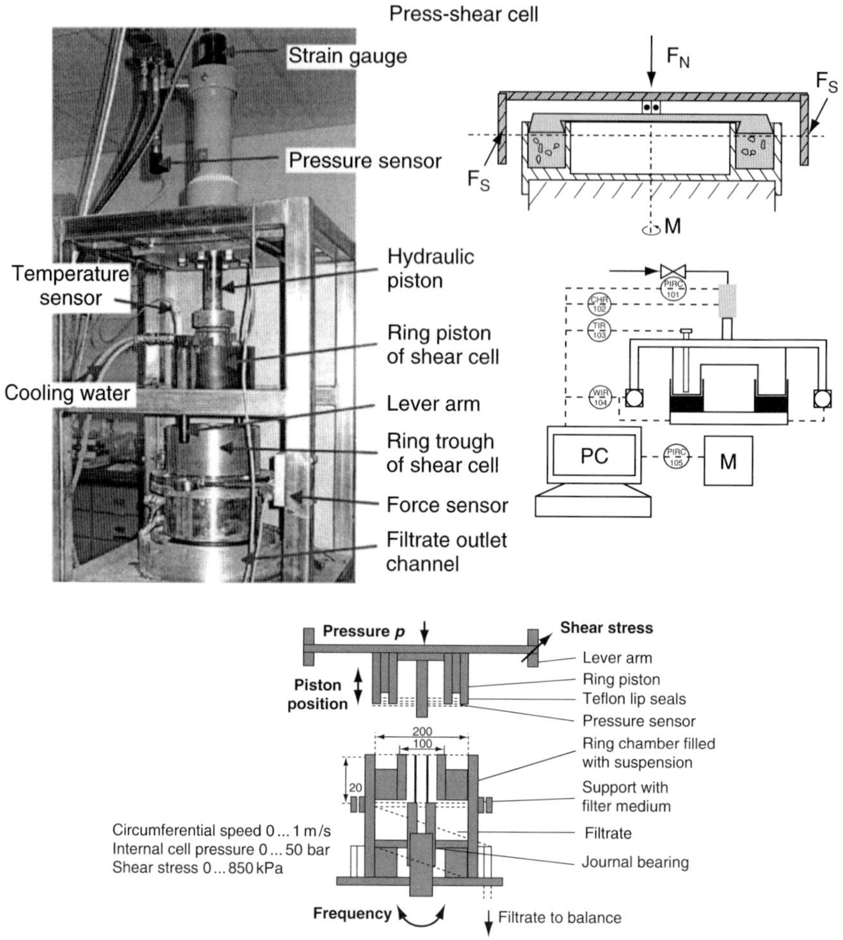

Figure 7.3 Press-shear cell developed by Reichmann and Tomas. (Reichmann and Tomas, 2001. Reprinted by permission of Elsevier.)

where P_A is the applied compression load; K, the ratio between the vertical stress and the radial stress; f, the wall friction coefficient; and c_w, the particle cohesive stress. The average value of the compressive load is

$$\overline{p}_s = \frac{P_A + (c_w/fK)}{4fK(L/D)} \left[1 - \exp(-4fK\, L/D)\right] - \frac{c_w}{fK} \tag{7.2.2}$$

More recently, Lu *et al.* (1998) presented an analysis of stress distribution within a C-P cell of cylindrical configuration taking into account that the surface friction coefficient and the stress ratio may not be constant. Using experimentally determined f and K values, stress distribution and average stresses of four different types of cakes (bentonite, $CaCO_3$, celite and kaolin) under different compression loads were calculated.

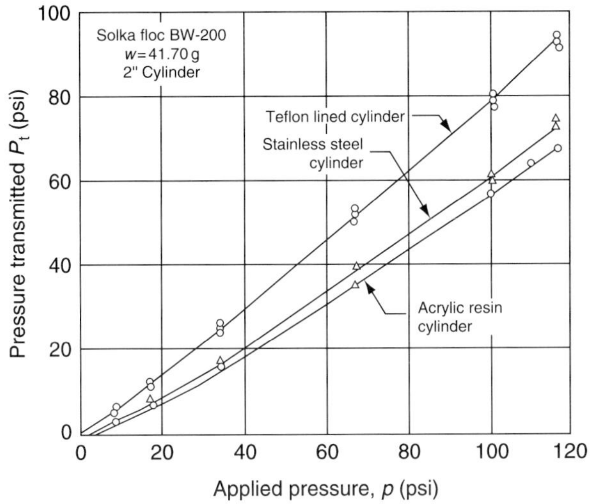

Figure 7.4 Transmitted vs. applied pressure of C-P cells constructed with different materials. (Tiller, Haynes and Lu, 1972. Reprinted by permission of The American Institute of Chemical Engineers.)

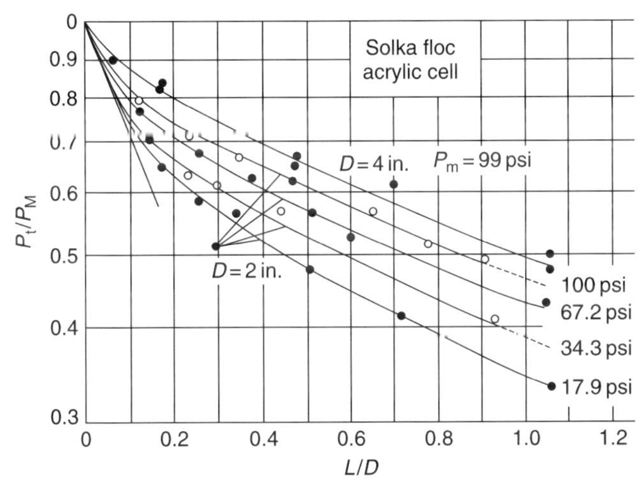

Figure 7.5 Fraction of applied load transmitted (P_A/P_t) as a function of cell geometry. (Tiller, Haynes and Lu, 1972. Reprinted by permission of The American Institute of Chemical Engineers.)

These results were compared with those estimated from Equation (7.2.2) as well as the logrithmic is mean value of P_A and P_t or

$$\bar{p}_s = \frac{P_A - P_t}{\ln \frac{P_A}{P_t}} \tag{7.2.3}$$

The estimated values according to Equation (7.2.3) were found to agree well with the more exact calculations.

Comparisons of the constitution relationships (ε_s vs. p_s and α vs. p_s) based on corrected and uncorrected compressive stress were also made. Since the average compression stress experienced by a cake was always less than the applied values, the relationships based on the uncorrected stress values tend to underestimate ε_s but overestimate α. The average difference ranged from being negligible up to 8%. In all cases, the difference was more significant at lower compression load.

7.3 DETERMINATION OF CAKE PROPERTIES FROM FILTRATION EXPERIMENTAL DATA

7.3.1 A general search-optimization procedure

Constitutive relationships of ε_s vs. p_s and k (or α) vs. p_s, in principle, can be obtained by matching experimental data with predictions based on appropriate analyses and assumed constitutive relationships. The problem may be treated as one of search and optimization. If A is a measurable quantity of cake filtration experiment (such as cake thickness or cumulative filtrate volume), an objective function Φ, may be defined as

$$\Phi = \sum_{j=1}^{N} [A_j^E - A_j^P]^2 \tag{7.3.1.1}$$

where the superscripts E and P denote experimental and predicted values, respectively. The subscript j refers to the j-th measurement (i.e. at $t = t_j$), and there are N measurements.

If the constitutive relationships are given by the power-law expression of the types of Equation (3.2.2) or

$$\varepsilon_s = \varepsilon_s^o \left(1 + \frac{p_s}{P_A}\right)^{\beta} \tag{7.3.1.2a}$$

$$k = k^o \left(1 + \frac{p_s}{P_A}\right)^{-\delta} \tag{7.3.1.2b}$$

$$\alpha = \alpha^o \left(1 + \frac{p_s}{P_A}\right)^{n} \tag{7.3.1.2c}$$

where

$$\alpha^o = (k^o \varepsilon_s^o \rho_s)^{-1} \tag{7.3.1.3}$$

$$n = -\beta + \delta \tag{7.3.1.4}$$

The relationships are given by the values of the coefficients and exponents of the above expression. The determination of these quantities is made on the minimization of the objective function.

Stamatakis (1990) and Stamatakis and Tien (1991) developed and applied a search-optimization method for determining cake constitutive relationships based on the numerical solution of Equations (3.4.9)–(3.4.13b) and (3.4.14a)–(3.4.14b) with $f' = -1$. Because of the relatively large number of parameters (namely, ε_s^o, α^o, P_A, β and n) to be determined, Stamatakis found it expedient to apply the complex method for the search. A procedure for estimating the ranges of the parameter values from relevant experimental data was also established by Stamatakis (1990).

The results of one sample determination is given here as an example. The determination was based on the experimental results of constant rate filtration of talc suspensions reported by Chase (1989). The experimental data are summarized in Table 7.4. The estimated ranges of parameter values, the starting parameter values and the results from the minimization of Φ using the complex method are listed in Table 7.5. As an indication of the validity of the search, the predicted cake thickness history from the solution of Equation (3.4.10)–(3.4.13c) using the parameter values obtained from the search was compared with the experimental values. The comparison is shown in Fig. 7.6.

The search-optimization method described above, generally speaking, is computational demanding. While it is certainly feasible, its use may not be practical from benefit-effort considerations and simpler methods may be preferred. Several such methods will be discussed in the following sections.

Table 7.4

Constant-rate filtration data of Chase (1989) 4% talc suspensions

$q_{\ell m} = 7.818 \times 10^{-4}$ m/s, $1/R_m = 5.044 \times 10^{-11}$ m, t < 179 s
$1/R_m = 5.044 \times 10^{-11}/[1 + 1.013 \times 10^{-2}(t - 179)]$, t > 179 s

Time (s)	Cake thickness (m)	Operating pressure (kPa)
100	0.004	24.9
126	0.006	29.4
153	0.007	34.3
179	0.009	44.3
205	0.010	55.9
232	0.012	71.6
258	0.013	88.9
284	0.015	111.7
311	0.016	133.6
337	0.018	154.5
363	0.022	173.5

Table 7.5

Search results of constitutive relationship parameters based on results of Table 4.2

(1) Estimated parameter ranges

$$0.55 \leq \varepsilon_s^o \leq 0.65$$
$$0.02 \leq \beta \leq 0.05$$
$$1.0 \times 10^3 \leq P_A \leq 3.0 \times 10^3 \text{ (Pa)}$$
$$5 \times 10^{-13} \leq k^o \leq 1 \times 10^{-11} \text{ (m}^2\text{)}$$
$$0.4 \leq \delta \leq 0.70$$

(2) Initial estimation and final parameter value

	Initial estimation	Search result
ε_s^o	0.6	0.6293
β	0.04	0.0287
P_A (P_s)	2.8×10^3	1.74×10^3
k^o (m^2)	5×10^{-12}	1×10^{-12}
δ	0.5	0.6365

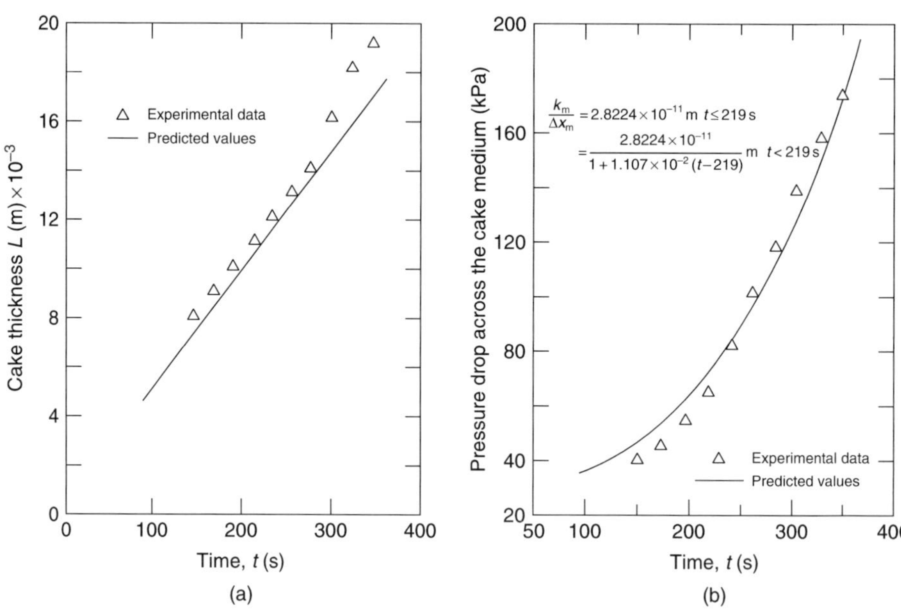

Figure 7.6 Comparison of predicted cake filtration performance with experiments of Chase, 1989. Predictions made using constitutive relationship parameters obtained from the search-optimization parameters based on the minimization of the objective function defined by Equation (7.3.1): (a) cake thickness vs. time; (b) Δp vs. time of $CaCO_3$. (Stamatakis and Tien, 1991. Reprinted by permission of Elsevier.)

7.3.2 Determination of ε_s vs. p_s from expression results

The results of ε_s vs. p_s can be obtained from batch filtration experiments if the experiments are extended beyond filtration into cake consolidation (or expression experiments). From the amount of the particles present in the test suspension, and the ultimate cake height achieved at equilibrium, the corresponding cake solidosity can be readily calculated and related to the pressure applied.

The results obtained are similar to what can be obtained in C-P measurements described previously. Problems encountered in C-P measurements in determining ε_s vs. p_s is, therefore, also present here, namely, the presence of the cell surface friction and the inhomogeneity of the cake formed. Assuming that these problems can be ignored or taken care of, a series of constant pressure expression experiments carried out under various operating pressure may yield results of the relationship of ε_s vs. p_s.

7.3.3 Determination of the average specific cake resistance based on the linear plot of t/V vs. V of constant-pressure filtration data

The simplest and most widely used method of determining specific cake resistance (or permeability) is based on the linear plot of t/V vs. V of constant-pressure filtration data. As discussed in Section 6.4.2, from Equations (2.2.2) and (2.2.3), one has

$$\mu s \rho \overline{(1-\overline{m}s)^{-1}(\alpha_{av})_{p_{s_m}}} \frac{V^2}{2} + \mu R_m V = p_o t \tag{7.3.3.1}$$

and

$$\overline{(1-\overline{m}s)^{-1}(\alpha_{av})_{p_{s_m}}} = \frac{2}{V^2} \int_0^V (1-\overline{m}s)^{-1} (\alpha_{av})_{p_{s_m}} \, dV \tag{7.3.3.2}$$

where μ, s and ρ denote, respectively, the filtrate viscosity, particle mass fraction of the suspension and the filtrate density. $\overline{m}$ is the wet to dry cake mass ratio and $[\alpha_{av}]_{p_{s_m}}$ is the average specific cake resistance defined as

$$[\alpha_{av}]_{p_{s_m}} = \frac{\Delta p_c / p_{s_m}}{\left[\int_0^{p_{s_m}} \left(\frac{1}{\alpha}\right)(-f')\mathrm{d}p_s \right] \Big/ p_{s_m}} \tag{7.3.3.3}$$

where Δp_c is the pressure drop across the cake and p_{s_m} is the compressive stress at the cake/medium interface. f' is defined as

$$f' = \frac{\mathrm{d}p_\ell}{\mathrm{d}p_s} \tag{7.3.3.4}$$

The quantity $\overline{(1-\overline{m}s)^{-1}(\alpha_{av})}$ is not a constant, but varies with time since both $\overline{m}$ and p_{s_m} are functions of time (or V). However, as stated previously, both quantities approach their ultimate values rather quickly; namely, $\overline{m}$ becomes a constant as the average cake solidosity reaches a constant (see Fig. 6.19) and p_{s_m} becomes the value corresponding to $p_\ell = 0$ (i.e. the value of p_ℓ at the cake/membrane interface if the medium resistance is negligible). This is shown in Fig. 6.22 that with the exception of the initial period, there exists a linear relationship between t/V vs. V of constant-pressure filtration data given by Equation (2.2.13) or

$$t/V = (1/2\,p_o)\,\mu s\rho\,[(1-\overline{m}s)^{-1}]_{\Delta p_c = p_o}][\alpha_{av}]_{\Delta p_c = p_o}\,\frac{V + \mu R_m}{p_o} \tag{7.3.3.5}$$

If the slope of the linear segment of t/V vs. V is denoted as B, the average specific cake resistance over $0 < p_s < p_{s_m}$ (corresponding to $0 < p_\ell < p_0$) with p_{s_m} corresponding to $p_\ell = 0$ is simply

$$[\alpha_{av}]_{\Delta p_s = p_o} = \frac{2Bp_o\,(1-\overline{m}s)_{\Delta p_c = p_o}}{\mu s\rho} \tag{7.3.3.6}$$

As described in the previous chapter (see Section 6.4.2), non-synchronization of time-filtrate volume data may result in errors in the plot of t/V vs. V. To avoid this possible error, a different procedure for obtaining $[\alpha_{av}]_{\Delta p_c = p_o}$ from filtration data may be used. From Equation (2.1.20), one has

$$\left(\frac{dV}{dt}\right)^{-1} = \mu s\rho\,[(1-\overline{m}s)^{-1}]_{\Delta p_c = p_o}[\alpha_{av}]_{\Delta p_c = p_o}(V/p_o) + \left(\frac{\mu R_m}{p_o}\right) \tag{7.3.3.7}$$

In other words, there exists a linear relationship between $(dV/dt)^{-1}$ vs. V. More important, an error in V due to non-synchronization has no effect on the slope of the line of $(dV/dt)^{-1}$ vs. V, which was pointed out by both Christensen and Dick (1985) and Shirato *et al.* (1987).

By applying the method described above to data of constant-pressure filtration carried out under different operating pressure, p_o, the results of $[\alpha_{av}]_{\Delta p_c = p_o}$ vs. p_o can be obtained. However, if one wishes to relate the values of $[\alpha_{av}]_{\Delta p_c = p_o}$ of a cake with the compressive stress, p_s, to which the cake is subject, the relationship between the filtration pressure p_ℓ and the compressive stress p_s must be known. According to the multiphase theory (see Section 3.2), there are a number of possibilities. The relationship that $dp_\ell + dp_s = 0$ or $f' = -1$ is often used although one cannot validate its use with any theoretical argument.

In a more recent study, Tien *et al.* (2001) proposed that one may identify the $p_\ell - p_s$ relationship for a given system by comparing the average specific cake resistance obtained from constant-pressure filtration data with the values calculated according to Equation (7.3.3.3) using different $p_\ell - p_s$ relationships and the α vs. p_s relationship based on C-P measurement. Some of the possible $p_\ell - p_s$ relationships are those given by Equations (3.2.17a)–(3.2.17d) or

$$f' = -1 \qquad \text{or Type 1 relationship} \qquad (7.3.3.8a)$$

$$f' = \frac{-1}{1-\varepsilon_s} \qquad \text{or Type 2 relationship} \qquad (7.3.3.8b)$$

$$f' = \frac{-\varepsilon_s}{1-\varepsilon_s} \qquad \text{or Type 3 relationship} \qquad (7.3.3.8c)$$

$$f' = -\frac{(1-\varepsilon_s^o)p_o - p_s}{(1-\varepsilon_s)^2}\frac{d\varepsilon_s}{dp_s} + \frac{\varepsilon_s}{1-\varepsilon_s} \qquad \text{or Type 4 relationship} \qquad (7.3.3.8d)$$

The comparisons are shown in Figs 7.7 (for $CaCO_3$) and 7.8 (for Kaolin cake). For $CaCO_3$ cakes, (α_{av}) calculated with Type 2 relationship gives the best agreement and for Kaolin cakes, the α_{av} obtained from constant-pressure filtration experiments are bracketed between the calculated (α_{av}) based on Type 1 and Type 3 relationships. These results are largely consistent with the comparisons shown in Figs 6.27 and 6.28. The $p_\ell - p_s$ relationship appears to be system specific.

7.3.4 Determination of cake material functions from batch filtration data based on the diffusional theory

The two material functions characterizing cake properties according to the diffusion theory of cake filtration are the filtration diffusivity $D(e)$ and the function $e(p_s)$. The function $e(p_s)$ can be obtained from the relationship of ε_s vs. p_s, the determination of

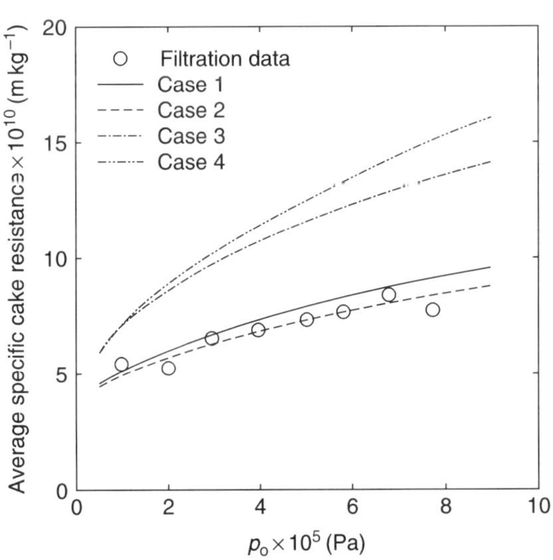

Figure 7.7 Comparison of average specific resistance vs. compressive stress obtained from constant-pressure filtration experiments with those from C-P measurements. (Tien, Teoh and Tan, 2001. Reprinted by permission of Elsevier.)

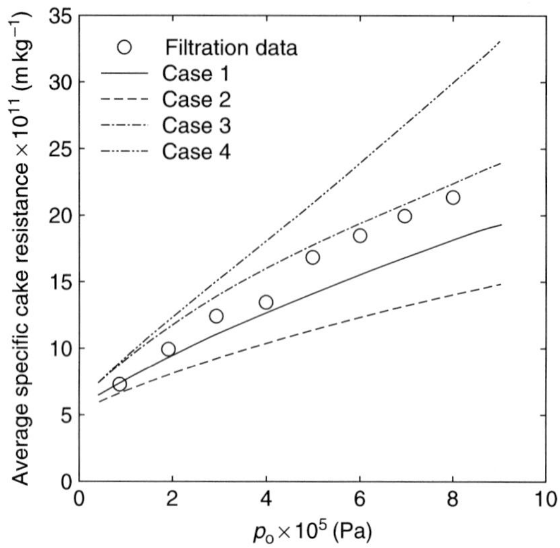

Figure 7.8 Comparison of average specific resistance vs. compressive stress of Kaolin cake obtained from constant-pressure cake filtration with those from C-P measurements. (Tien, Teoh and Tan, 2001. Reprinted by permission of Elsevier.)

which is described in Section 7.3.2. For the determination of $D(e)$, the method based on the void ratio profile may be applied (Smiles, 1987). From Equation (4.3.1.7), one has

$$D(e)\frac{\mathrm{d}e}{\mathrm{d}\lambda} = \frac{1}{2}\int_{e}^{e_o} \lambda \, \mathrm{d}e \tag{7.3.4.1}$$

where λ is the similarity variable defined as $m/t^{1/2}$, m is the material coordinate given as

$$m = \int_{o}^{x} \varepsilon_s \, \mathrm{d}x = \int_{o}^{x} \frac{\mathrm{d}x}{1+e} \tag{7.3.4.2}$$

and e_o is the void ratio of the suspension $[e_o = (1-\varepsilon_{s_o})/\varepsilon_{s_o}]$.

Accordingly if the solidosity profile ε_s vs. x at various times are known, the relationship of e vs. λ can be established from which $\mathrm{d}e/\mathrm{d}\lambda$ and the integral of Equation (7.3.3.1) can be readily determined, thus establishing the relationship of $D(e)$ vs. e.

As an example of applying this method, Fig. 7.9 gives the results of e vs. λ based on the dissection of filter cakes formed in constant-pressure expression (filtration followed by consolidation) of bentonite suspensions under a pressure of 63.7 kPa at three different temperatures. The results show that the temperature effect can be well accounted for by the factor of μ/μ^* where μ^* is the filtrate viscosity at 293 K. For large λ, the void ratio was nearly constant, corresponding the cake consolidation phase. The results of $D(e)$ vs. e estimated according to Equation (7.3.4.1) based on this profile are shown in Fig. 7.10.

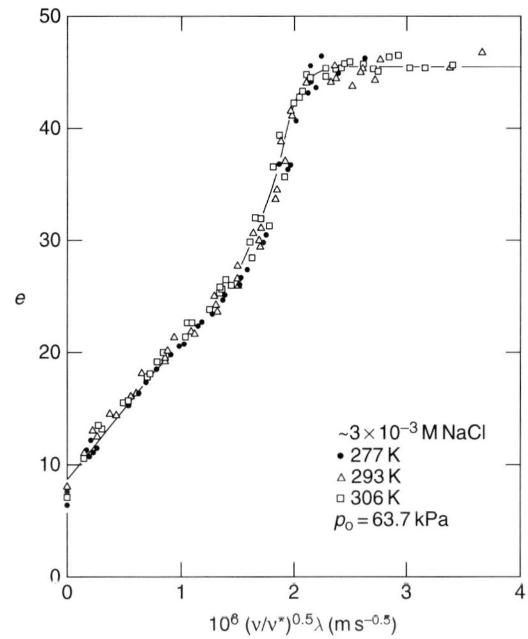

Figure 7.9 The void ratio profile (e vs. $\lambda = m/t^{1/2}$) of bentonite cakes based on dissection of filter cakes obtained from constant-pressure filtration ($p_o = 63.7$ kPa) at three different temperatures. (Smiles, 1986. Reprinted by permission of Elsevier.)

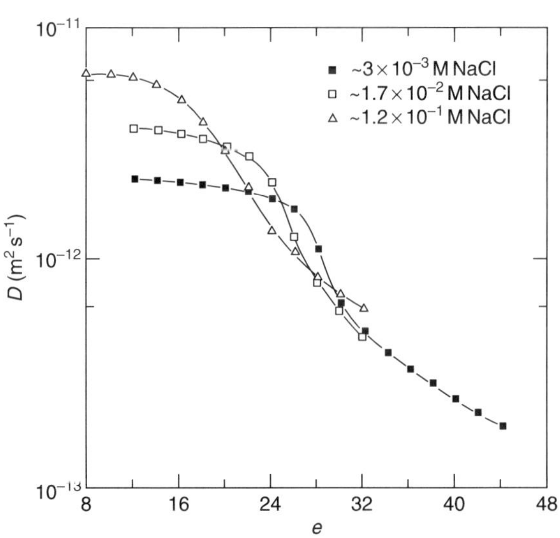

Figure 7.10 Results of $D(e)$ vs. e obtained from the void ratio profile of Fig. 7.9. (Smiles, 1986. Reprinted by permission of Elsevier.)

Also included in Fig. 7.10 are results obtained using the same suspension of different salt concentration.

The results shown in Fig. 7.10 are well defined and vary systematically with salt concentration. However, it is difficult to estimate the accuracy of the results since the method based on Equation (7.3.4.1) requires differentiation and integration of experimental data (Smiles, 1986). In particular, for the e vs. λ profile of Fig. 7.9, obtaining $d\lambda/de$ near the region with e approaching what appears to be a constant value was rather difficult.

7.3.5 Determination of filtration diffusivity from constant-pressure filtration data

Based on the diffusion theory of filtration, if the medium resistance is ignored, for constant pressure filtration, the cumulative filtrate volume, V, is given as [see Equation (4.3.1.10)]

$$V = St^{1/2} \tag{7.3.5.1}^*$$

where S is a function of the operating pressure, p_o, if the profile of e vs. λ is unique (see Fig. 7.9).

Using certain heuristic arguments advanced by Philip (1973), the filtration diffusivity, $D(e)$, can be expressed as

$$D(e) = \frac{1}{2(e(p_o) - e_o)} \left[\frac{dS^2}{de(p_o)} - \frac{rS^2}{e(p_o) - e_o} \right] \tag{7.3.5.2}$$

and r may assume values of zero, one or $2 - \dfrac{\pi}{2}$.

From a series of constant pressure filtration experiments, values of S corresponding to different operating pressures can be obtained by fitting the cumulative filtrate volume data V vs. t according to Equation (7.3.5.1). This information together with the constitutive relationship of ε_s (or e) vs. p_s and the p_s–p_ℓ relationship enables the establishment of the relationship of S vs. $e(p_o)$, With known values of r, $D(e)$ can be readily calculated from Equation (7.3.5.2).

Smiles and Harvey (1973) applied the procedure mentioned above to obtain the filtration diffusivity of Wyoming bentonite suspensions. A typical set of cumulative filtration volume data is shown in Fig. 7.11. The results of S vs. p_o and S vs. e are shown respectively in Figs 7.12 and 7.13 with the constitutive relationship of e vs. p_s given in Fig. 7.14. The results of D(e) vs. e are given in Fig. 7.15 with curves A, B and C corresponding to $r = 0$(A), $r = 1$ (B), and $r = 2 - (\pi/2)$ (C). Also included in this

---

*S used here differ from the S of Equation (4.3.1.10) by a factor of 2.

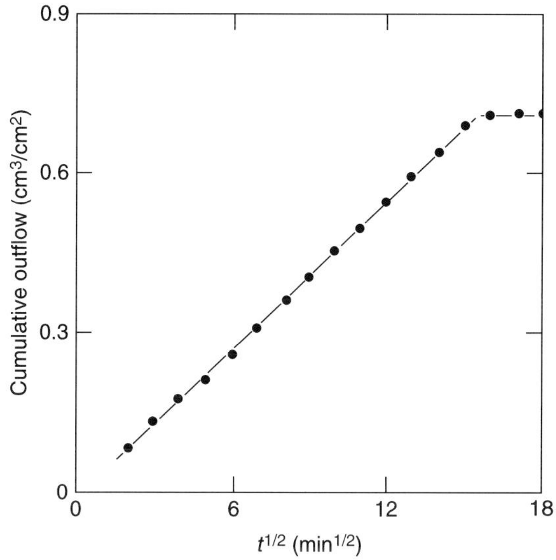

Figure 7.11 Experimental results of cumulative filtrate volume, V vs. $t^{1/2}$, constant pressure filtration of Wyoming bentonite suspension ($e_o = 63.5$ or $\varepsilon_{s_o} = 0.0155$). (Smiles and Harvey, 1973. Reprinted by permission of Lippincott, Williams and Wilkins.)

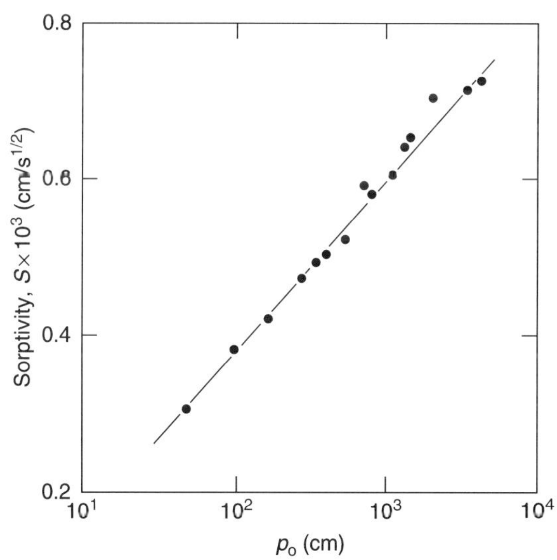

Figure 7.12 Results of S vs. p_o of Wyoming bentonite suspensions. (Smiles and Harvey, 1973. Reprinted by permission of Lippincott, Williams and Wilkins.)

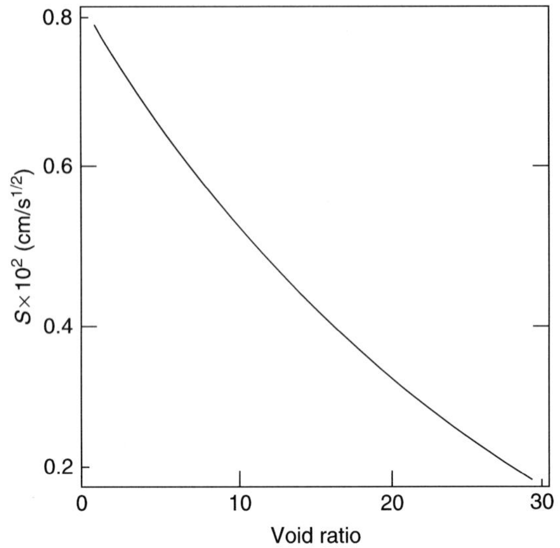

Figure 7.13 Results of S vs. $e(p_o)$ of Wyoming bentonite cake. (Smiles and Harvey, 1973. Reprinted by permission of Lippincott, Williams and Wilkins.)

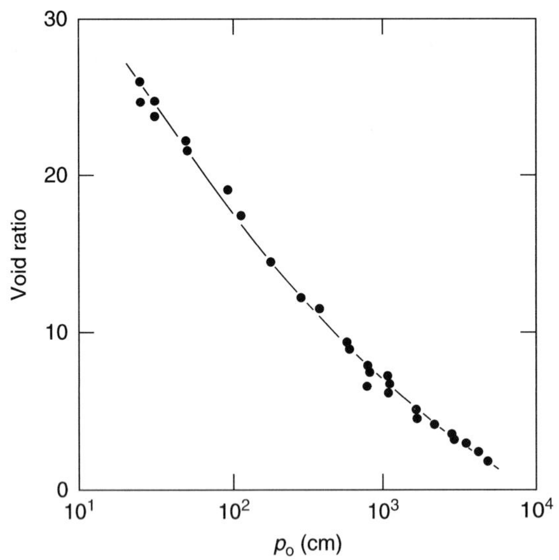

Figure 7.14 Results of $e(p_o)$ vs. p_o of Wyoming bentonite cake. (Smiles and Harvey, 1973. Reprinted by permission of Lippincott, Williams and Wilkins.)

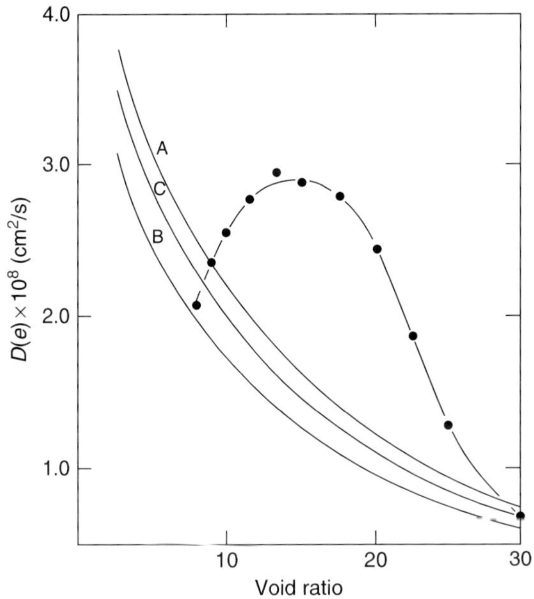

Figure 7.15 Results of $D(e)$ vs. e of Wyoming bentonite cake. A: $r = 0$; B: $r = 1$; C: $r = 2-(\pi/2)$; I: Obtained using the method based on the void profile (e vs. λ). (Smiles and Harvey, 1973. Reprinted by permission of Lippincott, Williams and Wilkins.)

figure are the results obtained from the method based on the void ratio profile (e vs. λ) discussed in the previous section. The lack of agreement between these two methods is apparent.

Equation (7.3.5.2) may also be written in terms of cake solidosity. Noting that $e = (1/\varepsilon_s) - 1$, Equation (7.3.5.2) becomes

$$2\left(\frac{1}{\varepsilon_{s_o}} - \frac{1}{\varepsilon_{s_{po}}}\right) D(e) = \varepsilon_{s_{po}}^2 \frac{dS^2}{d\,\varepsilon_{s_{po}}} - \frac{rS^2}{\dfrac{1}{\varepsilon_{s_o}} - \dfrac{1}{\varepsilon_{s_{po}}}} \tag{7.3.5.3}$$

If $r = 0$, and noticing $D(e)/\varepsilon_{s_{po}}^2 = [D]_{\ell we}$ [see Equation (7.1.6)], one has

$$[D]_{\ell we} = \frac{dS^2/d\varepsilon_{s_{po}}}{2\left(\dfrac{1}{\varepsilon_{s_i}} - \dfrac{1}{\varepsilon_{s_{po}}}\right)} \tag{7.3.5.4}$$

Equation (7.3.5.4) is the same as the physically derived (or re-derived) Smiles–Harvey's formula by Landman *et al.* (1999) (see their Equation (33)).* The same expression was also used by deKretsec *et al.* (2001) and Brown and Zukoski (2003) in their experimental studies.

It was apparently overlooked by Philip and others that Equation (7.3.5.4) can be readily obtained from the conventional theory of cake filtration. Referring to Equation (7.3.3.1), neglecting the medium resistance and that $\Delta p_o \cong p_o$, with $\overline{m}$ and $[\alpha_{av}]_{p_{s_m}}$ approaching their asymptotic values, V^2 can be seen to be proportional to t. In terms of Equation (7.3.3.1), S^2 is simply

$$S^2 = \frac{2}{\mu \rho s}(1-\overline{m}s)\int_{o}^{p_s} \varepsilon_s \rho_s k(-f')\mathrm{d}p_s \tag{7.3.5.5}$$

The differentiated form of the above expression is

$$\left(\frac{\mathrm{d}S^2}{\mathrm{d}\varepsilon_{s_{p_o}}}\right)\left(\frac{\partial \varepsilon_s}{\partial p_s}\right)_{p_o} = \frac{2}{\mu \rho s}(1-\overline{m}s)\rho_s[\varepsilon_s k(-f')]_{p_o}$$

and

$$\left(\frac{\mathrm{d}S^2}{\mathrm{d}\varepsilon_{s_{p_o}}}\right) = \frac{2}{s}\frac{\rho_s}{\rho}(1-\overline{m}s)\left[\varepsilon_s \frac{k}{\mu}(-f')\frac{\partial p_s}{\partial \varepsilon_s}\right]_{p_o} \tag{7.3.5.6}$$

$(1-\overline{m}s)/(\rho s)$ is the ratio V/w, namely the cumulative filtrate volume per unit cake mass. Therefore, the quantity

$$\frac{(1-\overline{m}s)\rho_s}{\rho s} = \frac{V}{w/\rho_s} \tag{7.3.5.7}$$

is the volume of cumulative filtrate per unit cake particle volume. From overall mass balance, one has

$$\frac{V + (w/\rho_s) + (w/\rho_s)\dfrac{1-\bar{\bar{\varepsilon}}_s}{\bar{\bar{\varepsilon}}_s}}{(w/\rho_s)} = 1/\varepsilon_{s_o} \tag{7.3.5.8}$$

The above expression may be simplified as

$$\frac{V}{(w/\rho_s)} = \frac{1}{\varepsilon_{s_o}} - \frac{1}{\bar{\bar{\varepsilon}}_s} \tag{7.3.5.9}$$

where $\bar{\bar{\varepsilon}}_s$ is the average cake solidosity.

* There is an error in Equation (7) of Landman *et al.* (1999). It should be

$$\frac{\mathrm{d}S^2}{\mathrm{d}\varepsilon_{s_{p_o}}} - \frac{1}{\varepsilon^2_{s_{p_o}}}\frac{rS^2}{\dfrac{1}{\varepsilon_{s_o}} - \dfrac{1}{\varepsilon_{s_p}}} = 2[D]_{\ell we}\left(\frac{1}{\varepsilon_{s_o}} - \frac{1}{\varepsilon_{s_{p_o}}}\right)$$

Equation (7.3.5.6) now becomes

$$\frac{dS^2}{d\varepsilon_{p_o}} = 2\frac{V}{(w/\rho_s)}\left[\varepsilon_s \frac{k}{\mu}(-f')\frac{dp_s}{d\varepsilon_s}\right]_{p_o} = 2\left[\varepsilon_s \frac{k}{\mu}(-f')\frac{dp_s}{d\varepsilon_s}\right]_{p_o}\left(\frac{1}{\varepsilon_{s_o}} - \frac{1}{\varepsilon_s}\right) \quad (7.3.5.10)$$

with $f' = -1$ (or $dp_\ell + dp_s = 0$) and the equivalence relationship of Equation (7.1.3b) and (7.1.6), one has

$$[D]_{\ell we} = \frac{1}{2}\frac{dS^2/dp_o}{(1/\varepsilon_{s_o}) - (1/\bar{\bar{\varepsilon}}_s)} \quad (7.3.5.11)$$

If filtration is extended to consolidation under mechanical equilibrium, neglecting the surface friction effect $\bar{\bar{\varepsilon}}_s = \varepsilon_{s_{p_o}}$, Equation (7.3.5.11) reduces to Equation (7.3.5.4) the expression given by Landman *et al.* (1999).

From what is shown above, it is clear that the method advanced by Landman *et al.* (1999) is actually a simplified version of the commonly applied method based on the linear plot of t/V vs. V data of constant pressure filtration data. Equation (7.3.5.1) on which the method of Landman *et al.* is based is the same as the parabolic law of constant pressure filtration with negligible medium resistance. The quantity "S^2" is merely the reciprocal of the slope of the t/V vs. V plot [i.e. Equation (7.3.5.5)]. Differentiating S^2 with $\varepsilon_{s_{p_o}}$ for the purpose of obtaining $[D]_{\ell we}$ or $[r]_{\ell we}$ is equivalent to the evaluation of the specific cake resistance from the differentiation of the average specific resistance. Furthermore, the method of Landman *et al.* is valid under the conditions of negigible resistance and the p_ℓ–p_s relationship being $dp_\ell + dp_s = 0$. The method, therefore, can hardly be described as one which "allows a mathematically exact determination of $r(\varepsilon_s)$" in the words of deKretsec *et al.* (2001).

There are two factors which warrant further considerations. First, differentiating experimental data invariably introduces errors. This disadvantage and the use of the additional assumption of $dp_\ell + dp_s = 0$ therefore make the method less attractive than the method of C-P measurement. Secondly, as a practical matter, if Equation (7.3.5.1) is sufficiently accurate to be used in evaluating cake properties from experimental data, it certainly can be used to predict filtration performance. For this purpose, the knowledge of S^2 (tentamount to α_{av}) should suffice. There is no need to carry out the differentiation.

7.3.6 Determination of cake properties based on stepped pressure, filtration data

This method which utilizes data obtained from experiments conducted with operating pressure increasing stepwise was first proposed by Murase *et al.* (1989a). In more recent years, the same method was also studied by Usher *et al.* (2001) and Tarleton (2004). Tarleton's study was conducted using his mechatronic apparatus and the results obtained were very similar to those of Murase *et al.* In the work of Usher *et al.*, an ad hoc model developed with intuitive reasoning without theoretical considerations was applied for data interpretation. It is, therefore, difficult to relate their results to those of other investigations. The following discussions are centered on the original work of Murase *et al.* (1989a).

Unlike the methods based on constant pressure filtration data (see Sections 7.3.2 and 7.3.4), which yields one value of the specific cake resistance (or filtration diffusivity) corresponding to the operating pressure p_o used, the method based on stepped pressure filtration experiments enables the determination of specific cake resistance corresponding to a number of operating pressure, thus reducing the effort required to establish the constitutive relationship.

The method of Murase *et al.* was formulated on the following considerations: From the conventional theory [i.e. Equation (2.1.20)], the instantaneous filtration rate is given as

$$\frac{dV}{dt} = \frac{p_o}{\mu (\alpha_{av})_{\Delta p_c} \dfrac{V\rho s}{1-\overline{m}s} + R_m} \tag{7.3.6.1}$$

The operating pressure p_o increases stepwise, or

$$\begin{array}{ll} p_o = (p_o)_1 & 0 < t = t_1 \\ p_o = (p_o)_2 & t_1 < t < t_2 \\ \cdot & \cdot \\ \cdot & \cdot \\ \cdot & \cdot \\ p_o = (p_o)_i & t_{i-1} < t < t_i \end{array}$$

Equation (7.3.6.1) may be rewritten as

$$\frac{dV}{dt} = \frac{(p_o)_i}{\mu (\alpha_{av})_{(\Delta p_c)_i} \dfrac{V\rho s}{1-\overline{m}s} + R_m} \quad \text{for} \quad t_{i-1} < t < t_i \tag{7.3.6.2}$$

where $(\Delta p_c)_i$ is the pressure drop across the cake during the period $t_{i-1} < t < t_i$.

As before, $(\alpha_{av})_{(\Delta p_c)_i}$ and $\overline{m}$ may be treated as constants if the medium resistance is ignored or $(\Delta p_c)_i \cong (p_o)_i$ and $\overline{m}$ may be taken to be the asymptotic value at $p_o = (p_o)_i$. Integration of (7.3.6.2) yields:

For $0 < t < t_1$

$$(\alpha_{av})_{(p_o)_1} \frac{\rho s}{\mu (1-\overline{m}s)_{(p_o)}} \frac{V^2}{2} + R_m V = \frac{(p_o)_1}{\mu} t \tag{7.3.6.3}$$

For $t_1 < t < t_2$

$$(\alpha_{av})_{(p_o)_2} \frac{\rho s}{\mu (1-\overline{m}s)_{(p_o)_2}} \frac{(V-V_1)^2}{2} + R_m (V-V_1) = \frac{(p_o)_2}{\mu} t \tag{7.3.6.4}$$

For $t_{i-1} < t < t_i$

$$(\alpha_{av})_{(p_o)_i} \frac{\rho s}{\mu(1-\overline{m}s)_{(p_o)_i}} \frac{(V-V_{i-1})^2}{2} + R_m(V-V_{i-1}) = \frac{(p_o)_i}{\mu} t \tag{7.3.6.5}$$

where V_{i-1} is the cumulative filtrate volume collected at the end of the period $t_{1-2} < t < t_{i-2}$. From the differentiated form of Equation (7.3.6.5), one has

$$\frac{(p_o)_i}{\mu}\frac{dt}{dV} = \frac{(p_o)_i}{\mu}\left(\frac{dV}{dt}\right)^{-1} = (\alpha_{av})_{(p_o)_i} \frac{\rho s}{(1-\overline{m}s)_{(p_o)_i}}(V-V_i) + R_m \tag{7.3.6.6}$$

The above derivation is based on the assumption that the cake structure (and therefore cake solidosity and permeability) depends only on the cake compressive stress. The cake structure changes instantaneously from one corresponding to $(p_o)_{i-1}$ to that of $(p_o)_i$ at the transition from the $(i-1)$th period to the i-th period.

If Equation (7.3.6.6) is valid, plotting experimental data obtained from stepped pressure experiment in the form of $(dV/dt)^{-1}$ vs. V should yield a number of linear segments with slopes equal to $[\mu/(p_o)_i](\alpha_{av})_{(p_o)_i}\rho s(1-\overline{m}s)^{-1}_{(p_o)_i}$ and discontinuities at V_i's. Furthermore, from the slopes of the respective linear segments of $(dV/dt)^{-1}$ vs. V, values of $(\alpha_{av})_{(p_o)_i}$ can be obtained. This was largely confirmed with the data obtained by Murase *et al.* (1989a) and the more recent results of Tarleton (2004). A set of results obtained by Murase *et al.* is shown in Fig. 7.16. A comparison of the results based on the stepped pressure experiments, constant pressure experiment and C-P measurements (with $-f' = 1$) is given in Fig. 7.17. Agreement of the results obtained from these three different methods is excellent.

The method proposed by Murase *et al.* is not without its problems. Tarleton (2004) reported that the results obtained may depend on the magnitude of the operating pressure increment as well as the duration of the individual period of operation (i.e. $t_i - t_{i-1}$). Physically speaking, the assumption of instantaneous cake response to pressure change being instantaneous can only be approximately correct. Larger pressure increments and shorter period of operation certainly will make this assumption less valid.

7.3.7 Determination of specific cake resistance based on instantaneous filtration rate data

A method of determining specific cake resistance based on instantaneous filtration rate was recently proposed by Teoh (2003). Similar to the work of Murase *et al.* (1989a), the development of this method was motivated by the desire of reducing the experimental effort required to establish the constitutive relationship. Recognizing that the method based on the linearity of t/V vs. t is widely accepted by industrial practitioners because of its simplicity, its replacement by any new method is unlikely if the necessary data acquisition and interpretation require elaborate facilities and/or are time-consuming.

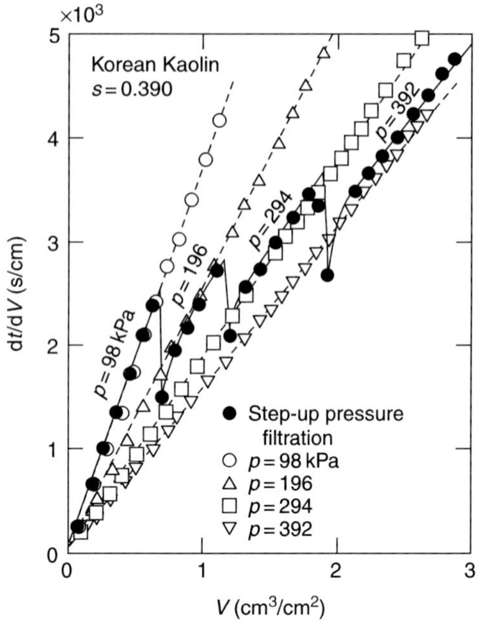

Figure 7.16 Relationship between reciprocal filtration rate vs. cumulative filtration volume obtained from filtration experiment carried out under operating pressure. Kaolin suspensions with $s = 0.36$. (Murase, Iritani, Cho and Shirato, 1989a. Reprinted by permission of The Society of Chemical Engineering, Japan.)

The basis of Teoh's method can be described as follows: Referring to Equation (2.1.16), the instantaneous filtration rate is

$$q_\ell = \frac{\mathrm{d}V}{\mathrm{d}t} = \frac{p_o}{\mu\left[(\alpha_{av})_{\Delta p_c}\dfrac{V\rho s}{(1-\overline{m}s)} + R_m\right]} \qquad (7.3.7.1)$$

The above expression may be rewritten as

$$[\alpha_{av}]_{\Delta p_c} = \frac{\Delta p_c}{\mu \dfrac{V\rho s}{(1-\overline{m}s)}\dfrac{\mathrm{d}V}{\mathrm{d}t}} \qquad (7.3.7.2)$$

and the pressure drop across the filter cake, Δp_c, is

$$\Delta p_c = p_o - \mu R_m \frac{\mathrm{d}V}{\mathrm{d}t} \qquad (7.3.7.3)$$

For constant-pressure filtration, Δp_c is zero initially and then increases monotonically with time and approaches asymptotically to p_o. The traditional method (i.e. that based on the linear plot of t/V vs. V) gives the values of $(\alpha_{av})_{\Delta p_c}$ with $\Delta p_c \to p_o$. However, with

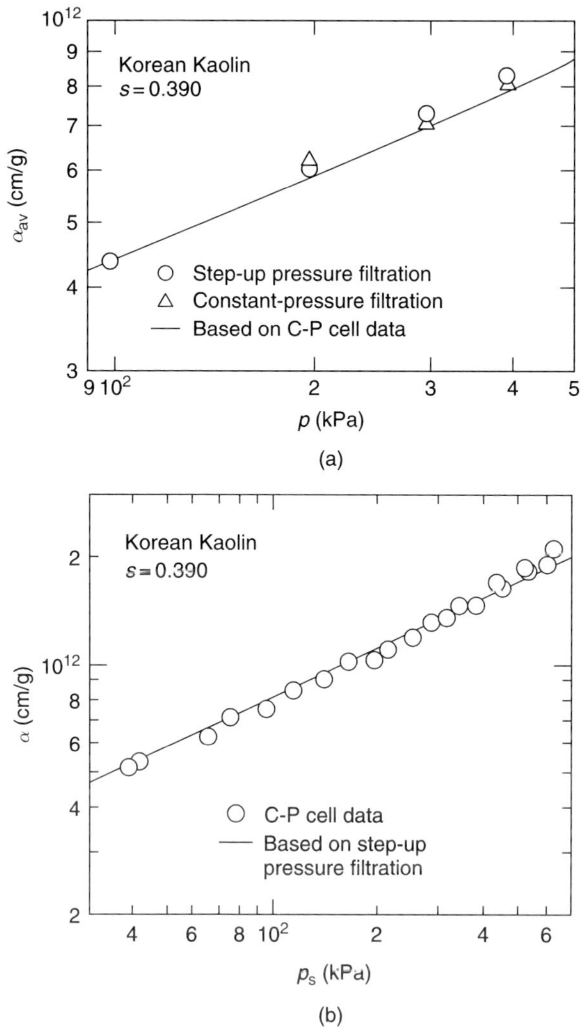

Figure 7.17 Comparison of results based on Fig. 7.16 and C-P measurements: (a) average specific cake resistance vs. p_s; (b) specific cake resistance vs. p_s; Korean Kaolin cake. (Murase, Iritani, Cho and Shirato, 1989a. Reprinted by permission of The Society of Chemical Engineering, Japan.)

the information of Δp_c vs. t, one should be able to obtain values of $(\alpha_{av})_{\Delta p_c}$ extending over Δp_c ranging from zero to nearly p_o. This is the basic principle of Teoh's method.

From Equation (7.3.7.2), $(\alpha_{av})_{\Delta p_c}$ may be expressed as

$$[\alpha_{av}]_{\Delta p_c} = \frac{\Delta p_c}{\mu \dfrac{V \rho s}{(1 - \overline{m} s)} \dfrac{dV}{dt}} \tag{7.3.7.4}$$

To utilize the above expression for the determination of $(\alpha_{av})_{\Delta p_c}$, values of $\Delta p_c, \overline{m}$ and $(\mathrm{d}V/\mathrm{d}t)$ at various times during the course of one constant-pressure filtration run are required. From the results of cumulative filtrate volume and cake thickness vs. time, these quantities may be obtained in the following manner:

(a) Numerical differentiation of the data of V vs. t gives the values of $\mathrm{d}V/\mathrm{d}t$ at various times.

(b) The wet to dry cake mass ratio volume $\overline{m}$ can be found according to the following expression

$$\overline{m} = 1 + \frac{\rho}{\rho_s}\frac{\rho_s L(1-s) - \rho v s}{\rho(V+L)s} \tag{7.3.7.5}$$ *

Thus from the V vs. t and L vs. t data, $\overline{m}$ at various times can be calculated.

(c) The pressure drop across the filter cake Δp_c can be obtained from Equation (7.3.7.3) if the value of R_m is known. Past investigators including Teoh (2003) and Meeten (2000) have shown that R_m may not be constant, but varies with time and the operating pressure (see Section 6.4.4). The initial value of R_m, $(R_m)_o$, can be found to be

$$(R_m)_o = \lim_{t \to o} R_m = \frac{1}{\mu} \lim_{t \to o} \frac{\Delta p_m}{(\mathrm{d}V/\mathrm{d}t)} = \frac{1}{\mu}\frac{p_o}{(\mathrm{d}V/\mathrm{d}t)_{t \to o}} \tag{7.3.7.6}$$

As cake thickness increases, $\Delta p_c \to p_o$ and the linearity of t/V vs. V is established. The ultimate medium resistance $(R_m)_\infty$ may be determined from the intercept of the t/V vs. V plot using the latter part of the $V_\infty t$ data. The question of which R_m value should be used will be discussed later.

Assuming that the values of $(\mathrm{d}V/\mathrm{d}t)$, $\overline{m}$ and R_m are known, the average specific cake resistance $[\alpha_{av}]_{p_{s_m}}$ can be found from the following expression

$$(\alpha_{av})_{p_{s_m}} = \frac{p_o - \mu R_m(\mathrm{d}V/\mathrm{d}t)}{\mu \frac{V\rho s}{(1-\overline{m}s)}\frac{\mathrm{d}V}{\mathrm{d}t}} \tag{7.3.7.7}$$

The value of p_{s_m} depends on the applied pressure p_o as well as the p_ℓ–p_s relationship. From Equations (3.2.12b), (3.2.13b), (3.2.14b), and (3.2.15b), p_{s_m} is given as

$$p_{s_m} = p_o - \Delta p_m \qquad \text{for Type 1 relationship} \tag{7.3.7.8}$$

$$\int_o^{p_{s_m}} \frac{\mathrm{d}p_s}{1-\varepsilon_s} = p_o - \Delta p_m \qquad \text{for Type 2 relationship} \tag{7.3.7.9}$$

*This expression may be obtained by combining Equations (6.3.3) and (6.4.1.1).

$$\int_{o}^{p_{s_m}} \frac{\varepsilon_s}{1-\varepsilon_s} dp_s = p_o - \Delta p_m \qquad \text{for Type 3 relationship} \qquad (7.3.7.10)$$

$$p_{s_m} = \frac{1-\varepsilon_s^o}{\varepsilon_{s_m}} p_o - \frac{1-\varepsilon_{s_m}}{\varepsilon_{s_m}} \Delta p_m \qquad \text{for Type 4 relationship} \qquad (7.3.7.11)$$

Before presenting some of the results obtained by Teoh, the difference due to using different R_m values will be first examined. Fig. 7.18 gives the results obtained based on constant-pressure filtration of 5% Kaolin suspension using $(R_m)_o$ and $(R_m)_\infty$ with Type 1 p_ℓ–p_s relationship. The general trends displayed, the degrees of data scattering and the ranges of p_s covered suggest strongly that $(R_m)_o$ should be used in calculating $(\alpha_{av})_{p_{s_m}}$. Accordingly, $(R_m)_o$ was used in Teoh's work.

The results of $(\alpha_{av})_{p_s}$ vs. p_s obtained according to Teoh's method are shown in Fig. 7.19 ($CaCO_3$ cakes) and Fig. 7.20 (Kaolin cakes). Also included in these figures are calculated values of α_{av} according to Equation (7.3.3.3) based on the results of C-P measurements and the various types of $p_\ell - p_s$ relationships. The $CaCO_3$ results are shown to be bracketed between those based on Type 1 and Type 2 relationships. The Kaolin results largely agree with those based on Type 3 relationships. These findings are the same as those of Tien *et al.* (2001) mentioned previously and consistent with the comparisons shown in Figs 6.27 and 6.28.

The results of Figs 7.19 and 7.20 seem to suggest the validity of Teoh's method. However, a full realization of the purpose in developing this method remains to be accomplished. The method was conceived and developed with the expectation of obtaining

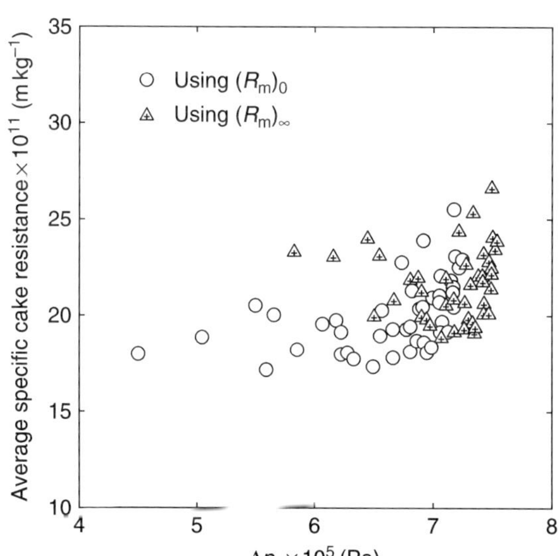

Figure 7.18 Average specific cake resistance based on instantaneous filtrate rate using two different medium resistance, constant-pressure cake filtration of Kaolin suspensions, $p_o = 800\,\text{kPa}$. (Teoh, 2003. Reprinted by permission of Dr S.-K. Teoh.)

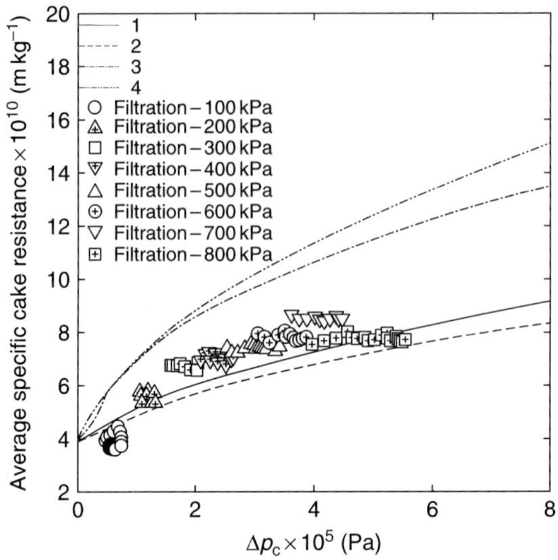

Figure 7.19 Comparison of the average cake specific resistance vs. p_s based on instantaneous filtration rate vs. C-P measurements, $CaCO_3$ cakes. 1. Type 1 $p_\ell - p_s$ relationship; 2. Type 2 $p_\ell - p_s$ relationship; 3. Type 3 $p_\ell - p_s$ relationship; 4. Type 4 $p_\ell - p_s$ relationship. (Teoh, 2003. Reprinted by permission of Dr S.-K. Teoh.)

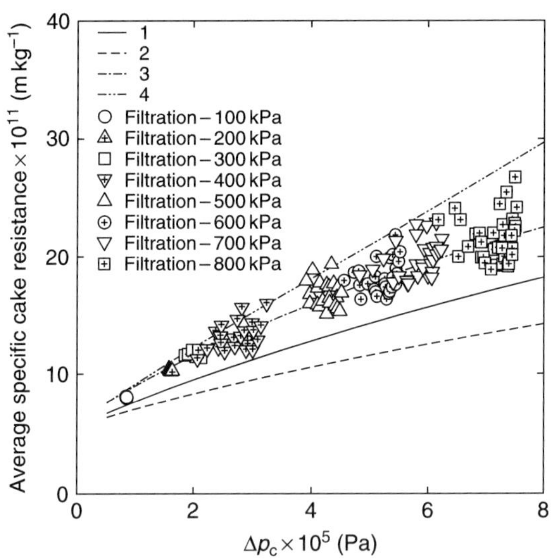

Figure 7.20 Comparison of the average cake specific resistance vs. p_s based on instantaneous filtration rate vs. C-P measurements, Kaolin cakes. 1. Type 1 $p_\ell - p_s$ relationship; 2. Type 2 $p_\ell - p_s$ relationship; 3. Type 3 $p_\ell - p_s$ relationship; 4. Type 4 $p_\ell - p_s$ relationship. (Teoh, 2003. Reprinted by permission of Dr S.-K. Teoh.)

values of (α_{av}) over a wide range of p_s values. As shown in these figures, based on the results of one experiment, several values of α_{av} were obtained but not covering the entire range of p_s values. The limitations were caused by the fact that accurate values of dV/dt were difficult to obtain at both the initial period (due to the inevitable start-up disturbance) and latter part of the experiment when filtration rate diminishes. This difficulty also underscores the problem of numerically differentiating experimental results in data treatment.

7.3.8 Determination of cake properties based on local measurements

Sedin *et al.* (2003) and Johansson *et al.* (2004) studied the determination of cake constitutive relationships based on measurements of solidosity and pressure profiles across filter cakes (for their determinations, see Section 6.3). The liquid-solid relative velocity inside a filter cake is given by Equation (3.1.4) or

$$\frac{q_\ell}{\varepsilon} - \frac{q_s}{\varepsilon_s} = -\frac{1}{\varepsilon}\frac{k}{\mu}\frac{\partial p_\ell}{\partial x} \tag{7.3.8.1}$$

which can be rearranged to give

$$k = \frac{\mu}{(\partial p_\ell/\partial x)}\left(q_\ell - \frac{1-\varepsilon_s}{\varepsilon_s} q_s\right) \tag{7.3.8.2}$$

Since the second term of the right hand side of the above expression is much less than q_ℓ, except in the region adjacent to the cake surface, and furthermore, if one assumes that

$$q_\ell \cong q_{\ell_m} = \frac{dV}{dt}$$

that is, q_ℓ may be replaced by the instantaneous filtration velocity, the local cake permeability becomes

$$k = \frac{\mu}{(\partial p_\ell/\partial x)}\frac{dV}{dt} \tag{7.3.8.3}$$

With the knowledge of V vs. t and the filtration pressure profiles at various times, the local values of k throughout a cake can be readily obtained. Furthermore, knowing the filtrate pressure profile and the $p_\ell - p_s$ relationship (i.e. Equations (7.3.3.8a)–(7.3.3.8d)), the compressive stress across the cake can be calculated. This information together with the results of the solidosity measurements provides the information of ε_s vs. p_s; k (or α) vs. p_s; and k (or α) vs. ε_s. A typical set of such results obtained by Johasson *et al.* (2004) of Kaolin suspensions assuming $f' = -1$ is shown in Fig. 7.21a,b. Also included in these figures are the results obtained from C-P measurements. The consistency of the results is excellent and the results also corroborate well with those of C-P measurements.

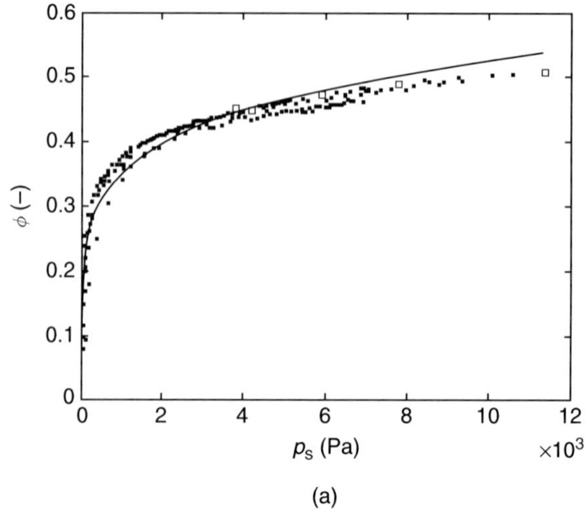

Figure 7.21a Relationship between ε_s vs. p_s. Kaolin cake at pH 2.5. • Local measurement results; □ results based on C-P measurements; - correlations of local measurement data according to Equation (7.3.2a); - - - correlation of C-P data according to Equation (7.3.2a) (Johansson *et al.*, 2004). (Courtesy of Professor H. Theilander).

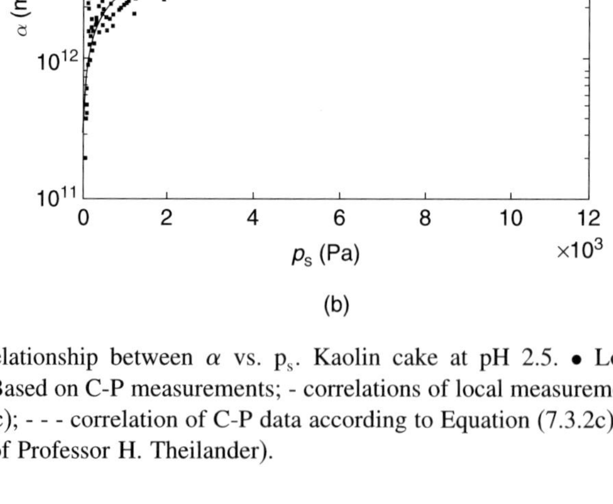

Figure 7.21b Relationship between α vs. p_s. Kaolin cake at pH 2.5. • Local measurement results; □ results Based on C-P measurements; - correlations of local measurement data according to Equation (7.3.2c); - - - correlation of C-P data according to Equation (7.3.2c) (Johansson *et al.*, 2004). (Courtesy of Professor H. Theilander).

7.3.9 Comparisons

The main differences among the various methods described above may be summarized as follows:

(a) Among these methods, only the C-P measurements yield results which relate cake properties (ε_s and k) with the cake compressive stress. The search-optimization procedure developed by Stamatakis and Tien (see Section 7.3.1), in principle, can be extended to include the determination of the $p_s - p_\ell$ relationship as part of the search. However, this will increase the number of parameters to be determined and may render the search impractical. The method of determining the void ratio e as a function of p_s based on consolidation data is actually a C-P measurement.
(b) With the results of ε_s vs. p_s known, one may estimate the relationship of k vs. p_s using the Kozeny-Carmen equation or other similar expressions. This kind of estimation is likely to introduce considerable error (for example, see Fig. 5.1).
(c) For the determination of k and its equivalent, all the methods discussed under Section 7.3 with the exception of the C-P measurements yield results of k vs. p_0 (as an approximation of the cake pressure drop). The constitutive relationships were often established based on a particular $p_s - p_\ell$ relationship; namely, $\mathrm{d}p_\ell + \mathrm{d}p_s = 0$ or $f' = -1$.
(d) Other than the C-P measurements, the results obtained can be readily used for scale-up or predictions based on the conventional theory without the knowledge of the $p_s - p_\ell$ relationship. However, they cannot be used for the more rigorous analysis (i.e. the solution of the volume-averaged equation of continuity without specifying the $p_s - p_\ell$ relationship).
(e) The methods discussed in Sections 7.3.3, 7.3.6 and 7.3.7 are based on the conventional theory. The method used for determining the filtration diffusivity described under Sections 7.3.4 and 7.3.5 are equivalent to the method based on the linear plot t/V vs. V with further assumption of negligible medium resistance.
(f) Both the stepped pressure method and the method based on instantaneous filtration rate have the potential time-saving advantage. However, they require further improvement for this potential to be realized.
(g) Selecting a particular method of determination depends largely on the purpose of the users. The decision inevitably involves a trade-off between the efforts to be expended and the benefit derived. The method based on the linear plot of t/V vs. V is the most widely used because of its simplicity and long history of usage. Its replacement by other methods is unlikely in the foreseeable future.

7.4 CAKE PROPERTY RESULTS AND CORRELATIONS

Over the years, a large number of experimental studies have been reported in the literature on the determination of filter cake properties of various systems. The variables which strongly influence the constitutive relationships include the particle size, size distribution, shape, and more importantly, the solution variables of the suspending liquid (ionic strength, pH and the presence of surface active agents). Grace (1953) reported that for $CaCO_3$ cakes formed at pH = 9.8, the specific cake resistance varied from $5.24 \times 10^{10}\,\mathrm{m\,kg^{-1}}$ ($p_s = 7$ kPa) to $1.314 \times 10^{11}\,\mathrm{m\,kg^{-1}}$ ($p_s = 698$ kPa). The corresponding values at pH = 10.3 were 1.483×10^{11} to $2.693 \times 10^{11}\,\mathrm{m\,kg^{-1}}$. Similar results for other types of cakes were also found by Laure *et al.* (2003).

Theoretical analysis of the interaction effects have been attempted by several investigators in recent years (McDonough *et al.*, 1992; Bowen and Jenner, 1995; Koenders and Wakeman, 1997a,b). None of the studies have yielded results which can be applied for predicting cake permeability (or its equivalents).

To account for the compressible behavior of filter cakes, power law expressions have been used to correlate data of specific systems. In addition to Equations (7.3.1.2a)–(7.3.1.2c), some of the commonly used simple expressions include

$$y = y_o p_s^{a} \tag{7.4.1}$$

$$y = y_1 + y_2 \,\ell n\, p_s \tag{7.4.2}$$

$$y = y_1 - \frac{a}{b + p_s} \tag{7.4.3}$$

where y stands for either ε_s or k (or α).

In contrast, the void ratio e $(= \varepsilon/\varepsilon_s)$ is often related to the compressive stress by the expression shown below (smiles, 1986),

$$\ell n(p_s) = a_1 + a_2 e \tag{7.4.4}$$

Based on the above expression and the definition of the filtration diffusivity given by Equation (7.1.3), with $f' = -1$, one has

$$D(e) = -2.303 a_2 \frac{e^3}{(1+e)^2} \exp\left[2.303(a_1 + a_2 e)\right] \tag{7.4.5}$$

The choice of which one of these expressions should be used to represent the constitutive relationship data depends mainly upon the fitness of the data set to a particular expression. However, there are differences among them. Equations (7.4.1), (7.4.2) and (7.4.4) do not give finite values of ε_s (or k, α) at $p_s = 0$. As stated before, the values of ε_s and k (or α) at the zero-stress state are required for the solution of the volume–averaged continuity equation (or the corresponding diffusion equation). In this sense, the values of ε_s^o, k^o (or α^o) of Equations (7.3.1.2a)–(7.3.1.2b) and $y_1 - a/b$ of Equation (7.4.4) are not merely fitting parameters but quantities with certain physical significance. In particular, ε_s^o is a threshold value which differentiates suspension from filter cake. In contrast, extraneous assumptions about the conditions prevailed at the cake/suspension interface must be introduced for those expressions which do not give the zero-stress values to be used in cake filtration analysis.

Generally speaking, Equation (7.4.1) was applied widely in earlier studies. However, using Equations (7.3.1.2a)–(7.3.1.2c) has become more common recently. Tabulation of the constitutive relationship data compiled by Shirato *et al.* (1987) according to Equation (7.4.1) are given in Tables 7.6 and 7.7.

Table 7.6

Constitutive relationships of porosity vs. compressive stress $\varepsilon = \varepsilon_o p_s^{-\lambda}$. Based on results compiled by Shirato *et al.* (1987)

Particle type	ε_o	λ	Range of p_s (10^5 Pa)
Almina	0.96	0.0102	0.07–689
	0.73	0.0047	0.82–4.89
Asbestos	1.06	0.017	0.52–6.89
Calcium carbonate	0.755	0.036	0.7–3.45
in distilled water	1.04	0.033	0.07–6.89
$Na_4P_2O_7$ solution	0.795	0.013	0.34–6.89
Carbon	0.845	0.021	0.82–7.58
Carbonyl iron	0.425	0	0.14–68.95
Celite	1.05	0.017	0.82–4.14
Cement	0.82	0.058	1.5–3.4
Clay (Mitsukuri Garirome)	1.23	0.0618	1.0–4.9
Copper oxide	0.505	0.021	0.82–5.49
Ferric oxide	1.12	0.037	70–345
Kaolin			
100%	0.88	0.045	0.1–6.89
$Al_3(SO_4)_3$ Solution	0.9	0.049	0.07–6.89
$Na_4P_2O_7$ Solution	0.79	0.031	0.07–6.89
Hong Kong Pink	1.0	0.047	0.01–9.8
Korean	1.03	0.0604	0.5–3.4
Shinmer	0.98	0.046	0.06–6.9
Limestone (crushed)	0.45	0.015	0.5–6.89
Magnesium carbonate	1.10	0.011	0.7–3.45
Silica (fine)	1.03	0.006	0.07–13.79
Suchar	0.965	0.0091	0.1–6.89
Super Cel			
Standard	0.96	0.0104	0.07–6.89
100% Hyflo	0.995	0.014	0.1–6.89
75% Hyflo, 25% Kaolin	1.16	0.02	0.1–6.89
50% Hyflo, 50% Kaolin	1.07	0.032	0.1–6.89
25% Hyflo, 75% Kaolin	1.00	0.044	0.1–6.89
Talc	1.39	0.054	0.07–6.89
TiO_2 R-100 (flocculated)	1.12	0.038	0.07–6.89
R-100	0.82	0.019	0.13–6.89
Tungsten (UT 238)	1.02	0.023	0.07–6.89
Zinc sulfide (A)	1.10	0.023	0.07–6.89
(B)	1.36	0.047	0.07–6.89

Table 7.7

Constitutive relationships of specific cake resistance vs. compressive stress, $\alpha = \alpha_o p_s^n$; Based on results of Shirato *et al.* (1987)

Particle type	α_o	n	Range of $p_s (\times 10^5$ Pa)
Alumina (Type A Linde) in			
0.01 mol dm^{-3} $Al_2(SO_4)_3$	2.37×10^8	0.3	0.07–6.89
Alumina hydroxide	3.32×10^{11}	0.34	1.73–6.84
$CaCO_3$ (superlight)			
Distilled water pH 9.8	8.93×10^9	0.20	0.07–6.89
(Superlight) 0.01 mol dm^{-3}			
$Na_4P_2O_7$ pH 10.3	4.69×10^{10}	0.13	0.07–6.89
Cement	2.20×10^{10}	0.298	1.0–8.8
Chalk (flocculated)	4.37×10^{10}	0.14	
Clay			
Colloidal	7.43×10^{11}	0.16	1.73–6.89
Hara Gairome	1.44×10^{11}	0.612	1.0–8.8
Mica	6.45×10^9	0.37	1.73–6.89
Mitsukusi Gairome	7.82×10^7	0.669	1.0–8.8
Darco (G-60) 0.01 mol dm^{-3} $Na_4P_2O_7$	2.07×10^8	0.51	0.07–6.89
Ferric oxide	2.59×10^{11}	0.39	1.73–6.89
Ignition pink slurry	4.20×10^9	0.563	1.0–8.8
Kaolin 0.01 mol dm^{-3} $Al_2(SO_4)_3$			
Hong Kong Pink	4.79×10^{10}	0.27	0.07–6.89
50% Hong Kong Pink	6.48×10^9	0.485	1.0–8.8
50% Mitenkuri Gairome	6.47×10^9	0.605	1.0–8.8
Magnesium Hydroxide	1.35×10^{10}	0.47	1.73–6.89
Silen (fine)	3.70×10^6	1.28	0.07–6.89
Solka Floc (BW 200) 0–1 mol dm^{-3} $Al_2(SO_4)_3$	2.22×10^3	1.01	0.07–6.89
Super-Cel (standard) distilled water pH 7.38	7.46×10^9	0.17	0.07–6.89
Talc 0.01 mol dm^{-3} $Al_2(SO_4)_3$	7.05×10^8	0.51	0.07–6.89
TiO_2			
R-110 0.01 mol dm^{-3} HCl pH 3.45	9.29×10^{11}	0.058	0.07–6.89
R-110 distilled water pH 3.8	1.27×10^{10}	0.32	0.07–6.89
ZnS			
Precipitated hot pH 9.1	2.11×10^8	0.69	0.07–6.89
Precipated cold pH 9.07	1.48×10^8	0.92	0.07–6.89

7.5 GENERALIZED CONSTITUTIVE RELATIONSHIP

A complete constitutive relationship should, in principle, give values of ε_s and k (or α) extending over both low and high compressive stress values. Experimentally this is difficult to accomplish. C-P measurements do not yield accurate results for $p_s \leq 10$ kPa. Shirato *et al.* (1983) suggested the use of sedimentation test for determining ε_s and k (or α) at low values of p_s.

Consider batch sedimentation of suspension of finite volume of H_o (m^3/m^2) with a known amount of suspended particles (M, volume of particle per unit cross-sectional area, m^3/m^2). The sediment height upon completion of sedimentation is H_∞, and is a function of M.

If M is increased by a quantity of δM, the corresponding increment of H_∞ being δH_∞, the solidosity of the sediment segment of height δH_∞ immediately above the bottom of the sediment chamber ε_s by definition

$$(\delta H_\infty)(\varepsilon_s) = \delta M \tag{7.5.1}$$

In the limit, one has

$$(\varepsilon_s)_{\text{bottom}} = \frac{dM}{dH_\infty} \tag{7.5.2}$$

The sediment at the bottom of the sedimentation chamber is subject to a compressive stress $p_s|_{\text{bottom}}$ equal to

$$p_s|_{\text{bottom}} = (\rho_s - \rho)Mg \tag{7.5.3}$$

If sedimentation measurements were made with different M values, from the results of H_∞ vs. M, the derivative dM/dH_∞ can be readily estimated. The relationship between ε_s vs. p_s can be established according to Equations (7.5.2) and (7.5.3).

A simplification of the procedure mentioned above was suggested by Shirato *et al.* (1983). Assuming that the experimental data of H_∞ vs. M can be approximated, or

$$H_\infty = \bar{a}M^{\bar{b}} \tag{7.5.4}$$

and the relationship of ε_s vs. p_s is of the type of Equation (7.4.1), or

$$\varepsilon_s = Bp_s^\beta \tag{7.5.5}$$

A simple manipulation of Equations (7.5.2)–(7.5.5) yields

$$\frac{1}{\bar{a}\bar{b}}M^{1-b} = B[(\rho_s - \rho)Mg]^\beta$$

or

$$\beta = 1 - b \tag{7.5.6}$$

$$B = \{ab[(\rho_s - \rho)g]^{1-b}\}^{-1} \tag{7.5.7}$$

Shirato *et al.* (1983) conducted sedimentation experiments using three types of suspensions (zinc oxide, ferric oxide and Mitsukuri-Gairome clay) and applied the procedure mentioned above to obtain the constitutive relationship of ε_s vs. p_s. The sedimentation data of H_s vs. M are shown in Fig. 7.22. It is obvious that the data can be adequately represented by Equations (7.5.4). With the coefficient and parameter of Equation (7.5.5) evaluated according to Equations (7.5.6) and (7.5.7), the constitutive relationships of ε_s vs. p_s of the three systems are shown in Fig. 7.23.

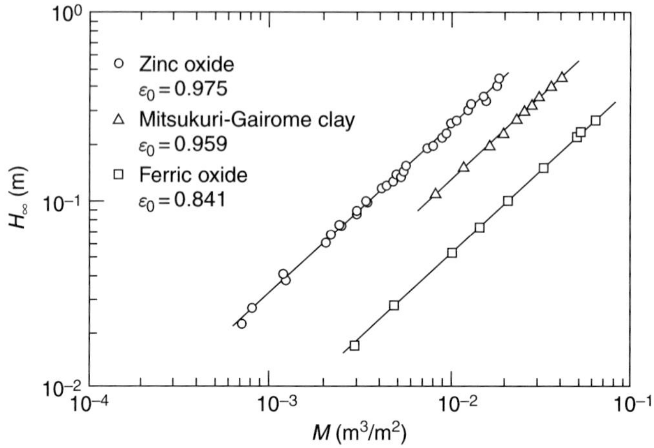

Figure 7.22 Relationship between the ultimate height and the total particles present, (H_∞ vs. M) zinc oxide, Mitsukuri-Gairome clay and ferric oxide suspensions. (Shirato, Murase, Eritani and Hayashi, 1983. Reprinted by permission of Elsevier.)

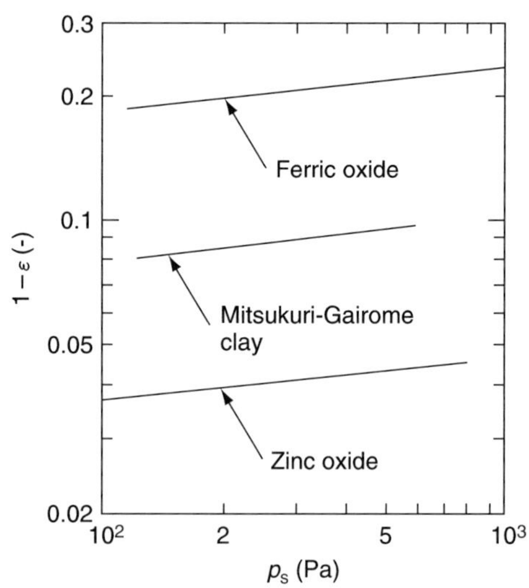

Figure 7.23 Relationship between ε_s and p_s on data of Fig. 7.22. (Shirato, Murase, Eritani and Hayashi, 1983. Reprinted by permission of Elsevier.)

From sedimentation results, one may also establish the relationship of α vs p_s. The generalized Darcy's law gives

$$\frac{q_{\ell s}}{1-\varepsilon_s} = \frac{q_\ell}{1-\varepsilon_s} - \frac{q_s}{\varepsilon_s} = \frac{k}{(1-\varepsilon_s)\mu}\frac{\partial p_\ell}{\partial z}$$

or

$$q_{\ell s} = \frac{k}{\mu}\frac{\partial p_\ell}{\partial z} \tag{7.5.8}$$

Assuming $f' = -1$, one has

$$\frac{\partial p_\ell}{\partial z} + (\rho_s - \rho) g \varepsilon_s + \frac{\partial p_s}{\partial z} = 0 \tag{7.5.9}$$

Initially, $\partial p_s / \partial z = 0$ throughout the suspension phase, or

$$\left(\frac{\partial p_\ell}{\partial z}\right)_{t=o} + (\rho_s - \rho) g \varepsilon_s = 0 \tag{7.5.10}$$

In terms of the material coordinate, m is defined as

$$m = \int_o^z \varepsilon_s \, \mathrm{d}z \tag{7.5.11}$$

$$\left(\frac{\partial p_\ell}{\partial m}\right)_{t=o} = -(\rho_s - \rho) g \tag{7.5.12}$$

The quantity $(q_{\ell s})$ may be taken as the initial settling velocity $|\mathrm{d}H/\mathrm{d}t|$ with H being the height of the suspension.

Combining Equations (7.5.8) and (7.5.11), with $|q_{\ell s}| = |\mathrm{d}H/\mathrm{d}t|$, one has

$$\alpha = \frac{1}{k \varepsilon_s \rho_s} = \frac{(\rho_s - \rho) g}{\mu \rho_s v_o} \tag{7.5.13}$$

where v_o is $|\mathrm{d}H/\mathrm{d}t|_{t=o}$.

The sedimentation data indicate that $\mathrm{d}H/\mathrm{d}t$ is a function only of the particle volume fraction as shown in Fig. 7.24. From sedimentation results obtained using suspension of different particle volume fractions, the relationship between α and ε_s can be established. The result of zinc oxide suspensions are shown in Fig. 7.25. Included in this figure are results obtained using C-P cell measurements at higher compressive stress values. These results suggest that dependence of α and ε_s at higher values of ε_s differs from that at lower values of ε_s. Extrapolation of data obtained from one range of p_s to another is likely to introduce significant errors. This observation was substantiated by a later study of Murase *et al.* (1989b). In their work, Murase *et al.* determined the constitutive relationships of two systems using different methods including C-P measurements, centrifugal test, sedimentation and constant-rate compression test. The general trends shown in Fig. 7.26 (ε_s *vs.* p_s) and Fig. 7.27 (α *vs.* ε_s) are in agreement with those of Fig. 7.25.

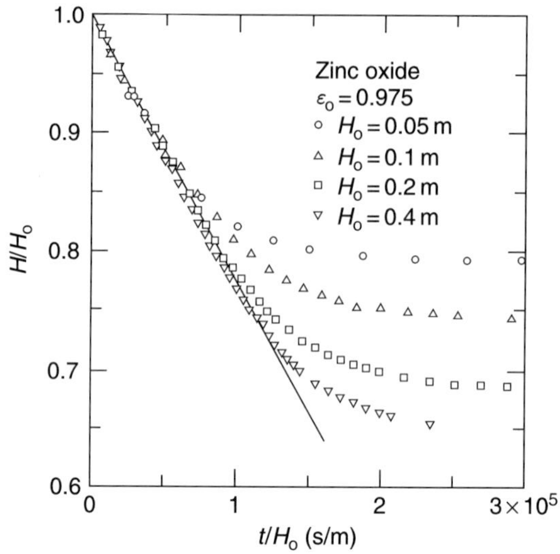

Figure 7.24 Sedimentation curve of zinc oxide suspension. (Shirato, Murase, Eritani and Hayashi, 1983. Reprinted by permission of Elsevier.)

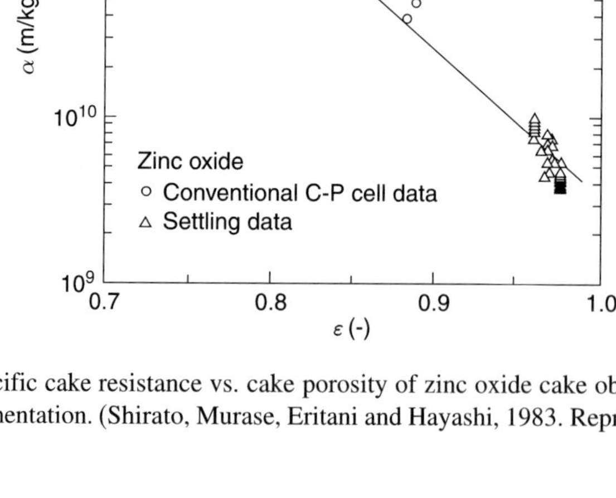

Figure 7.25 Specific cake resistance vs. cake porosity of zinc oxide cake obtained by C-P measurement and sedimentation. (Shirato, Murase, Eritani and Hayashi, 1983. Reprinted by permission of Elsevier.)

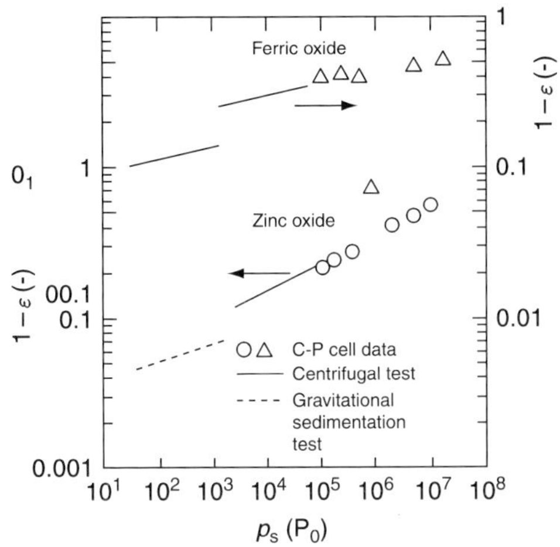

Figure 7.26 Constitutive relationship of ε_s $(= 1 - \varepsilon)$ vs. p_s of zinc oxide and ferric oxide cakes determined by different methods (Murase *et al.*, 1989b). (Murase, Iwata, Adachi, Gmaehowski and Shirato, 1989. Reprinted by permission of The Society of Chemical Engineering, Japan.)

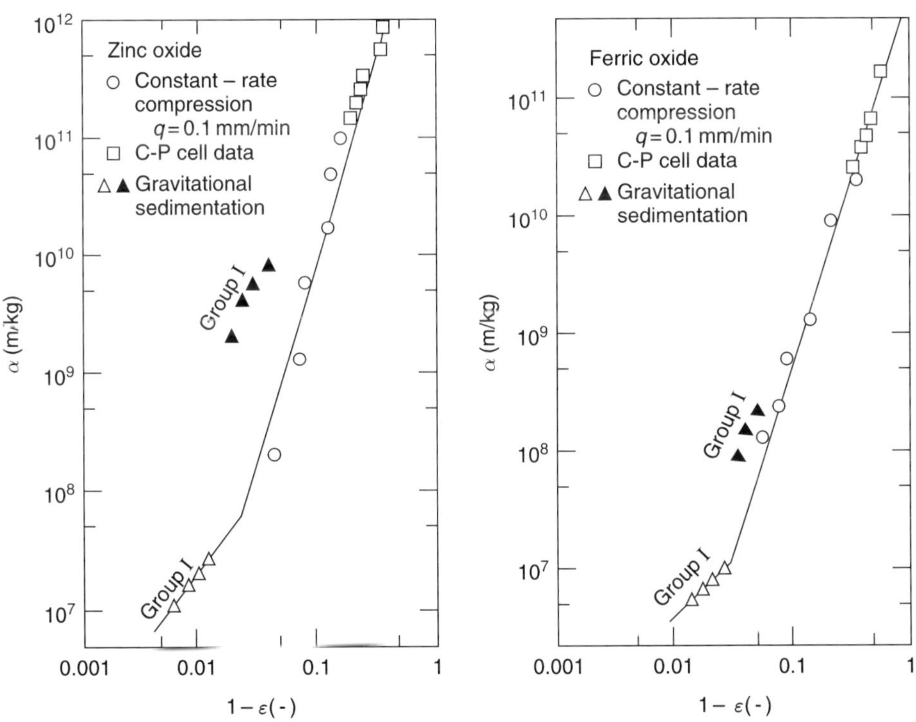

Figure 7.27 α vs. ε_s $(= 1 - \varepsilon)$ relationship of zinc oxide and ferric oxide cakes determined by different methods (Murase *et al.*, 1989b). (Murase, Iwata, Adachi, Gmaehowski and Shirato, 1989. Reprinted by permission of The Society of Chemical Engineering, Japan.)

REFERENCES

Bowen, W.J. and Jenner, F., *Chem. Eng. Sci.*, **50**, 1707 (1995).

Brown, L.A. and Zukoski, C.F., *AIChE J.*, **49**, 362 (2003).

Chase, G.G, "Continuum Analysis of Constant Rate Cake Filtration", Ph.D. Thesis, University of Akron (1989).

Christensen, G.L. and Dick, R.I., *J. Environ. Eng. Div.*, ASCE, **111**, 258 (1985).

deKretsec, R.G., Usher S.P., Scales, P.J. and Boger, D.V., *AIChE J.*, **47**, 758, (2001).

Donald, M.B. and Hunneman, R.D., *Trans. I. Chem. E.*, **1**, 97 (1923).

Grace, H.P., *Chem. Eng. Prog.*, **44**, 303 (1953).

Johansson, C., Sedin, P. and Theilander, H., "Determination of Local Filtration Properties", *9th World Filtration Congress*, New Orleans, Louisiana (2004).

Koenders, M.A. and Wakeman, R.J., *Trans. Inst. Chem. E.*, **75**, 310 (1997a).

Koenders, M.A. and Wakeman, R.J., *AIChE J.*, **43**, 946 (1997b).

Landman, K.A. and Russel, W.B., *Phys. Fluids A*, **5**, 550 (1993).

Landman, K.A. and White, L.R., *AIChE J.*, **43**, 3147 (1997).

Landman, K.A., Sirakoff, C. and White, L.R., *Phys. Fluids A*, **3**, 1495 (1991).

Landman, K.A., White, L.R. and Eberl, M., *AIChE J.*, **41**, 1867 (1995).

Landman, K.A., Stankovich, J.M. and White, L.R., *AIChE J.*, **45**, 1875 (1999).

Laure, O., Vorobiev, E., Vu, C. and Durand, B., *Sep. Pur. Tech.*, **31**, 177 (2003).

Lu, W.-M., Huang, Y.-P. and Hwang, K.-J., *Powder Tech.*, **97**, 16 (1998).

McDonough, R.M., Welsch, K., Fell, C.J.O. and Fane, A.G., *J. Membrane Sci.*, **72**, 197 (1992).

Meeten, G.H., *Chem. Eng. Sci.*, **55**, 1755 (2000).

Murase, T., Iritani, E., Cho, J.H. and Shirato, M., *J. Chem. Eng. Japan*, **22**, 373 (1989a).

Murase, T., Iwata, M., Adachi, T., Gmaehowski, L. and Shirato, M., *J. Chem. Eng. Japan*, **22**, 378 (1989b).

Philip, J.R., *Soil Sci.*, **116**, 328 (1973).

Reichmann, B. and Tomas, J., "Expression Dynamics of Fluid Particle Suspension", *4th Italian Conference on Chemical and Process Engineering*, Florence, Italy (1999).

Reichmann, B. and Tomas, J., *Powder Tech.*, **121**, 182 (2001).

Ruth, B.F., *Ind. Eng. Chem.*, **38**, 564 (1946).

Sedin, P., Johansson, C. and Theilander, H., *Trans. I Chem E*, **81**, A 1393 (2003).

Shirato, M., Aragaki, T., Mori, R. and Sawamato, K., *J. Chem. Eng. Japan*, **1**, 86 (1968).

Shirato, M., Murase, T., Iritani, E. and Hayashi, N., *Filtration and Separation*, 404 (1983).

Shirato, M., Murase,T., Iritani, E., Tiller, F.M. and Alciatore, A.F., "Filtration in the Chemical Process Industry", in *Filtration: Principles and Practices*, M.J. Matteson and C. Orr, eds, Marcel Dekker, New York (1987).

Smiles, D.E., "Principles of Constant-Pressure Filtration", in *Encyclopedia of Fluid Mechanics*, Vol. 5, N.P. Chermisinoff, ed., Gulf Publishing, Houston, TX (1986).

Smiles, D.E. and Harvey, A.G., *Soil Sci.*, **116**, 391 (1973).

Stamatakis, K., "Analysis of Cake Formation in Solid-Liquid separation", Ph.D. Dissertation, Syracuse University (1990).

Stamatakis, K. and Tien, C., *Chem. Eng. Sci.*, **46**, 1917 (1991).

Tan, R.B.H., He, D. and Tien, C., *Multifunction Test Cell for Cake Filtration Studies*, Singapore Patent No. 9804082-7 (1998).

Tarleton, E.S., "Determination of Cake Structure During Stepped and Variable Pressure Filtration", *9th World Filtration Congress*, New Orleans, Louisiana (2004).

Teoh, S.K., "Studies in Filter Cake Characterization and Modeling", Ph.D. Thesis, National University of Singapore (2003).
Tien, C., Teoh, S.K. and Tan, R.B.H., *Chem. Eng. Sci.*, **56**, 5361 (2001).
Tiller, F.M. and Lu, W.-M., *AIChE J.*, **18**, 569 (1972).
Tiller, F.M., Haynes, S. and Lu, W.-M., *AIChE J.*, **18**, 13 (1972).
Usher, S.P., deKretsec, R.G. and Scales, P.J., *AIChE J.*, **47**, 1561 (2001).

PART III

Application and Extension of Cake Filtration Theories

Cake formation and growth is present in a number of solid/fluid separation processes. Analyses of these processes require the incorporation of the theories discussed in Part I with information relevant to the particular process studied. Three such examples are presented and discussed in this section.

–8–

INCORPORATION OF CAKE FILTRATION PRINCIPLES TO THE ANALYSIS OF SOLID/FLUID SEPARATION PROCESSES

Notation

a_p	particle radius (m)
a	half distance between two parallel plates (approximating membrane module) (m)
c	particle concentration of feed gas stream (kg/m^3)
D	diffusivity (m^2 s^{-1})
D_{BM}	Brownian diffusivity (m^2 s^{-1})
D_{max}	deceleration (m s^{-2})
D_s	shear-induced diffusivity (m^2 s^{-1})
d_p	particle diameter (m)
e^i	$i = 1, 2$, void ratio defined as $\varepsilon^i/\varepsilon_s^i$
e^2_∞	ultimate void ratio of aggregate
F_{adh}	adhesion force (Pa)
F_{disl}	dislodgement from (Pa)
F_p	force along the cross-flow diameter (N)
F_q	force along the filtrate flow direction (N)
F_{q_i}	$i = 1, 2, 3$, forces given by Equations (8.2.2.10a)–(8.2.2.10c) (N)
F_{q_4}	particle-surface interaction force (N)
$f(F_{adh})$	probability density function of the adhesion force
f_1, f_2	hydrodynamic retardation correction factor [see Equations (8.2.2.8) and (8.2.2.10a)] (-)
$f_i(w_j)$	fraction of bag surface covered with cake mass of w_j in the i-th cycle of operation (-)
f'	defined as $\mathrm{d}p_\ell/\mathrm{d}p_s$ (-)
h	protrusion height (m)
h_c	critical protrusion height (m)

h_{max}	maximum protrusion height (m)
J_B	particle back diffusion flux ($m\,s^{-1}$)
K	rate constant of Equation (8.3.14) ($m\,s^{-1}\,Pa^{-1}$)
k	Boltzman's constant 1.3803×10^{-23} ($J\,K^{-1}$) or mass transfer coefficient ($m\,s^{-1}$)
k_1, k_2	quantities defined by Equations (8.1.2.2a) and (8.1.2.2b)
L	longitudinal length (m)
m	defined by Equation (8.3.6a) (m)
$\overline{m}$	wet to dry cake mass ratio (-)
N	number of particle types (-)
n	number of bags (-)
p_ℓ	filtrate pressure (Pa)
p_o	operating pressure (Pa)
p_s	compressive press (Pa)
p_{s_m}	value of p_s at cake/membrane interface (Pa)
Q	gas throughput rate defined by Equation (8.1.3.1) ($m\,s^{-1}$)
Q_{av}	a quantity defined by Equation (8.1.3.1) ($m\,s^{-1}$)
q_ℓ	filtrate velocity ($m\,s^{-1}$)
q_{ℓ_m}	instantaneous filtration rate ($m\,s^{-1}$)
$(q_{\ell_m})_c$	critical filtration velocity (flux) ($m\,s^{-1}$)
$(\overline{q}_{\ell_m})_c$	average value of $(q_{\ell_m})_c$ defined by Equation (8.2.3.3)
R	interface flux of Equation (8.3.2a) or (8.3.2b) ($m\,s^{-1}$)
R_m	membrane resistance (m^{-1})
s	particle mass fraction ratio (-)
T	absolute temperature K, or cycle time (s)
t	time (s)
$\overline{t}$	time defined by Equation (8.3.6b) (s)
t_1	time when the steady state is attained (s)
u_L	lift velocity ($m\,s^{-1}$)
w	cake mass per unit surface area ($kg\,m^{-2}$)
w_s	cross-flow velocity (m)
x	longitudinal distance (m)
x_i	volume fraction of suspended particles of diameter d_{p_i} (-)
y_i	volume fraction of cake particles with diameter d_{p_i} (-)

Greek letters

α_{av}	average specific cake resistance defined by Equation (8.3.1.2)
β	fraction of cake removal or fraction of particle flux being deposited (-)
γ	adhesion probability – or a quantity defined by Equation (8.1.3.2) (-)
δ	film thickness (m)
Δp	pressure drop (Pa)
Δp_c	pressure drop across filter cake (Pa)

Δp_m	pressure drop across membrane (Pa)
ε_s	cake solidosity (-)
ε_{s_o}	volume of particle fraction of the suspension to be filtered (-)
ε_s^o	value of ε_s at the zero-stress state (-)
$\varepsilon_{s_{max}}$	maximum value of ε_s (-)
$(\varepsilon_s)_i$	value of ε_s at cake surface (-)
$(\varepsilon_s)_w$	value of ε_s at membrane surface (-)
μ	filtrate viscosity (Pa s)
$\overline{\mu}$	friction coefficient (-)
ν	gas velocity (m s^{-1})
ν_i	instantaneous gas velocity over bag covered with cake mass of w_i (m s^{-1})
$\overline{\nu}$	average gas velocity (m s^{-1})
ρ	filtrate density (kg m^{-3})
ρ_s	particle density (kg m^{-3})
τ_w	shear stress at wall (Pa)
ϕ	particle volume fraction of suspension (-)
ω	shear rate (s^{-1})

Superscript

i	macroscopic description ($i = 1$) or microscopic description ($i = 2$)

The principles of cake filtration and the experimental methods used for its study are discussed in the preceding chapters. Besides cake filtration and expression, cake formation, growth and/or consolidation are also present in several other solid/fluid separation operations. As an example, for gas emission control, fabric filtration is often applied. In fabric filtration, particle-laden gas is passed through filter bags such that filter cakes are formed outside bag surfaces. Because of the large quantities of gases to be treated and to insure bags' repetitive use, filter bags require periodic cleaning. A study of fabric filtration, therefore, must be based on the principles of cake filtration and that of filter cleaning. The present chapter is devoted to the analyses and discussions of such problems.

8.1 ANALYSIS OF PULSE JET FABRIC FILTRATION

Fabric filtration is widely used for gas emission control. In fabric filtration, dust-laden gaseous streams are passed through filter bags with the formation of cakes outside filter bags (see Fig. 8.1).

The nature of dust deposit (i.e. cakes) built-up, the limitation of cake built-up and other process requirements necessitate that each filter bag of a given system be operated cyclically; namely, repeated filtration followed by bag cleaning. Analysis of fabric

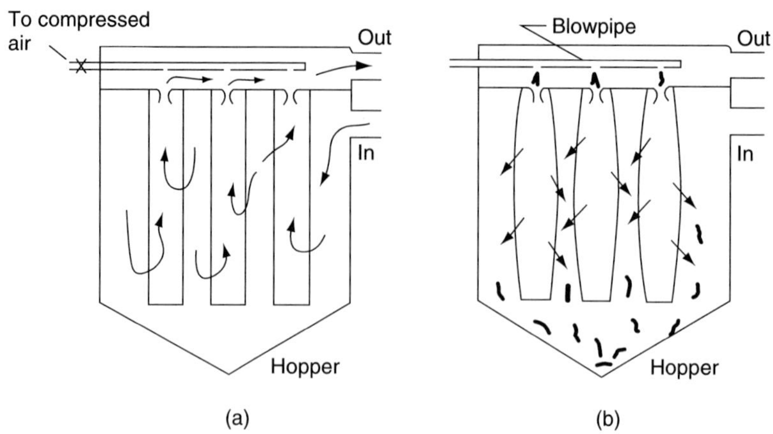

Figure 8.1 Schematic diagram of pulse jet fabric filtration system: (a) filtration; (b) filter cleaning. (Ju, Chiu and Tien, 2001. Reprinted by permission of Elsevier.)

filtration is, therefore, made by combining cake filtration theories with the relevant information of the cleaning process.

Depending upon the method applied for filter bag cleaning, fabric filters may be classified into the following categories: cleaning by shaking, cleaning by reverse gas flow and pulse air jet cleaning. In the present section, an account of the analysis of pulse air jet fabric filtration based on the recent work of Ju *et al.* (2000, 2001) is presented.

The formation and growth of cake in fabric filtration proceeds in the same manner as pressure filtration of liquid suspensions. The relevant results presented in the previous chapters can be readily applied. Similar to filtration of liquid suspensions, the built-up of cake deposits outside a filter bag increases the bag's resistance to gas flow. Thus, from both energy consumption and operational point of view, the cake must be removed periodically. This need coupled with the requirement that a large volume of gas be treated in most applications leads to the use of multibag systems. Proper design of fabric filtration systems, therefore, involves optimally scheduling a number of bags in their operations in order to handle a fixed volume of gases continuously.

8.1.1 Filter bag cleaning by pulse air jet

As filter bag cleaning is an integral part of fabric filtration, a description of filter cleaning process and its treatment is therefore in order. If the particle/filter bag adhesion force F_{adh}, extends over a range of values, it may be characterized by its probability function $f(F_{adh})$. For a bag covered with a cake of fixed thickness (or mass of cake per unit bag surface area, w), applying a dislodgement force F_{disl} to such a bag results in the removal of a fraction of the cake, β. The fraction β can be expressed as

$$\beta = \frac{\int_{0}^{F_{disl}} f(F_{adh})\, \mathrm{d}F_{adh}}{\int_{0}^{\infty} f(F_{ahd})\, \mathrm{d}F_{adh}} \tag{8.1.1.1}$$

For pulse jet cleaning, the dislodgement force F_{disl} is given as

$$F_{disl} = wD_{max} \tag{8.1.1.2}$$

where w is the cake mass per unit bag surface area and D_{max}, the deceleration caused by air jet injected into a bag during cleaning. An empirical expression relating D_{max} with the overpressure of the air jet during cleaning was given by DeRavin *et al.* (1998).

8.1.2 Individual bag performance

For a single bag of a fabric filtration system, it undergoes a cyclic operation: filtration followed by bag cleaning. The filtration phase extends over a period of $T(s)$. For convenience, $t=0$ is taken to be the beginning of the filtration phase. From Equation (2.1.16), the pressure drop–filtration velocity relationship may be written as

$$\Delta p = \left(k_1 + k_2 \int_{0}^{t} cv\, \mathrm{d}t \right) v \qquad t \in [0, T] \tag{8.1.2.1}$$

where

$$k_1 = \mu R_m \tag{8.1.2.2a}$$

$$k_2 = \mu [\alpha_{av}]_{\Delta p_c} \tag{8.1.2.2b}$$

$$\int_{0}^{t} cv\, \mathrm{d}t = w \tag{8.1.2.2c}$$

and the liquid velocity used previously is now replaced by v (gas velocity), Δp is the total pressure drop, or $\Delta p_c + \Delta p_m$, and c is the particle concentration of the feed gas stream. It is assumed that all particles are retained at the bag surface.

One may assume that the operation begins with a fresh bag. At $t=T$, the bag is covered with a cake of uniform thickness. After cleaning, part of the bag surface is free of cake. Therefore, at the beginning of the second cycle of operation, one part of the bag is clean and the remainder part is covered with cake with thickness equal to the thickness of the cake formed at the end of the first cycle. Since the feed gas pressure is fixed, local filtration velocity is determined by the local cake thickness. At the end of the second cycle, the cake thickness over the part of the bag surface which was free of cake at the beginning of the second cycle is the same as the cake thickness at the end of the first cycle. The other part is covered with the cake of greater thickness.

Given the assumption used in describing filter cleaning, i.e. Equation (8.1.1.1), a filter surface may be described by a sequence of functions $\{f(w_i)\}$ where $f(w_i)$ is the fraction of the bag surface covered with cake of mass w_i. With this notation, we may develop expressions for individual bag filtration performance as follows.

First cycle

At the beginning, $[f_1(w_o)]_{in} = 1$ with $w_o = 0$. The subscripts "1" denote the first cycle and "in", the initial condition. Let v_o denote the instantaneous filtration velocity over area covered with cake of mass w_o initially, the area-averaged filtration velocity of the bag $\bar{v}_1$ is

$$\bar{v}_1 = v_o [f_1(w_o)]_{in} = v_o \tag{8.1.2.3}$$

and v_o can be found from Equation (8.1.2.1), or

$$\Delta p = \left(k_1 + k_2 \int_o^t c v_o \mathrm{d}t \right) v_o \tag{8.1.2.4}$$

At the end of the first cycle, the bag is covered with cake of thickness w_1 given as

$$w_1 = \int_o^T c\, v_o \mathrm{d}t \tag{8.1.2.5}$$

Second cycle

The state of the bag surface at the start of the second cycle is given as

$$[f_2(w_o)]_{in} = \beta_1 \tag{8.1.2.6a}$$

$$[f_2(w_1)]_{in} = 1 - \beta_1 \tag{8.1.2.6b}$$

where β_1 is the fraction of cake of mass w_1 removed.

During filtration, the instantaneous velocities through these two parts of the surface are v_o and v_1. v_o can be obtained from Equation (8.1.2.4). Applying Equation (8.1.2.4), v_1 can be found from the following expression:

$$\Delta p = \left(k_1 + k_2 \left(w_1 + \int_o^t c v_1 \, \mathrm{d}t \right) \right) v_1 \tag{8.1.2.7}$$

and the area-averaged filtration velocity during the second cycle, $\bar{v}_2$ is

$$\bar{v}_2 = v_o [f_2(w_o)]_{in} + v_1 [f_2(w_1)]_{in} \tag{8.1.2.8}$$

At the end of the second cycle, the state of the bag surface is

$$[f_2(w_1)]_f = \beta_1 \tag{8.1.2.9}$$

$$[f_2(w_2)]_f = 1 - \beta_1 \tag{8.1.2.10}$$

where the subscript f denotes the final state and w_2 is given as

$$w_2 = w_1 + \int_o^T c v_1 \, dt \tag{8.1.2.11}$$

Third cycle

The fractions of removal of cakes with mass w_1 and w_2 and β_1 and β_2 which can be calculated from Equation (8.1.1.1). Thus, the state of the bag surface at the beginning of the third cycle is

$$[f_3(w_o)]_{in} = 1 - \beta_1(1 - \beta_1) - (1 - \beta_1)(1 - \beta_1)$$

fraction of surface with no cake (8.1.2.12a)

$[f_3(w_1)]_{in} = \beta_1(1 - \beta_1)$ fraction of surface with cake of mass w_1 (8.1.2.12b)

$[f_3(w_2)]_{in} = (1 - \beta_1)(1 - \beta_2)$ fraction of surface with cake of mass w_2 (8.1.2.12c)

During the third cycle, the instantaneous filtration velocities over the three parts of surface are v_o, v_1 and v_2, respectively. v_o and v_1 are given as before, and v_2 according to Equation (2.1.16) is

$$\Delta p = \left(k_1 + k_2 \left(w_2 + \int_o^t c v_2 dt \right) \right) v_2 \tag{8.1.2.13}$$

The area-averaged filtration velocity of the third cycle $\bar{v}_3$ is given as

$$\bar{v}_3 = v_o [f_3(w_o)]_{in} + v_1 [f_3(w_1)]_{in} + v_2 [f_3(w_2)]_{in} \tag{8.1.2.14}$$

At the end of the third cycle, the state of the bag surface is given as

$$[f_3(w_1)]_f = 1 - \beta_1(1 - \beta_1) - (1 - \beta_1)(1 - \beta_2) \tag{8.1.2.15a}$$

$$[f_3(w_2)]_f = \beta_1(1 - \beta_1) \tag{8.1.2.15b}$$

$$[f_3(w_3)]_f = (1 - \beta_1)(1 - \beta_2) \tag{8.1.2.15c}$$

Table 8.1

Equations describing the state of cake coverage of filter bag surface at the beginning of the i-th cycle and the area-averaged filtration rate expression

(1) The initial and final cake coverage states are related by the expression

$$[f_i(\omega_{j-1})]_{\text{in}} = [f_i(\omega_j)]_{\text{f}} \quad j = 1, 2, \ldots, i$$

(2) The final cake coverage state of the $(i-1)$th cycle and the initial state of the i-th cycle are related by the expression

$$[f_i(\omega_{j-1})]_{\text{in}} = [f_{i-j}(\omega_j)](1-\beta_j)$$

$$[f_i(\omega_{\text{o}})]_{\text{in}} = 1 - \sum_{i=1}^{i-1}(f_i(\omega_j)]_{\text{in}}$$

(3) The local instantaneous velocities of the i-th cycle, $v_{\text{o}}, v_1, \ldots, v_{i-1}$ can be found from

$$\Delta p = \left[k_1 + k_2\left(\omega_{j-1} + \int_{\text{o}}^{\text{t}} cv_{j-1}\text{d}t\right)\right]v_{i-1}$$

(4) The cake mass at the end of the i-th cycle, $\omega_1, \omega_2, \ldots, \omega_i$ are

$$\omega_j = w_{j-1} + \int_{\text{o}}^{\text{T}} cv_{j-1}\,\text{d}t \quad j = 1, 2, \ldots, i$$

(5) The area averaged filtration rate of the i-th cycle, $\bar{v}_i$ is given as

$$\bar{v}_i = \sum_{j=0}^{i-1} = v_j[f_i(\omega_j)]_{\text{in}}$$

and

$$w_3 = w_2 + \int_{\text{o}}^{\text{T}} cv_2\text{d}t \tag{8.1.2.15d}$$

By generalizing the above results, expressions for any arbitrary cycle can be obtained. These expressions are given in Table 8.1.

As expected, the surface condition of the filter bag approaches a steady state as the number of cycles increases, namely $[f_i(w_j)]_{\text{in}}$ and therefore $[f_i(w_j)]_{\text{f}}$ reach their respective ultimate values as $i \to \infty$. With the establishment of a steady state surface condition, the filtration performance also reaches its steady state. In other words, the relationship of $\bar{v}_i$ vs. t becomes independent of the cycle number as i increases. This behavior is confirmed through a number of sample calculations some of which are shown in Table 8.2.

8.1.3 Multiple bag operation

Under the condition of constant pressure and with the establishment of a steady-state surface condition, the gas throughput of a bag $\bar{v}_\infty$ is a function of time. To meet the

Table 8.2

Sample calculation results demonstrating the achievement of the steady state in a cyclic operation of a pulse jet air filter bag[a] filtration velocity at the end of the cycle ($m\,s^{-1}$)

P_a (kPa)	0.6	1.2	1.8	2.4	3.0
first cycle	1.0698×10^{-4}	1.8192×10^{-2}	2.1000×10^{-4}	2.3200×10^{-2}	2.5034×10^{-2}
second cycle	0.6492×10^{-2}	1.7835×10^{-2}	1.9861×10^{-2}	1.9468×10^{-2}	1.9550×10^{-2}
tenth cycle	0.9820×10^{-2}	1.7861×10^{-2}	2.0041×10^{-1}	2.0284×10^{-2}	1.7557×10^{-2}
twenty seventh cycle	0.9821×10^{-2}	1.7861×10^{-2}	2.0041×10^{-1}	2.0284×10^{-2}	1.7703×10^{-2}
Stray state	0.9821×10^{-2}	1.7861×10^{-2}	2.0041×10^{-1}	2.0284×10^{-2}	1.7703×10^{-2}

[a]Calculations were made with $T = 120\,s$, $c = 0.0275\,kg/m^3$, $k_1 = 10\,kPa\,s\,m^{-1}$, $k_2 = 300\,kPa\,s\,m\,kg^{-1}$, for $p_s < 0.099\,kPa$, $k_2 = 300 + 400\,(p_s - 0.099)^{0.5}$ for $p_s > 0.099\,kPa$.

requirement that the value of the gas to be treated is fixed, a system consisting of a number of bags operating according to a particular schedule may be employed. This is shown in the example given below.

Consider a system consisting of n bags and assume that the operation has reached its steady state. At any arbitrary time, these n bags have been in operation for various times since their cleanings. A sequence of values of $\{t_i\}$ may be used for its description. For example, for a system with five bags, $\{t_s\} = \{0, 12, 30, 50, 90\}$ denotes a state at which the first bag is in a state of just being cleaned and the second, third, fourth and fifth bags have been in operation for 12, 30, 50–90 seconds, respectively. The total gas throughput at time t is

$$Q = \sum_{i=1}^{n} \overline{v}_\infty (t + t_i) \tag{8.1.3.1}$$

since $\overline{v}_\infty$ is a function of time, Q is also a function of time. To meet the condition that the feed volume is constant, $\{t_i\}$ and n should be so specified that the variation of Q with time is held to a minimum. The degree of this time variation of Q can be seen from the value of γ defined as

$$\gamma = \frac{\max(Q) - \min(Q)}{Q_{av}} \tag{8.1.3.2}$$

and

$$Q_{av} = \frac{\int_{o}^{T} Q\,dt}{T} \tag{8.1.3.3}$$

Intuitively, one may select $\{t_i\}$ such that at $t_1 = 0$, the times of filtration experienced by each of the n bags of a system are

$$t_i = (i-1)\frac{T}{n} \qquad n = 1, 2, \ldots, n \tag{8.1.3.4}$$

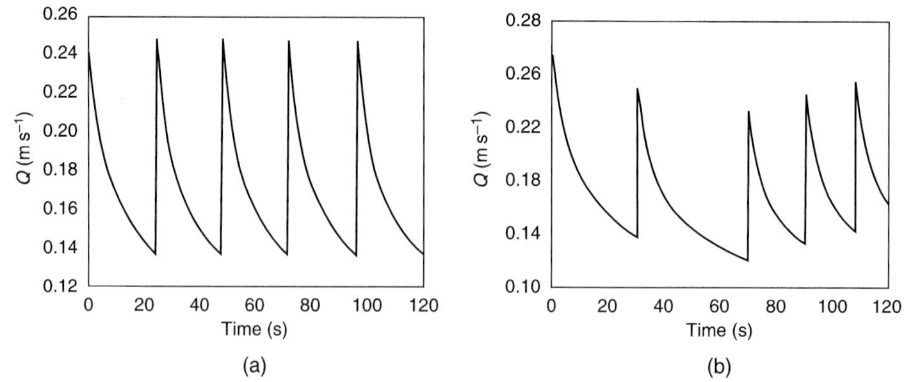

Figure 8.2 Q vs. time for two five-bag systems arranged as: (a) $\{0, 24, 48, 72, 96\}$; (b) $\{0, 12, 30, 50, 90\}$. (Ju, Chiu and Tien, 2000. Reprinted by permission of The Air and Waste Management Association.)

In Fig. 8.2, the values of Q vs. t from two sets of $\{t_i\}$, one according to Equation (8.1.1.22) and the other which is chosen arbitrarily, are shown. It is clear that the arrangement according to Equation (8.1.3.4) gives better results in reducing the variation of Q.

On physical grounds, it can be seen readily that increasing the value of n is effective in reducing the time dependence of Q. The results of Q vs. t corresponding to two systems consisting of 120 and 240 bags arranged according to Equation (8.1.3.4) are given in Fig. 8.3. For the 120-bag system, Q varies approximately from 3.99 to 4.085 ms (or a γ of 0.0228) and the corresponding values for 240-bag system are 8.036 to 8.15 ms (or a γ of 0.0086). Practically speaking, the constancy of Q is achieved in both cases.

8.2 CROSS-FLOW MEMBRANE FILTRATION

Cross-flow membrane filtration is one application of the so-called membrane technology, which has become popular and important in recent years. Membrane technology for separation is extremely versatile and can be applied to both homogeneous mixtures (gas and liquid) as well as heterogeneous mixtures (i.e. suspensions) based on certain physical and/or chemical characteristics of the membrane used. Accordingly, cross-flow filtration is commonly divided into two categories: microfiltration for removing particles of size ranging from 10^{-1} to $10^2\,\mu$m and ultrafiltration for particles (or macromoleculars) with size up to 100 nm. Separation is achieved by the sieving effect and the membrane used functions as a sieve. The rejected (or removed) particles form cakes on the upstream side of the membrane.

Unlike regular cake filtration discussed before, in cross-flow filtration, the feed flow and the filtrate flow are not in the same direction as mentioned previously (see Section 1.3, Chapter 1). While the consequence of the cake formation is the same as

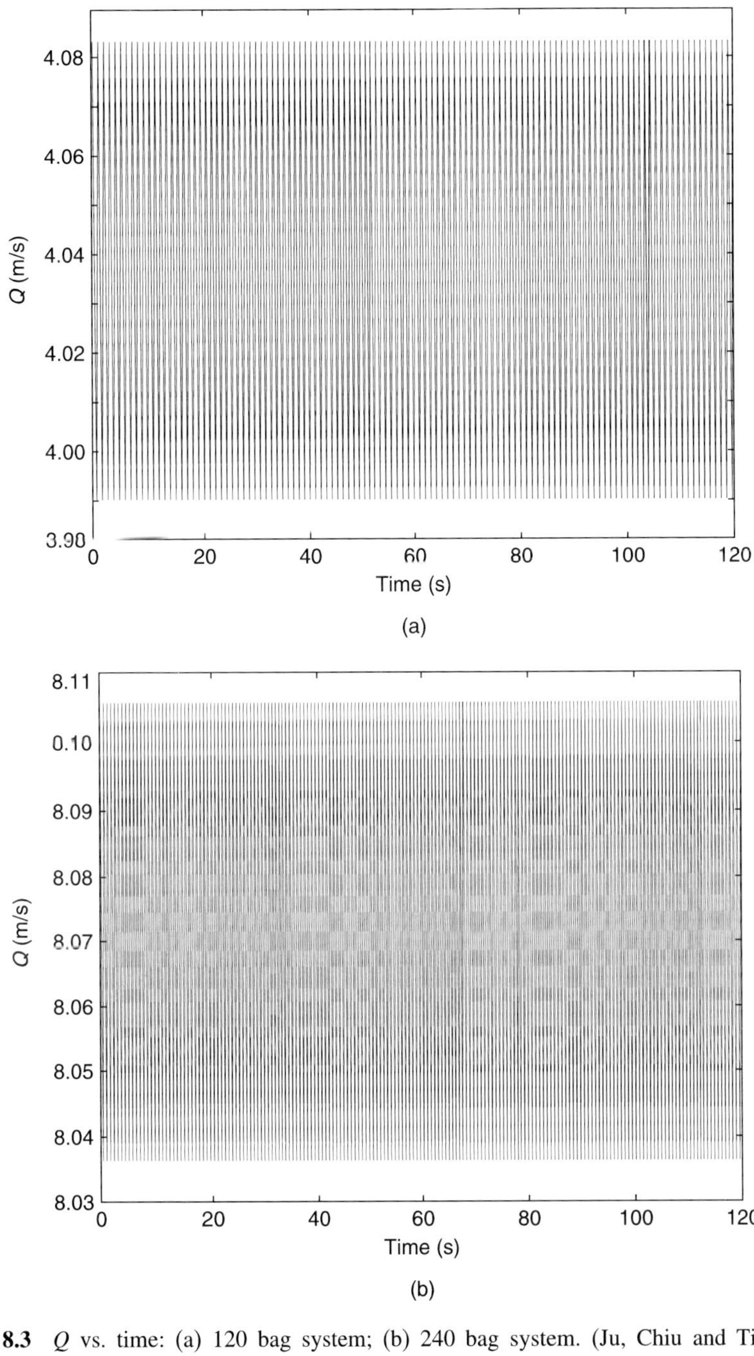

Figure 8.3 Q vs. time: (a) 120 bag system; (b) 240 bag system. (Ju, Chiu and Tien, 2000. Reprinted by permission of The Air and Waste Management Association.)

that of regular cake filtration, namely, an increase of the total resistance to filtrate flow, the manner and the extent of cake formation in cross-flow filtration is quite different from the discussions given previously. This difference represents the main issues to be discussed below.

8.2.1 Features of cross-flow membrane filtration

Based on experimental observations, the three main features of cross-flow membrane filtration are:

(1) There exists a threshold filtration velocity (critical flux) below which the phenomenon of flux declination with time is absent. The critical flux hypothesis was first suggested by Howell and coworkers (Howell, 1995; Field *et al.*, 1995). Experimentally, it was shown that in constant rate filtration, if the filtration rate is below the critical flux value, the operating pressure (or the transmembrane pressure, TMP) necessary to maintain the operation does not increase with time or there is no membrane fouling. If one assumes that membrane fouling is caused mainly by particle deposition at membrane surface, this implies that cake formation does not take place in cross-flow filtration if the filtration rate is less than the critical flux value.[1]

The presence of the critical flux can be seen clearly from the results of Kwon *et al.* (2000) as shown in Fig. 8.4a,b. The cross-flow filtration cell used by these investigators consisted of nine identical channels with the dimension of 6 cm (length) × 0.6 cm (width) × 0.036 cm (height). The experiments were conducted using suspensions of particles of uniform size and a commercial PVDF (polyvinyl defluoride) membrane (Millipore catalogue No. GVLPOMS10) under the constant rate condition. The required transmembrane pressure was recorded as a function of time.

The results shown in Fig. 8.4a (for $d_p = 0.46\,\mu m$) include four groups of data corresponding to different filtration rates. At the extreme left of the figure, the value of $66\,\ell\,m^{-2}h^{-1}$ refers to the filtration rate. Over a period of nearly one hour, the TMP remained constant, indicating that there was no cake formation. The same situation was observed when the filtration rate was later set at a higher value ($88\,\ell\,m^{-2}h^{-1}$). However, when the filtration rate was increased further (i.e. $91\,\ell\,m^{-2}h^{-1}$), the required TMP was found to increase with time. The same type of behavior was also found for larger particles ($d_p = 3.2\,\mu m$) of Fig. 8.4b.

(2) There is a preferential deposition of smaller particles in cross-flow filtration of polydisperse suspensions. A number of investigators (Fischer and Raasch, 1986;

[1] There are two forms of critical flux, a strong form and a weak form depending upon whether the TMP required to maintain the constant rate operation (below the critical flux value) is the same or greater than that required to maintain the flow of particle-free liquid streams at the same rate.

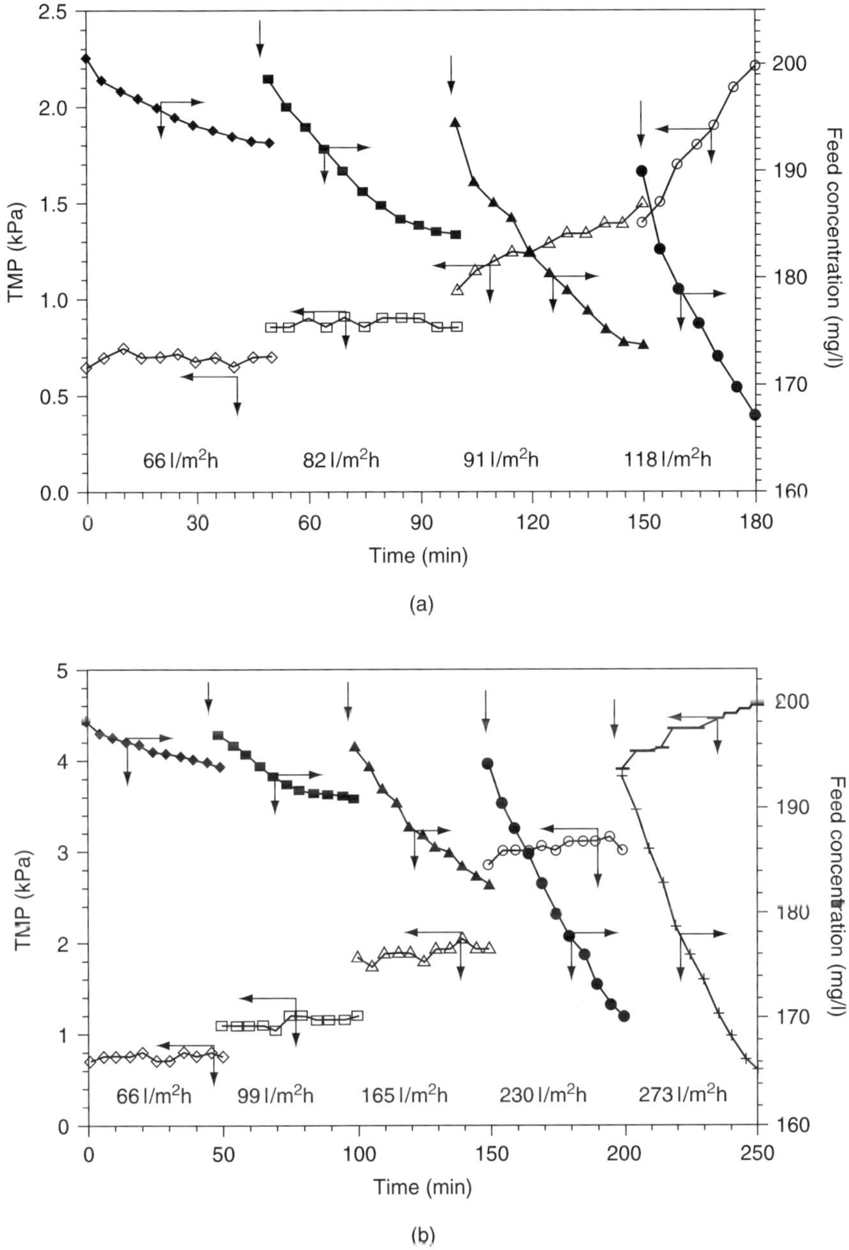

Figure 8.4 Critical velocity determined from constant-rate filtration experiment by Kwon *et al.*: (a) $d_p = 0.46\,\mu m$; (b) $d_p = 3.2\,\mu m$. (Kwon, Vigneswaran, Fane and Ben Aim, 2000. Reprinted by permission of Elsevier.)

Lu and Ju, 1989; Meier *et al.*, 2002) have demonstrated conclusively that in cross-flow filtration, there is a selective deposition of finer particles. A set of data obtained by Meier *et al.* is shown in Fig. 8.5 in which the particle size distribution of the feed suspension and those of the cakes formed at different positions are shown (for example, data points designate as pos *x/y* in this figure refers to cake formed between *x* and *y* mm down stream). The experiments were performed using a quartz powder suspension (mean particle diameter of 3 μm) with a feed concentration of 0.5% (by weight) and a flow duct module made of 0.4 μm rated flat sheet nylon membrane under a cross velocity of $2\,\mathrm{m\,s^{-1}}$ and filtration rate of $1450\,\ell\,\mathrm{m^{-2}\,h^{-1}}$. The size-selective effect was more pronounced initially but reached a constant state further downstream. The results of Meier *et al.* demonstrate selective deposition under all conditions. In fact, these investigators suggested that with selective deposition phenomena, it is feasible to develop a new method of wet classification of ultrafine particle. Some experimental data were presented to substantiate their claim.

The experimental observations of Meier *et al.* provide the necessary base for formulating the so-called adhesion model in the analysis of cross-flow filtration. This will be discussed later in this chapter.

(3) In terms of filtration performance, under the constant-pressure condition, filtration rate declines with time, but the rate of declination diminishes with time. For many cases, the results seem to suggest an attainment of a steady state. A typical example of this behavior obtained by Ould-Dris *et al.* (2000) under different

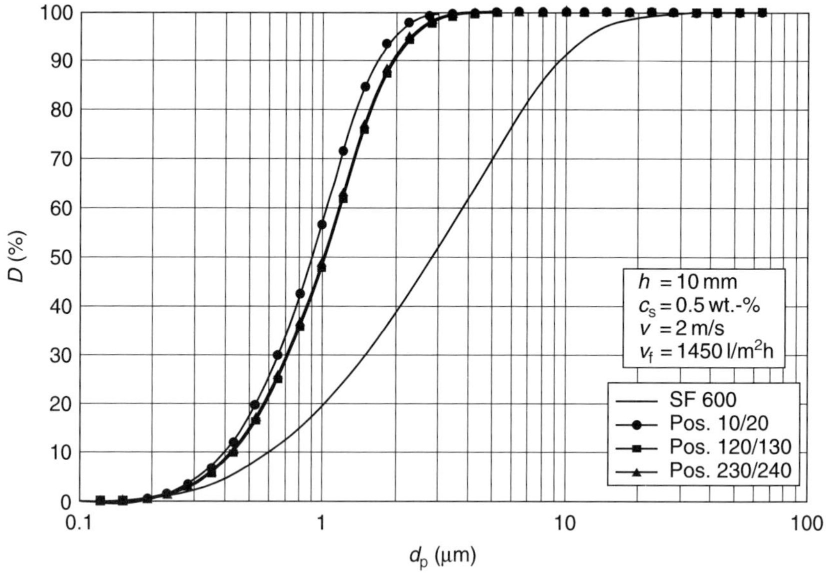

Figure 8.5 Particle size distributions of cakes formed at different longitudinal positions obtained by Meier *et al.* (Meier, Klein and Kottke, 2002. Reprinted by permission of Elsevier.)

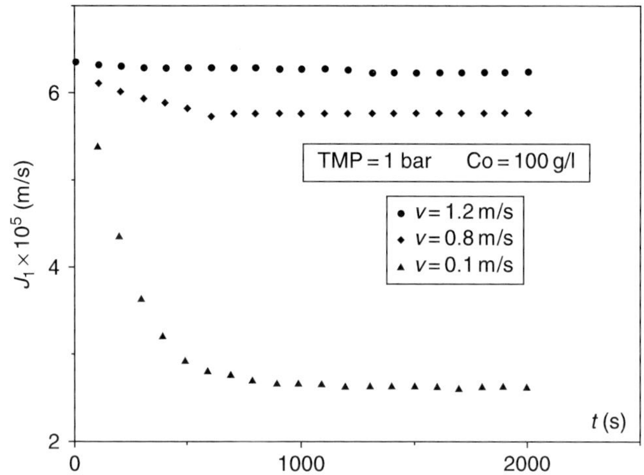

Figure 8.6 Filtration rate vs. time data of Ould-Dris *et al.*, cross-flow filtration of $CaCO_3$ suspensions; $d_p = 4.65\,\mu m$; polysulfone membrane with pore diameter of $0.1\,\mu m$. (Ould-Dris, Jaffrin, Si-Hassen and Neggaz, 2000. Reprinted by permission of Elsevier.)

operating pressures is shown in Fig. 8.6. The decrease of filtration rate with time is commonly considered as a consequence of membrane fouling. For cross-flow filtration, membrane fouling may be attributed mainly (if not totally) to cake formation at membrane surface. As any rate process, the rate of filtration may be expressed as

$$\text{cross-flow filtration rate} = \text{driving force/resistance} \tag{8.2.1.1}$$

The driving force, of course, is the operating pressure (transmembrane pressure), or the pressure difference between the suspension pressure and the pressure of the permeate side. The resistance may be assumed to be consisted of two parts, the membrane resistance and the resistance due to the cake formed. If one assumes that the membrane resistance remains constant,[2] cake resistance can be expected to increase as cake thickness increases. Accordingly, one may expect a decrease of the filtration rate with time as cake thickness increases.

There are two problems with the above argument. First, what is the explanation for the attainment of a steady state of the filtration rate? Does this imply that cake growth ceases once the steady state is attained? If this is indeed the case, what are the factors causing the cessation of cake growth?

Second, the magnitude of cake resistance may be approximately proportional to the cake mass. However, for a given cross-flow filtration, if one assumes that all

[2] A more complete discussion on membrane medium resistance is given in Section 6.5.5 of Chapter 6.

the particles transported to the membrane surface associated with the filtrate flow result in cake formation, the cake resistance would be so large and would give a filtration rate at least one order of magnitude less than experimental values (the so-called flux paradox, Green and Belfort, 1980). Based on the observed filtration rate, only a fraction of the particles transported by convection become deposited and form filter cakes. What are the explanations?

8.2.2 Particle depolarization

To resolve the "flux paradox" mentioned above and for proper analysis of cross-flow filtration, various hypotheses aimed at explaining and quantifying the reduction of particle accumulation and deposition from that based on particle flux due to filtrate flow have been proposed in the past. These hypotheses, broadly speaking, may be grouped into three categories: those based on the back transport of particles (opposite to the direction of filtrate flow), those based on particle lateral migration and those based on the failure of particle adhesion. Their descriptions are presented below.

(1) *Back diffusion of particles*
With increased particle concentration at membrane surface, for sufficiently small particles, particle diffusion away from the surface due to the Brownian diffusion could be significant in a manner similar to the transport of solute species in reverse osmosis. The Brownian diffusivity D_{BM} according to the Stokes–Einstein relationship is

$$D_{BM} = \frac{kT}{6\pi\mu\, a_p} \tag{8.2.2.1}$$

where k is the Boltzman constant, T, the absolute temperature, a_p, particle radius and μ, the filtrate viscosity.

Back diffusion may also be caused by particle movement due to mutually induced velocity fields present in shear flow of relatively concentrated suspensions. This type of particle migration, termed "induced shear diffusion" was first observed by Eckstein *et al.* (1977) and applied by Zydney and Colton (1986) in membrane filtration. Based on the data of Eckstein *et al.* (Fig. 8.7), D_s may be approximated by the following empirical expressions:

$$\frac{D_s}{a_p^2\omega} = 0.1\varepsilon_s \quad \text{for} \quad 0 < \varepsilon_s < 0.2 \tag{8.2.2.2a}$$

$$\frac{D_s}{a_p^2\omega} = 0.03 \quad \text{for} \quad 0.2 < \varepsilon_s < 0.5 \tag{8.2.2.2b}$$

where ε_s is the particle volume function of the suspension and ω is the stress rate.

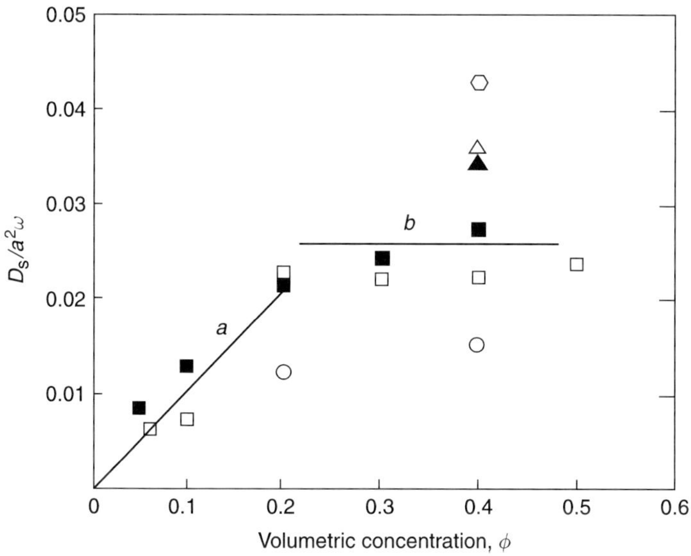

	■	▲	□	△	⬡	○
Particle	Disk	Disk	Sphere	Sphere	Sphere	Sphere
a (cm)	0.16	0.16	0.16	0.16	0.16	0.05>
ω (s^{-1})	10	1	10	1	0.4	10
$a^2\omega$ ($cm^2\, s^{-1}$)	0.256	0.0256	0.256	0.0256	0.0102	0.025

Figure 8.7 Shear-induced diffusivity as a function of particle volume function. (Eckstein, Bailey and Shapiro, 1977. Reprinted by permission of Cambridge Press.)

A somewhat different correlation of D_s was later proposed by Leighton and Acrivos (1987). $D_s/a_p^2\omega$ was found to be

$$\frac{D_s}{a_p^2\omega} = 0.5\varepsilon_s^2(1+0.09\,e^{7\varepsilon_s}) \tag{8.2.2.3}$$

Both Eckstein *et al.* and Leighton and Acrivos obtained their data using relatively large particles (a_p greater than 10^3 μm in the former study and approximately 300 μm in the latter), and the differences between these two sets of data are rather substantial (several fold) for ε_s greater than 0.2.

The difference between using D_{BM} and using D_s in assessing particle depolarization is significant. D_{BM} is a function of particle size and increases with the decrease in a_p. If the Brownian diffusion is the mechanism for particle back diffusion, this implies that back diffusion is more important for smaller particles, which is contrary to the observed preferential deposition of finer particles. Furthermore, since D_{BM} is constant for particles of a given size, the extent of back diffusion based on D_{BM} should be insensitive to the flow condition. This, of course, is in contrast to experimental observations that particle depolarization (and therefore the critical filtration velocity) is enhanced with the increase of the cross velocity. In contrast,

D_s is shown to increase with the increase in both a_p and the cross-flow velocity (or ω). It is, therefore, no surprise that particle transport based on the shear-induced diffusion hypothesis has been more useful in analyzing cross-flow filtration (Li *et al.*, 2000; Ripperger and Altmann, 2002).

(2) *Lateral migration of particles*

Altena and Belfort (1984) first suggested that particle polarization may be reduced due to the lift velocity experienced by particles. For a neutrally buoyant particle in slow two-dimensional shear flow, the lift velocity, u_L, according to Vasseur and Cox (1976), is given as

$$u_L = \frac{61\rho}{576\mu}\omega^2 a_p^3 \tag{8.2.2.4a}$$

A different expression which takes into account for relatively high flow rate and porous boundary (i.e. cake surface) is also available (Drew *et al.*, 1991):

$$u_L = 0.577\frac{\rho a_p^2 \omega^2}{16\mu} \tag{8.2.2.4b}$$

As u_L is directly proportional to $a_p^3 \omega^2$, particle depolarization due to lateral migration becomes significant for lag particles and under greater cross velocity, which is in accord with available experimental data. However, the extent of depolarization due to particle lateral migration is always less than experimental results (see Section 8.3 of this chapter).

(3) *Particle Adhesion*

A simple and direct explanation of particle depolarization can be made on the basis that only a fraction of the convectively transported particles become deposited and form cakes, with the balance being swept away due to cross-flow. This simple idea was used by several investigators in their studies of particle deposition in cross-flow filtration (Fischer and Raasch, 1986; Lu and Ju, 1989, 1993; Stamatakis and Tien, 1993; Altmann and Ripperger, 1997). Generally speaking, these studies determined the extent of deposition based on mechanical equilibrium conditions of particles in contact with a surface.

Following Stamatakis and Tien (1993), consider a particle of diameter d_p coming into contact with a surface against a protrusion of height h (Fig. 8.8). If F_p and F_q respectively denote the forces acting on the particle along the directions of the cross-flow and the filtration flow, the condition of particle deposition (that is, the particle remains static and does not roll over the protrusions) is

$$F_q\left[\left(\frac{d_p}{2}\right)^2 - \left(\frac{d_p}{2} - h\right)^2\right]^{1/2} \geq F_p\left[\frac{d_p}{2} - h\right] \tag{8.2.2.5}$$

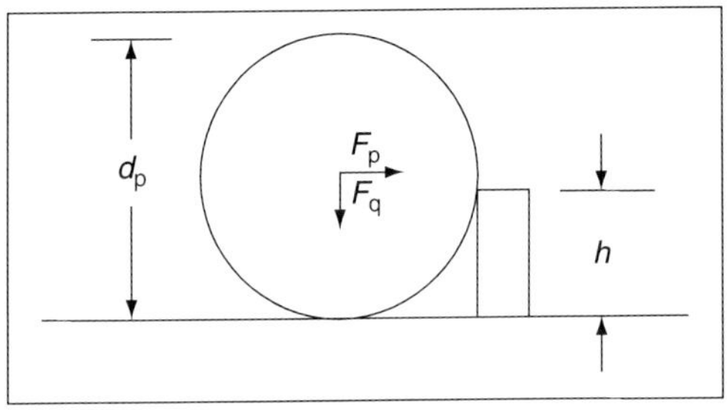

Figure 8.8 Deposition according to the adhesion hypothesis.

The condition of deposition expressed in terms of the required height of the protrusion, or the critical height, h_c, is

$$h_c \geq \left(1 - \frac{1}{\sqrt{(F_p/F_q)^2 + 1}}\right)\frac{d_p}{2} \tag{8.2.2.6}$$

The protrusion height may be used to characterize the surface condition. Alternatively, one may describe the surface condition by the friction coefficient between the surface and the particle. The condition of deposition is

$$F_p \leq \overline{\mu} F_q \tag{8.2.2.7}$$

where $\overline{\mu}$ is the friction coefficient consisting of a rolling and a gliding part (Altmann and Ripperger, 1997).

For estimating the force terms of Equation (8.2.2.5), or (8.2.2.6), the expression of the drag force exerted on a single sphere in a shear-flow field given by O'Neill (1968) may be used to calculate F_p, or

$$F_p = 3\pi\mu\, d_p \frac{\tau_w}{\mu}\frac{d_p}{2} f_1 \tag{8.2.2.8}$$

where τ_w is the wall shear stress and f_1, the hydrodynamic retardation correction factor.

The force along the direction of filtrate flow F_q may be expressed as

$$F_q = F_{q_1} + F_{q_2} + F_{q_3} + F_{q_4} \tag{8.2.2.9}$$

where F_{q_i}, $i = 1, 2, 3, 4$, denotes respectively the drag force due to filtrate flow, the net buoyant force, the lift force and the surface–particle interaction force. As approximations, they may be estimated as

$$F_{q_1} = 3\pi\mu d_p q_{\ell_m} f_2 \quad (8.2.2.10a)$$

$$F_{q_2} = (\pi/6)\, d_p^3 (\rho_s - \rho) \quad (8.2.2.10b)$$

$$F_{q_3} = 3\pi\mu\, u_L \quad (8.2.2.10c)$$

and f_2 is the hydrodynamic retardation correction factor.

The particle–surface interaction force, F_{q_4} may be assumed to be consisting of the London-van der Waals force and the double layer force. Expression of these forces can be found in any standard colloid chemistry text.[3]

Applying either of the two adhesion criteria [i.e. Equations (8.2.2.6) or (8.2.2.7)] requires the knowledge of the protrusion height or the value of the friction coefficient. The friction coefficient can be expected to be system specific. Halow (1973) showed that for relatively large particles (20–5000 μm), $\overline{\mu}$ varies from 0.06 to unity. This uncertainty of $\overline{\mu}$ values renders the use of Equation (8.2.2.8) questionable.

Equation (8.2.2.6) indicates that deposition (adhesion) takes place if the protrusion height exceeds a critical value. For a typical system (Table 8.3) and using the force expressions given above but ignoring F_{q_4}, values of h_c as a function of particle diameter at various operating conditions are shown in Fig. 8.9a,b. For a given filtration rate, h_c increases with the increase of the cross velocity. The critical protrusion height, for a given cross velocity, however, increases with the decrease of the filtration rate.

As the protrusion height is used to characterize the surface condition, for a clean membrane (medium), its surface roughness depends upon the method used for its preparation. In principle, the surface roughness can be determined by scanning membrane samples with AFM (atomic force microscopy) instruments. The modular height (which

Table 8.3

Conditions used to obtain the critical protrusion height results shown in Fig. 8.9a,b

Parameter	Value
ε_{s_o}	0.002
d_p	1 μm
ρ_ℓ	1,000 kg m^{-3}
ρ_s	2,600 kg m^{-3}
μ	0.001 Pa s
$\varepsilon_s = \varepsilon_s^o$	0.25
$k = k^o$	1.0×10^{-14} m^2
R_m	10^{10} m

[3] Calculation of these forces requires the knowledge of the particle–surface separation distance. This distance is often assumed to be 4 nm.

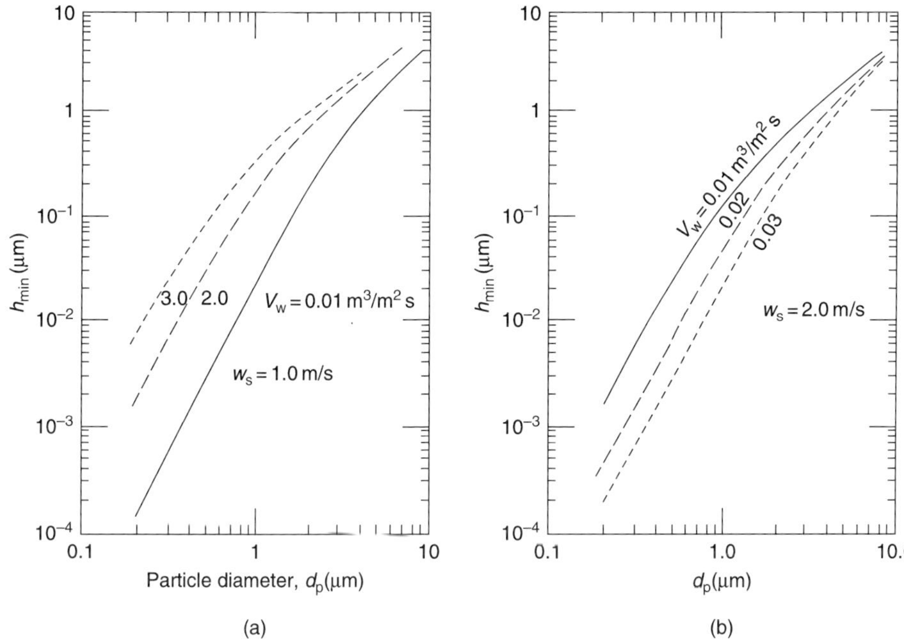

Figure 8.9 Critical protrusion height as a function of particle diameter. (Stamatakis and Tien, 1993. Reprinted by permission of The American Institute of Chemical Engineers.)

may be taken as protrusion) observed varied considerably depending upon the membrane type and even size of the area scanned (Khulbe *et al.*, 1997; Chung, 2004, private communication).

Once a cake is formed, the surface to be considered is that of the cake/suspension interface. Since a cake is composed of deposited suspended particles, its surface roughness can be expected to be determined largely by the size of particles as well as the manner of their depositions. The protrusion height, therefore, can be expected to cover a range of values. Assuming that h is a continuous random variable with a continuous distribution function, Stamatakis and Tien (1993) argued that the probability of a particle of diameter d_p to be deposited equals the probability of the presence of a local protrusion with a height equal to or greater than h_c, $P(h \geq h_c)$, or

$$\gamma = P(h \geq h_c) = 1 - P(h \leq h_c) \tag{8.2.2.11}$$

where γ is the probability for the particles to be deposited (or adhesion probability). Furthermore, if one assumes that the height distribution for a uniform distribution over $0 < h < h_{max}$, where h_{max} is the maximum protrusion height, Equation (8.2.2.11) becomes

$$P(h \leq h_c) = \frac{h_c}{h_{max}} \tag{8.2.2.12}$$

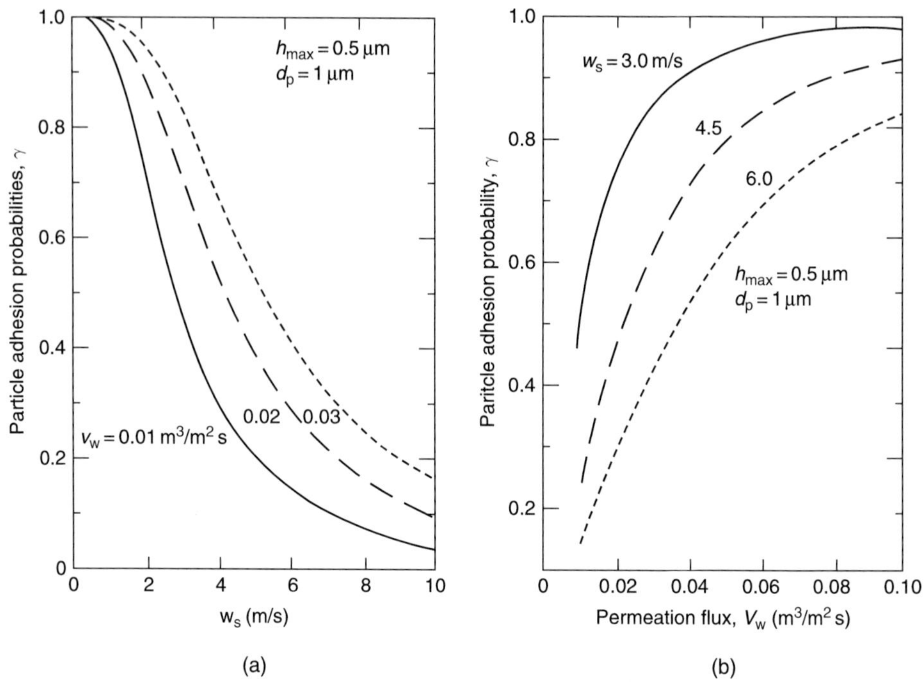

Figure 8.10 (a) Adhesion probability vs. cross-flow velocity; (b) adhesion probability vs. filtration rate. (Stamatakis and Tien, 1993. Reprinted by permission of The American Institute of Chemical Engineers.)

Combining Equations (8.2.2.6), (8.2.2.11) and (8.2.2.12), γ is found to be

$$\gamma = 1 - \left(1 - \frac{1}{\sqrt{(F_p/F_q)^2 + 1}}\right)\frac{d_p}{2h_{max}} \tag{8.2.2.13}$$

Figure 8.10a,b shows the results of γ vs. w_s (and γ vs. q_{ℓ_m}) obtained from Stamatakis and Tien (1993) for the filtration of suspensions of 1 μm particles.

8.2.3 Estimation of the critical filtration velocity

The depolarization hypotheses discussed in the preceding section can be readily applied to estimate the critical filtration velocity (or critical flux) of cross-flow filtration. Estimation can be made as follows.

(1) *Based on the back-diffusion hypothesis*
A schematic representation depicting back diffusion of accumulated particles away from membrane surface is given in Fig. 8.11. If the critical filtration velocity

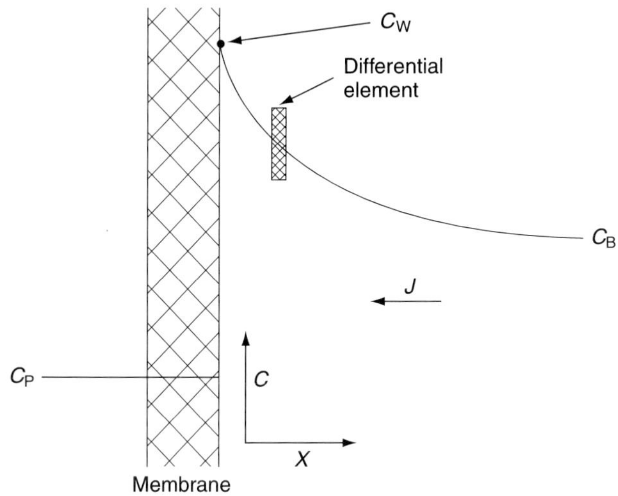

Figure 8.11 Schematic representation of back diffusion for depolarization.

(critical flux) is denoted as $(q_{\ell_m})_c$, the commencement of cake formation requires a balance of the convectively transported particle by the back diffusion particle flux or

$$-D\frac{\partial \varepsilon_s}{\partial x} + (q_{\ell_m})_c \cdot \varepsilon_s = 0 \tag{8.2.3.1}$$

Integrating the above expression over the thickness of the accumulated particle layer, δ, the result is

$$\left|(q_{\ell_m})_c\right| = \frac{D}{\delta} \ln \frac{(\varepsilon_s)_w}{\varepsilon_{s_o}} \tag{8.2.3.2a}$$

where $(\varepsilon_s)_w$ and $(\varepsilon_s)_o$ are the particle volume fractions at the membrane surface and of the feed, respectively. The absolute value of $(q_{\ell_m})_c$ is considered here since q_{ℓ_m} is negative according to the direction x (away from membrane surface).

The quantity D/δ may be interpreted as the transfer coefficient according to the film theory. For a shear flow field, the Leveque expression of the local transfer coefficient is

$$k(x) = 0.538\left(\frac{D^2\omega}{x}\right)^{1/3} \tag{8.2.3.2b}$$

where x is the longitudinal distance.

Substituting k of Equation (8.2.3.2b) for D/δ into Equation (8.2.3.2a), after integrating over a length of L, the average critical flux $(\overline{q}_{\ell_m})_c$ is found to be

$$(\overline{q}_{\ell_m})_c = \frac{1}{L}\int_o^L (q_{\ell_m})_c \, dx = 0.807\left(\frac{D^2\omega}{L}\right)^{1/3} \ell n \frac{(\varepsilon_s)_w}{(\varepsilon_s)_o} \tag{8.2.3.3}$$

With the knowledge of $(\varepsilon_s)_w$, the above equation can be used to calculate the average critical filtration velocity by substituting the appropriate diffusivity expression – namely, Equation (8.2.2.1) if the Brownian diffusion is the dominant mechanism, or Equations (8.2.2.2a), (8.2.2.2b) or (8.2.2.3) if the shear-induced diffusion is operative – into Equation (8.2.3.3). The value of the particle volume fraction at the membrane surface, $(\varepsilon_s)_w$, however, is generally unknown. Some investigators (for example, Hong *et al.*, 1997), assume $(\varepsilon_s)_w$ to be the same as the maximum packing density or $(\varepsilon_s)_w = 0.64$. This assumption is probably questionable since even under pressure as high as 10^3 kPa, no such highly packed cake has been observed experimentally.[4] In contrast, Li *et al.* (2000) assumed $(\varepsilon_s)_w$ to be 0.2 in accordance with their experimentally determined $(\overline{q}_{\ell_m})_c$ value.

Experimental determination of the critical filtration velocity in cross-flow membrane filtration was undertaken by Li *et al.* (2000). These investigators also compared their experimental results with predictions based on Equation (8.2.3.3). If the shear-induced diffusivity expression of Equation (8.2.2.2b) is used, one has

$$(\overline{q}_{\ell_m})_c = 0.078\,\omega\left(\frac{a_p^4}{L}\right)^{1/3} \ell n \frac{(\varepsilon_s)_w}{(\varepsilon_s)_o} \tag{8.2.3.4}$$

Comparisons between the experimentally determined $(\overline{q}_{\ell_m})_c$ and predictions based on Equation (8.2.3.4) with $(\varepsilon_s)_w = 0.2$ are shown in Fig. 8.12. For the four types of particles used in the experiments, Equation (8.2.3.4) underestimates for small particles ($d_p = 3\,\mu$m) but overestimates for large particles ($d_p = 11\,\mu$m). Similar but different expressions of the shear-induced diffusivity were also used for comparisons. The same kind of agreement shown in Fig. 8.12 was also observed.

Li *et al.* (2000) also attempted to establish an empirical expression of D_s based on their $(\overline{q}_{\ell_m})_c$ data. Assuming that $D_s/(a_p^2\omega)$ is a linear function of ε_s, the following correlation was obtained:

$$\frac{D_s}{a_p^2\omega} = 0.31(\varepsilon_s)_w \tag{8.2.3.5}$$

It is clear that Equation (8.2.3.5) gives higher values of D_s than either Equations (8.2.2.2a) or (8.2.2.2b).

[4] This can be seen from the compiled cake solidosity data (Table 7.6) given in Chapter 7.

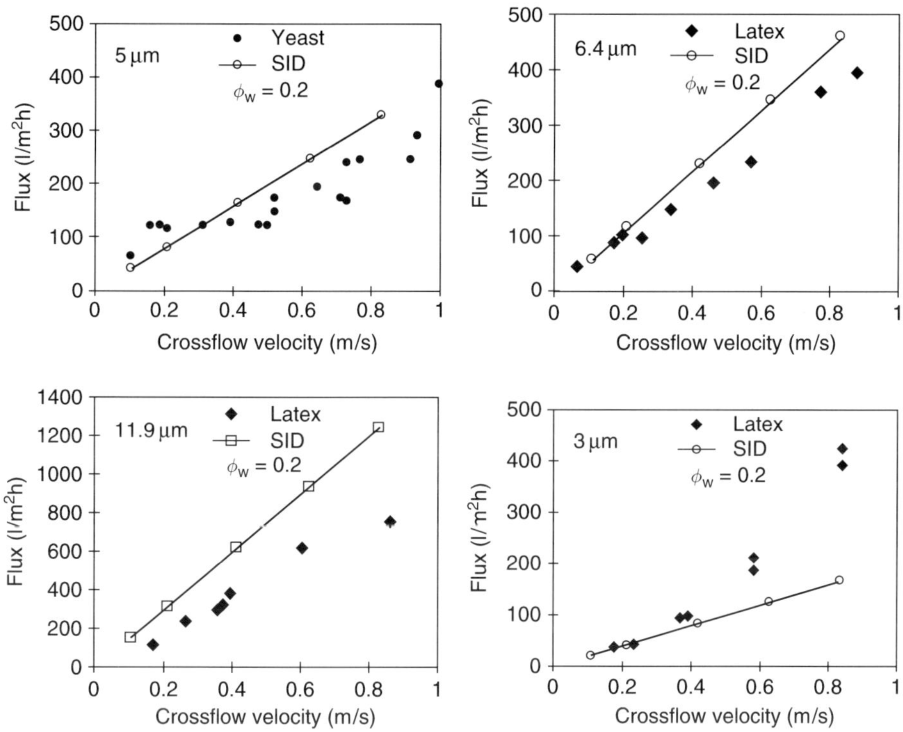

Figure 8.12 Comparisons of experimentally determined $(\bar{q}_{\ell_m})_c$ with predictions based on back diffusion hypothesis with $D = D_s$. (Li, Fane, Coster and Vigneswaran, 2000. Reprinted by permission of Elsevier.)

(2) *Based on particle lateral migration*

Based on simple mechanics, one may argue that there is no cake formation if the filtration velocity is less than or equal to the lift velocity u_L. The critical filtration velocity is, therefore, equal to the lift velocity (in magnitude, but opposite in direction). Accordingly $(\bar{q}_{\ell_m})_c$ can be estimated according to Equation (8.2.2.5).

Comparisons between the predicted $(\bar{q}_{\ell_m})_c$ based on Equation (8.2.2.5) and experiments are shown in Fig. 8.13. It is clear that the difference is rather significant (by order of magnitude). One may therefore conclude that the lateral migration effect is unimportant in particle depolarization in cross-flow filtration.

(3) *Based on particle adhesion hypothesis*

Equation (8.2.2.6), in principle, can be used to estimate the critical filtration velocity. For a simple case where F_{q_1} is the dominant term of $F_{\bar{q}}$, Equation (8.2.2.6) becomes

$$\frac{f_2}{f_1}\frac{(\bar{q}_{\ell_m})_c}{1.5 w_s}\frac{a}{d_p} = \sqrt{\frac{1}{\left(1 + 2\dfrac{h_c}{d_p}\right)^2 - 1}}$$

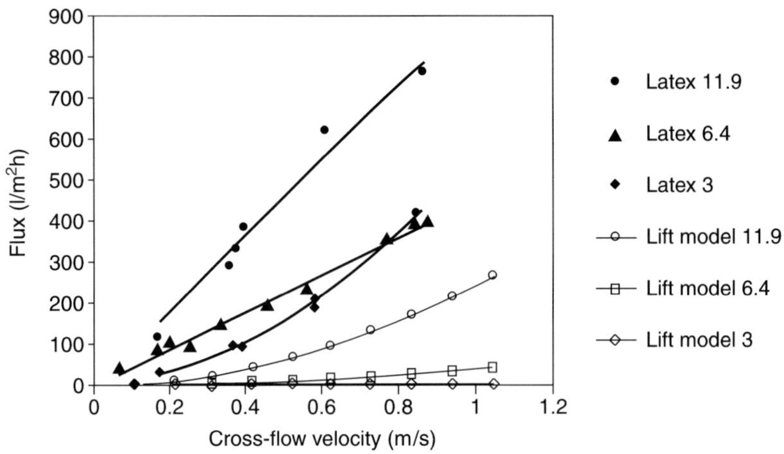

Figure 8.13 Comparisons of experimentally determined $(\overline{q}_{\ell_m})_c$ with predictions based on the lift velocity hypothesis. (Li, Fane, Coster and Vigneswaran, 2000. Reprinted by permission of Elsevier.)

or

$$(\overline{q}_{\ell_m})_c = \frac{f_1}{f_2}(1.5\,w_s)\left(\frac{d_p}{a}\right)\sqrt{\frac{1}{\left(1+2\,\dfrac{h_c}{d_p}\right)^2 - 1}} \qquad (8.2.3.6)$$

where w_s is the cross-flow velocity. The geometry of the membrane module is assumed to be that of two parallel plates, and a is the half distance between the plates.

Since h_c is generally not known, assessing the feasibility of applying the particle adhesion hypothesis of estimating $(\overline{q}_{\ell_m})_c$ can only be made indirectly (Tien, 2004). As an example, using the critical filtration velocity data obtained by Kwon *et al.* (2000) (Table 8.4), the required h_c values were determined according to Equation (8.2.3.6) (Tien, 2004) and included in this table. Based on the value of h_c for particles of $d_p = 0.46\,\mu\text{m}$, the critical filtration velocities for the other three particles ($d_p = 0.82, 3.2, 11.2\,\mu\text{m}$) were then calculated according to Equation (8.2.3.6). The estimated $(\overline{q}_{\ell_m})_c$ was found to be of the same order of magnitude of the experimental values.

The calculated h_c values listed in Table 8.4 are comparable to the modular height values reported by some investigators (Zhang *et al.*, 2003). It is also known that modular height results vary with the size of the scanned area (Chung *et al.*, 2002). Because of the uncertainties associated with h_c values, using Equation (8.2.3.6) for estimating $(\overline{q}_{\ell_m})_c$ can only yield approximate results.

Table 8.4

Experimentally determined values of the critical velocity, the requisite critical protrusion height and the predicted critical velocity

Particle diameter (μm)	Critical velocity[a] ($m\,s^{-1}$)	Critical protrusion height[b] (μm)	Predicted critical velocity[c] ($m\,s^{-1}$)
0.46	2.4082×10^{-5}	0.1495	–
0.82	3.0972×10^{-5}	0.1682	3.8481×10^{-5}
3.2	5.4862×10^{-5}	0.64	1.4949×10^{-4}
11.2	6.9617×10^{-5}	4.78	3.6728×10^{-4}

[a] From Kwon *et al.* (2000).
[b] According to Equation (8.2.3.6).
[c] Estimated with $h^* = 0.1495\,\mu$m.

8.2.4 Preferential deposition of finer particles

Based on the adhesion hypothesis, the size distribution of cake particles can be readily determined. Let x_i denote the volume fraction of particles with diameter d_{p_i} in the total population of particles and γ_i, the corresponding adhesion probability. There the volume fraction of particles of diameter d_{p_i} in the cake formed y_i is

$$y_i = \frac{x_i \gamma_i}{\sum_{j=1}^{N} x_j \gamma_j} \tag{8.2.4.1}$$

assuming that there are N types of particles present.

Equation (8.2.4.1) was used by Stamatakis and Tien (1993) to demonstrate the extent of preferential deposition of finer particles in cross-flow filtration. The results of one of these sample calculations are shown in Fig. 8.14. The suspended particles of the feed were assumed to follow a normal distribution with a mean particle size of $5\,\mu$m and a standard deviation of $1\,\mu$m. The results shown in Fig. 8.14 were obtained using $h_{max} = 2.5\,\mu$m, $w_s = 3\,m\,s^{-1}$ and two different filtration velocities. Preferential deposition of finer particles is clearly shown and the extent of preferential deposition increases with the decrease of the filtration velocity. The results are in accord with experimental observation as shown in Fig. 8.5.

8.2.5 Cross-flow filtration performances

Similar to regular cake filtration, a rigorous analysis of cross-flow filtration can be made through the solution of the volume-averaged continuity equations. With cross-flow filtration being two dimensional, the presence of one more spatial variable complicates

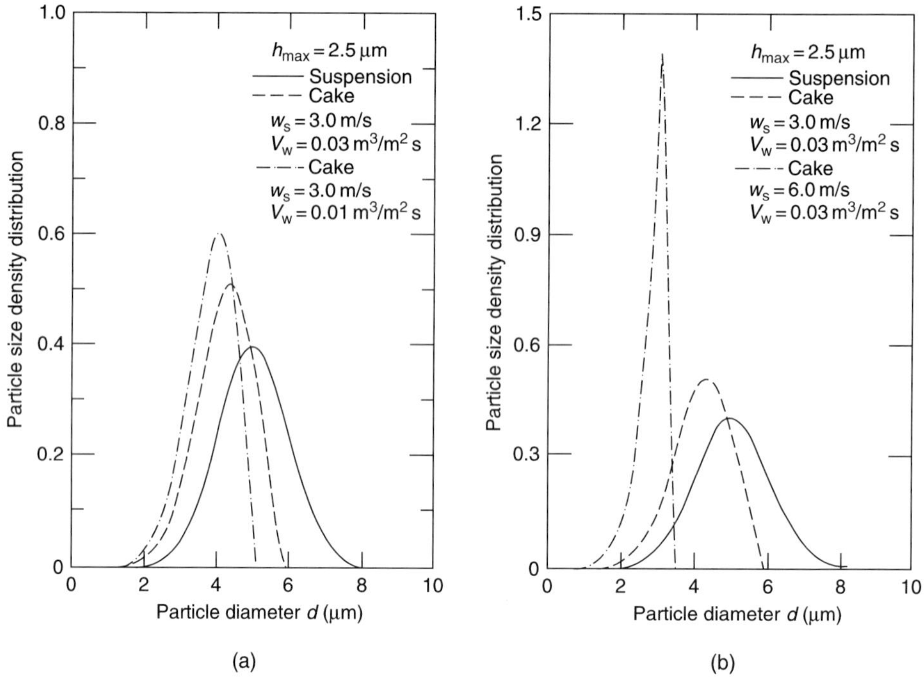

Figure 8.14 Preferential deposition of finer particles in cross-flow filtration according to the adhesion hypothesis. (Stamatakis and Tien, 1993. Reprinted by permission of The American Institute of Chemical Engineers.)

the solution of the continuity equations. Furthermore, the additional problems of properly formulating a two-dimensional constitutive relationship and the difficulty of evaluating the constitutive relationship parameters make this approach impractical.

Among the large number of studies on the analysis of cross-flow filtration performance of the past two decades, most of these studies were based on the conventional cake filtration theory with various degrees of modifications. The main effort has been aimed at incorporating depolarization in describing cake growth. In addition to those hypotheses described above, Romero and Davis (1988, 1990) gave a more involved analysis of cross-flow filtration. In their studies, depolarization was assumed to be that of back diffusion due to the shear-induced diffusion with further assumptions that the concentrated particle layer is composed of a stagnant region (filter cake) and a mobile part. Considerable computational efforts were used to carry out the analysis. More importantly, their analysis introduced a number of extraneous parameters (such as the particle volume fraction at the membrane surface, the critical axial distances and more importantly, cake resistance parameters) which cannot be determined independently. It is, therefore, questionable whether the Romero–Davis analysis can be used in applications.

A simple and practical procedure of predicting cross-flow filtration performance can be made by modifying the conventional filtration theory to account for the presence of particle depolarization in cake formation. Beginning with Equation (2.1.16), the instantaneous filtration velocity is

$$q_{\ell_m} = q_\ell = \frac{p_o}{\mu\left[w(\alpha_{av})_{p_{s_m}} + R_m\right]} \tag{8.2.5.1}$$

and

$$(\alpha_{av})_{p_{s_m}} = \frac{\Delta p_c / p_{s_m}}{\left[\int\limits_o^{p_{s_m}} \left(\frac{1}{\alpha}\right)(-f')\mathrm{d}p_s\right] / p_{s_m}} \tag{8.2.5.2}$$

where α is the local specific cake resistance; p_o, the operating pressure (or TMP); and p_{s_m}, the cake compressive stress at the membrane/cake interface. f' denotes the relationship between p_ℓ and p_s, or

$$f' = \frac{\mathrm{d}p_\ell}{\mathrm{d}p_s} \tag{8.2.5.3}$$

If w is the amount of cake particles per unit membrane area, at a given position, w increases with time. To account for the fact that only a fraction of the convectively transported particle becomes deposited, by overall mass balance, one has

$$s = \frac{\mathrm{d}w/\beta}{(\mathrm{d}w/\beta) + \mathrm{d}[(\overline{m}-1)w] + \rho \mathrm{d}V} \tag{8.2.5.4}$$

where s is the particle mass fraction of feed, $\overline{m}$ is the wet to dry cake mass ratio and β is the fraction of the deposited particles.

After rearrangement, Equation (8.2.5.4) becomes

$$\mathrm{d}V = \frac{1 - s[(m-1)\beta + 1]}{\beta s \rho}\mathrm{d}w + \frac{w}{\beta s \rho}\mathrm{d}(\overline{m}-1) \tag{8.2.5.5}$$

The wet to dry cake mass ratio remains essentially constant except in the initial period (see Fig. 6.19). If the last term of Equation (8.2.5.5) is neglected, the integrated form of Equation (8.2.5.5) is

$$w = s\rho \int\limits_o^t \frac{\beta q_{\ell_m} \mathrm{d}t}{1 - s[(\overline{m}-1)\beta + 1]} \tag{8.2.5.6}$$

Substituting Equations (8.2.5.6) into (8.2.5.1) yields

$$q_{\ell_m} = \frac{p_o}{\mu s \rho (\alpha_{av})_{p_{s_m}} \int\limits_o^t \frac{\beta q_{\ell_m} \mathrm{d}t}{1 - s[(\overline{m}-1)\beta + 1]} + \mu R_m} \tag{8.2.5.7}$$

Note that the above expression reduces to Equation (2.1.20) with $\beta = 1$. Simplified versions of Equation (8.2.5.7) have been used by some investigators in the past (Stamatakis and Tien, 1993; Song and Elimelech, 1995; Hong *et al.*, 1997; Bowen *et al.*, 2001).

Rewrite Equation (8.2.5.7) as

$$\frac{p_o}{q_{\ell_m}} = \mu s \rho (\alpha_{av})_{p_{s_m}} \int_o^t \frac{\beta q_{\ell_m} dt}{1 - s[(\overline{m} - 1)\beta + 1]} + \mu R_m \tag{8.2.5.8}$$

If both $(\alpha_{av})_{p_{s_m}}$ and R_m are independent of time, for constant pressure filtration, differentiating the above expression yields

$$\frac{dq_{\ell_m}}{dt} = -\frac{(\alpha_{av})_{p_{s_m}} \mu \rho s \beta q_{\ell_m}^3}{p_o\{1 - s[(\overline{m} - 1)\beta + 1]\}} \tag{8.2.5.9}$$

which is the same expression used by Hong *et al.* (1997).[5]

Equation (8.2.5.9) may be rearranged to give

$$\frac{dq_{\ell_m}}{dt} = -\frac{(\alpha_{av})_{p_{s_m}} \mu \rho s \beta q_{\ell_m}^2}{1 - s\{1 - s[(\overline{m} - 1)\beta + 1]\}} \frac{q_{\ell_m}}{p_o} \tag{8.2.5.10}$$

Combining Equations (8.2.5.10) and (8.2.5.7), one has

$$\frac{dq_{\ell_m}}{dt} = \left\{ \frac{q_{\ell_m}^2}{1 - s[(\overline{m} - 1)\beta + 1]} \right\} \frac{1}{\dfrac{1}{\beta} \displaystyle\int_0^t \frac{\beta q_{\ell_m} dt}{1 - s[(\overline{m} - 1)\beta + 1]} + \dfrac{R_m}{[\alpha_{av}]_{p_{s_m}} s \rho \beta}} \tag{8.2.5.11}$$

On account of $(\overline{m} - 1)\beta \ll 1$ (since $\overline{m} - 1$ is of the order of 1 and β is less than unity), $1 - s[(\overline{m} - 1)\beta + 1] \cong 1$, the above expression reduces to

$$\frac{dq_{\ell_m}}{dt} = \frac{-q_{\ell_m}^2}{\dfrac{1}{\beta} \displaystyle\int_0^t \beta q_{\ell_m} dt + \dfrac{1 - s}{[\alpha_{av}]_{p_{s_m}} \beta s \rho} R_m} \tag{8.2.5.12}$$

If β is assumed to be a constant (independent of time), the above expression becomes the same as that of Bowen *et al.* (2001).[6]

---

[5] Equation (15) of Hong *et al.* can be shown to be the same of Equation (8.2.5.9) with $\beta = 1$. Equation (15) of Hong *et al.* was for incompressible cakes, or $\overline{\overline{\varepsilon}}_s$ and $(\alpha_{av})_p$ both being constant and $\alpha = (k \overline{\overline{\varepsilon}}_s \rho_s)^{-1}$ with $k = (2a_p^2/9\overline{\overline{\varepsilon}}_s)[1/A_s(\overline{\overline{\varepsilon}}_s)]$ and $\overline{\overline{\varepsilon}}_s = (\varepsilon_s)_{max}$.

[6] Equation (12) of Bowen *et al.*, after correcting its typographical errors, is the same as Equation (8.2.5.12) if β is constant.

8.2.6 Application of Equation (8.2.5.7)

The parameters present in Equation (8.2.5.7) are: the average specific cake resistance, $[\alpha_{av}]_{p_{s_m}}$; the fraction of the convectively transported particle flux to the cake/suspension interface being deposited, β; the wet to dry cake mass ratio, $\overline{m}$; and the medium resistance R_m. If these quantities at various times are known, the instantaneous filtration rate can be readily calculated to give the filtration performance.

Estimations of these quantities can be made as follows.

(1) The average specific cake resistance is defined by Equation (8.2.5.2). $(\alpha_{av})_{p_{s_m}}$ can be calculated if the constitutive relationship α vs. p_s and the value of p_{s_m} are specified. p_{s_m}, in turn, can be found from the pressure drop across the cake Δp_c, which is given as

$$\Delta p_c = p_o - \Delta p_m = p_o - q_{\ell_m} \mu R_m \tag{8.2.6.1}$$

(2) For the estimation of β, if the particle back diffusion effect is negligible, β is the same as the adhesion probability γ, which can be estimated according to the Equation (8.2.2.13). If the back diffusion effect is significant, the net particle flux is $[q_{\ell_m} s/(1-s)] - J_B$ where J_B is the back diffusion flux. β now becomes

$$\beta = \frac{\{[q_{\ell_m} s/(1-s)] - J_B\}\gamma}{q_{\ell_m} s/(1-s)} = \left[1 - \frac{J_B(1-s)}{q_{\ell_m} s}\right]\gamma \tag{8.2.6.2}$$

J_B can be estimated from mass transfer considerations. For laminar flow and assuming that the Leveque approximate holds, J_B over a distance of $\overline{L}$ is given as (Chen and Wiley, 2002)

$$J_B = 0.807 \left(\frac{D^2 \omega_w}{\overline{L}}\right)^{1/3} \ell n \frac{(\varepsilon_s)_i}{\varepsilon_{s_o}} \tag{8.2.6.3}$$

where D is the diffusivity responsible for back diffusion (either the Brownian diffusivity or the shear-induced diffusivity), $(\varepsilon_s)_i$ is the value of ε_s at the cake/suspension interface.

It is clear that for estimating β [i.e. Equation (8.3.2.2)], the value of ε_s at the cake/suspension interface ε_{s_i} must be known. Some of the previous investigators have taken $(\varepsilon_s)_i$ to be $(\varepsilon_s)_{max}$. This is, however, not a reasonable assumption as stated previously. Alternatively, since the compressive stress at the cake/suspension interface is commonly assumed to be zero, $(\varepsilon_s)_i$ may be assumed to be ε_s^o, or the solidosity at the zero-shear state. However, using ε_s^o instead of $(\varepsilon_s)_{max}$ for $(\varepsilon_s)_i$, in most cases, will lead to a significant reduction of the extent of the back diffusion.

(3) As discussed before, the wet to dry cake mass ratio $\overline{m}$ is a function of the cake structure. $\overline{m}$ therefore is a function of time. However, as shown in Fig. 6.19, for practical purposes, $\overline{m}$ is largely a constant. Furthermore, with the use of Equation (8.2.5.12) as an approximation of (8.2.5.7), $\overline{m}$ no longer is present in the expression of q_{ℓ_m} and its estimation becomes unnecessary.

(4) R_m is commonly assumed to be constant and equal to the intrinsic membrane resistance (i.e. that determined using particle free-solvent). Generally speaking, R_m increases due to either internal clogging or surface clogging or both. While the former is determined by the relative particle/pore size and can be neglected if the membrane used has sufficiently small pores, the latter is caused by the formation and presence of filter cake over the membrane. As shown in Chapter 6 (see Section 6.4.4), R_m clearly is not a constant. For regular cake filtration, as the cake resistance is dominant except during the initial period, complications arising from the non-constancy of R_m, therefore, does not present serious problems. In contrast, cake formed in cross-flow filtration is often thin (in the order of mm), neglecting the increase of R_m due to cake presence may not be justifiable. Implication of ignoring the change of R_m will be further discussed in the next section.

Equation (8.2.5.7), or (8.2.5.12), can be applied to estimate the steady-state filtration rate. The attainment of a steady-state operation requires that the integral of Equation (8.2.5.7) reaches a constant value as time increases. This condition is met if the integrand of the integral vanishes for $t, t \geq t_1$ or

$$\beta = \gamma = 0 \quad \text{for} \quad t > t_1 \tag{8.2.6.4}$$

As shown in Fig. 8.8, the condition of particle deposition is a balance between the moment caused by the tangential force, F_p and that due to the normal force F_q. Since the cross-flow velocity is kept constant, F_p can be regarded constant. In contrast, F_q decreases due to the decrease of the filtration velocity. The condition of no deposition commences when the critical protrusion height equals h_{max}. With $h_c = d_p/2$, Equation (8.2.2.6) can be used to estimate the steady-state filtration velocity.

Equation (8.2.5.7) (or more precisely its various simplified versions) can be used for the prediction of filtration performance. Assuming that R_m remains constant (and equal to its intrinsic value), the cake is incompressible and there is no particle depolarization ($\beta = 1$), Hong *et al.* obtained good agreement between prediction based on the integrated form of Equation (8.2.5.9) using the permeability expression given by Happel (1958) for estimating the specific cake resistance with the assumption that the cake is densely packed to $(\varepsilon_s)_{max}$ (see Fig. 8.15). Equation (8.2.5.7) was also used for data fitting. Tien and Stamatakis showed that filtration data of silica suspensions can be described by Equation (8.2.5.7) over a wide range of conditions with consistent fitting parameters (i.e. specific cake resistance). More recently, Bowen *et al.* obtained good results by fitting their data with an expression similar to Equation (8.2.5.11).

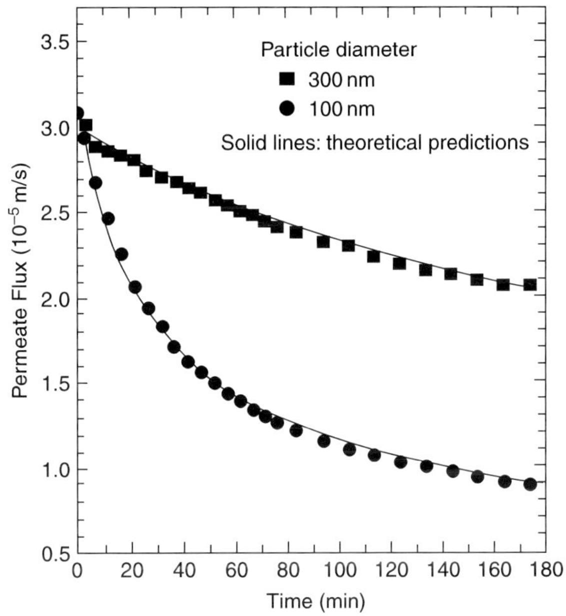

Figure 8.15 Comparison of cross-flow filtration performance q_{ℓ_m} vs. time reported by Song and Elimelech. (Hong, Faibish and Elimelech, 1997. Reprinted by permission of Elsevier.)

8.2.7 Limitations of Equation (8.2.5.7)

The main result presented above [i.e. Equation (8.2.5.7)] is not without its limitation. Accurate determination of the parameter present in Equation (8.2.5.7) is not always possible. For example, determining the extent of back diffusion requires the knowledge of $(\varepsilon_s)_i$ and the mass transfer coefficient. However, accuracy of the available correlations of the shear-induced diffusivity is not high and the value of $(\varepsilon_s)_i$ is really not known. Similarly, there are considerable uncertainties of the adhesion probability expression of Equation (8.2.2.13). As the force terms of their expression can only be approximately estimated and without direct observation, only a guess of the magnitude of h_{max} can be made. Furthermore, to apply the adhesion depolarization hypothesis for two-dimensional calculations, one must have the knowledge of the population of the protrusions at the cake/suspension interface and its distribution. And they can be expected to be functions of time.

The validity of an analysis of a physical process is often based on comparisons of experiments and predictions based on the analysis. Because of the nature of Equation (8.2.5.7), caution should be exercised in assessing the significance of any observed agreement between results from Equation (8.2.5.7) and experiments. For constant pressure filtration, the filtration rate is reduced by the increase in cake resistance or that of the medium resistance. An error in underestimating R_m can be compensated by overestimating the cake resistance. Thus, the effect of overlooking the effect of membrane surface clogging can be nullified by assigning higher values of cake resistance (such as using higher cake solidosity value or ignoring any particle depolarization). The good agreement obtained by Hong *et al.* as shown in Fig. 8.15 could be a case of such a trade-off, and agreement obtained does not necessarily validate the proposed model.

8.3 DEWATERING AND FILTRATION OF SUSPENSIONS OF POROUS ENTITIES

As a final example, the problem of dewatering (or filtration) of suspensions of porous and permeable entities will be discussed. By porous and permeable entities, we refer to entities such as fiber flocs or particle aggregates. In dewatering (or filtering) suspensions of such entities, liquid exchange between the suspended matter and suspending liquid must be considered. The exchange may also alter the dimension of the suspended matter which must be considered in the analysis. Examples of this kind of operation include the formation of paper mat and filtration of suspensions of aggregates of particles of extreme small size (for example, the so-called "nanoparticle"). In spite of its practical significance, its study has not yet appeared in the literature. In the following sections, attempts will be made to explore and speculate the possible approach which may be applied for its investigation.

For either filtration or expression of suspensions of porous and permeable entities, the cake formed (and dewatered) consists of a number of flocs or aggregates as shown in Fig. 8.16. Macroscopically, the cake may be described by its porosity ε^1 or solidosity ε_s^1 where the superscript "1" denotes the macroscopic description. At the same time, the aggregates may be viewed as an assembly of particles or fibers (for the case of fiber flocs). The structure of the aggregate may be described by the aggregate porosity ε^2 and solidosity ε_s^2 where the superscript "2" denotes the microscopic aggregate structure. By definition, one has

$$\varepsilon^1 + \varepsilon_s^1 = 1 \tag{8.3.1a}$$

$$\varepsilon^2 + \varepsilon_s^2 = 1 \tag{8.3.1b}$$

On an overall basis, the cake structure composition is:

Volume fraction of macroscopic void: ε^1
Volume fraction of microscopic void: $\varepsilon_s^1 \varepsilon^2$
Volume fraction of solid: $\varepsilon_s^1 \varepsilon_s^2$
The sum of the fractions is, of course, unity.

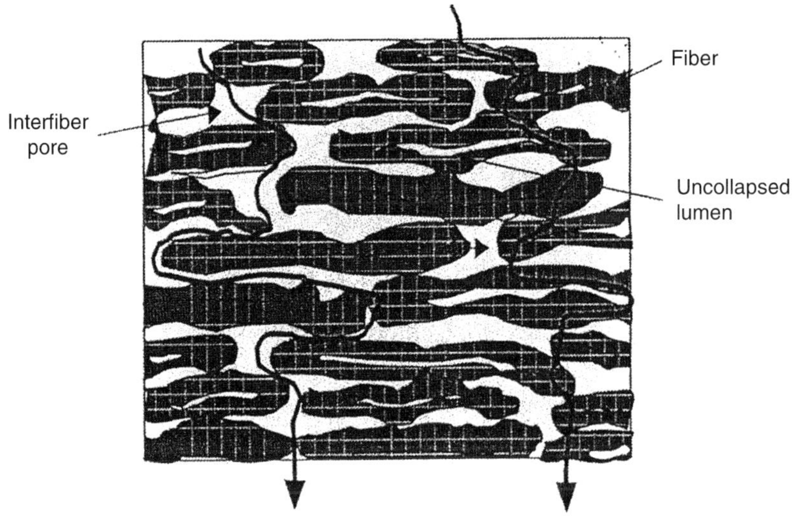

Figure 8.16 Schematic representation of cake formed with permeable particle aggregates.

Macroscopically speaking, the one-dimensional, volume-averaged conservation equations are:

$$\frac{\partial \varepsilon^1}{\partial t} + \frac{\partial q_\ell^1}{\partial x} = R \tag{8.3.2a}$$

$$\frac{\partial \varepsilon_s^1}{\partial t} + \frac{\partial q_s^1}{\partial x} = -R \tag{8.3.2b}$$

where q_ℓ and q_s are the liquid and aggregate velocities, respectively. R is the liquid flux entering into the macroscopic void from the aggregates on a volume basis.

The liquid-aggregate relative velocity $q_{\ell s}^1$ is given by Darcy's law as before or

$$\frac{q_{\ell s}^1}{\varepsilon^1} = \frac{q_\ell^1}{\varepsilon^1} - \frac{q_s^1}{\varepsilon_s^1} = -\frac{k^1}{\varepsilon^1 \mu}\frac{\partial p_\ell^1}{\partial x} \tag{8.3.3}$$

where k, p_ℓ and μ are the permeability, liquid pressure and liquid viscosity, respectively. Similar equations can be written from the aggregate phase. They are

$$\frac{\partial(\varepsilon_s^1 \varepsilon^2)}{\partial t} + \frac{\partial(\varepsilon_s^1 q_\ell^2)}{\partial x} = -R \tag{8.3.4a}$$

$$\frac{\partial(\varepsilon_s^1 \varepsilon_s^2)}{\partial t} + \frac{\partial(\varepsilon_s^1 q_s^2)}{\partial x} = 0 \tag{8.3.4b}$$

and

$$\frac{q_{\ell s}^2}{\varepsilon^2} = \frac{q_\ell^2}{\varepsilon^2} - \frac{q_s^2}{\varepsilon_s^2} = \frac{k^2}{\varepsilon^2 \mu}\frac{\partial p_\ell^2}{\partial x} \tag{8.3.5}$$

It may be more convenient to use the material coordinates instead of the spatial coordinates. If m is defined as

$$dm = \varepsilon_s^1 \varepsilon_s^2 \, dx - q_s^1 \varepsilon_s^2 \, dt \tag{8.3.6a}$$

$$d\bar{t} = dt \tag{8.3.6b}$$

In terms of the new independent variables, m and $\bar{t}$, from Equations (8.3.2a), (8.3.2b) and (8.3.3), one has

$$\frac{\partial e^1}{\partial \bar{t}} + \frac{1}{1+e^2}\frac{\partial}{\partial m}\left[-\frac{1}{1+e^2}\frac{k^1}{\mu}\varepsilon_s^1 \frac{\partial p_\ell^1}{\partial m}\right] = (1+e^1)^2 R \tag{8.3.7a}$$

and

$$e^1 = \frac{\varepsilon^1}{\varepsilon_s^1} \tag{8.3.7b}$$

Similarly, from Equations (8.3.4a), (8.3.4b) and (8.3.5), one has

$$\frac{\partial e^2}{\partial t} + \frac{\partial}{\partial m}\left[-\left(\frac{1}{1+e'}\right)^2 \frac{k^2}{\mu}\varepsilon_s^2 \frac{\partial p_\ell^2}{\partial m}\right] - (q_s^1 \varepsilon_s^2 - q_s^2 \varepsilon_s^1)\frac{\partial e^2}{\partial m}$$
$$= -(1+e^1)(1+e^2)R \tag{8.3.8a}$$

and

$$e^2 = \frac{\varepsilon^2}{\varepsilon_s^2} \tag{8.3.8b}$$

According to previous discussions (see Chapter 4), the quantity, $(k^i/\mu)\varepsilon_s^1 \dfrac{\partial p_\ell^i}{\partial m}$, $i = 1, 2$ may be written as

$$\frac{k^i}{\mu}\varepsilon_s^i \frac{\partial p_\ell^i}{\partial m} = \frac{k^i}{\mu(1+e^i)}\left(\frac{\partial p_\ell^i}{\partial p_s^i}\right)\frac{1}{(\partial e^i/\partial p_s^i)}\frac{\partial e^i}{\partial m} = D^i(e^i)\frac{\partial e^i}{\partial m} \tag{8.3.9}$$

Equations (8.3.7) and (8.3.8) now become

$$\frac{\partial e^1}{\partial \bar{t}} - \frac{1}{1+e^2}\frac{\partial}{\partial m}\left[\frac{1}{1+e^2}D^1(e^1)\frac{\partial e^1}{\partial m}\right] = (1+e^1)R \tag{8.3.10}$$

$$\frac{\partial e^2}{\partial \bar{t}} - \frac{1}{\partial m}\left[\left(\frac{1}{1+e^1}\right)^2 D^2(e^2)\frac{\partial e^2}{\partial m}\right] - \left(\frac{q_s^1}{1+e^2} - \frac{q_s^2}{1+e^1}\right)\frac{\partial e^2}{\partial m}$$
$$= -(1+e^1)(1+e^2)R \tag{8.3.11}$$

To complete the description, the relationships between p_ℓ^i and p_s^i, $i = 1, 2$, need to be specified. The various $p_\ell - p_s$ relationship discussed previously are applicable. For simplification, one may assume[7]

$$p_\ell^1 + p_s^1 = p_t \tag{8.3.12}$$[7]

where p_t is the applied pressure. For the aggregate (or floc) phase, one may assume

$$p_s^1 = p_\ell^2 + p_s^2 \tag{8.3.13}$$

It is also necessary to specify the flux expression R. Intuitively, one expects that liquid flow out of the micro void to the macro void results from the pressure difference. Accordingly, the simple expression

$$R = K\,(p_\ell^2 - p_\ell^1) \tag{8.3.14}$$

One may further modify the above expression by requiring that

$$R = 0 \quad \text{if} \quad p_\ell^2 < p_\ell^1 \tag{8.3.15}$$

In other words, if the micro void dimension is sufficiently small, even with $p_\ell^1 > p_\ell^2$, there would be no liquid flow into the micro void.

R may also be assumed to be

$$R = K(e^2 - e_\infty^2) \tag{8.3.16}$$

where e_∞^2 is the value of e corresponding to the irreducible moisture content of the aggregates (or floc). This expression assumes that liquid may be expelled out of the aggregate if excess liquid is present. The rate constant K may further be assumed to be a function of p_s^1. In other words, the flux depends upon the extent of compression to which the aggregates are subject.

The conservation equations given above may also be further simplified. For example, if q_s^1 and q_s^2 are negligible, Equation (8.3.11) then becomes

$$\frac{\partial e^2}{\partial \bar{t}} - \frac{\partial}{\partial m}\left[\left(\frac{1}{1+e^1}\right)^2 D^2(e^2)\,\frac{\partial e^2}{\partial m}\right] = -(1+e^1)(1+e^2)\,R \tag{8.3.17}$$

Furthermore, if one assumes that $D^2(e^2)$ is insignificant or there is only insignificant liquid flow within the aggregates, (8.3.17) may be further reduced to give

$$\frac{\partial e^2}{\partial \bar{t}} = -(1+e^1)(1+e^2)\,R \tag{8.3.18}$$

[7] This is tantamount to the relationship of $\partial p_\ell / \partial p_s = -1$.

As a possible topic of future investigation, one may wish to obtain solutions of Equations (8.3.10) and (8.3.11) [or (8.3.17), (8.3.18)] with appropriate initial and boundary conditions (see discussion in Chapter 4) and different expressions for D^1, D^2 and R and validate the analysis with experiments. Equally important, based on the analysis results, one may attempt to establish procedures with which the various relevant systems and rate parameters (D^1, D^2 and K) can be estimated from experiments.

REFERENCES

Altena, F.W. and Belfort, G., *Chem. Eng. Sci.*, **36**, 393 (1984).

Altmann, J. and Ripperger, S., *J. Membrane Sci.*, **124**, 119 (1997).

Bowen, W.R., Yousef, H.N.S. and Colvo, J.F., *Sep. Pur. Tech.*, **24**, 297 (2001).

Chen, V. and Wiley, D.E., "Particle Deposition in Membrane Systems", in *Transport Processes in Bubbles, Drops, and Particles* 2nd edn, D. DeKee and R.P. Chhabra, Editors, Taylor and Francis, New York (2002).

Chung, T.S., Qin, J.J., Huan, A. and Toh, K.C., *J. Membrane Sci.*, **196**, 251 (2002).

DeRavin, M., Humphries, W. and Postle, R., *Filtration and Separation*, p. 201 (1998).

Drew, D.A., Schonberg, J.A. and Belfort, G., *Chem. Eng. Sci.*, **46**, 3219 (1991).

Eckstein, E.C., Bailey, D.G. and Shapiro, A.H., *J. Fluid Mech.*, **79**, 1911 (1977).

Field, R.W., Wu, D., Howell, J.A. and Gupta, B.B., *J. Membrane Sci.*, **100**, 259 (1995).

Fischer, E. and Raasch, J., "Model Tests of Particle Deposition at the Filter-Medium in Cross-Flow Filtration", *4th World Filtration Congress Proc. Part II*, pp. 9.9–9.16 (1986).

Green, G. and Belfort, G., *Desalination*, **35**, 129 (1980).

Halow, J.S., *Chem. Eng. Sci.*, **28**, 1 (1973).

Happel, J., *AIChE., J.*, **4**, 197 (1950).

Hong, S., Faibish, R.S. and Elimelech, M., *J. Colloid Interface Sci.*, **196**, 267 (1997).

Howell, J.A., *J. Membrane Sci.*, **107**, 165 (1995).

Ju, J., Chiu, M.S. and Tien, C., *J. Air and Waste Management Asso.*, **50**, 600 (2000).

Ju, J., Chiu, M.S. and Tien, C., *Powder Tech.*, **118**, 79 (2001).

Khulbe, K.C., Matsuura, T., Lamarchi, G. and Kim, H.J., *J. Membrane Sci.*, **135**, 211 (1997).

Kwon, D.Y., Vigneswaran, S., Fane, A.G. and Ben Aim, R., *Sep. Pur. Tech.*, **19**, 169 (2000).

Li, H., Fane, A.G., Coster, H.G.L. and Vigneswaran, S., *J. Membrane Sci.*, **172**, 135 (2000).

Leighton, D. and Acrivos, A., *J. Fluid Mech.*, **181**, 415 (1987).

Lu, W.-M. and Ju, S-C., *Sep. Sci. Tech.*, **24**, 517 (1989).

Lu, W.-M. and Ju, S.-C., *Chem. Eng. Sci.*, **48**, 863 (1993).

Meier, J., Klein, G.-M. and Kottke, V., *Sep. Pur. Tech.*, **26**, 43 (2002).

O'Neill, M.E., *Chem. Eng. Sci.*, **23**, 1387 (1968).

Ould-Dris, A., Jaffrin, M.Y., Si-Hassen, D. and Neggaz, Y., *J. Membrane Sci.*, **211**, 227 (2000).

Ripperger, S. and Altmann, J., *Sep. Pur. Tech.*, **26**, 19 (2002).

Romero, C. and Davis, R.H., *J. Membrane Sci.*, **39**, 157 (1988).

Romero, C. and Davis, R.H., *Chem. Eng. Sci.*, **45**, 13 (1990).

Song, L. and Elimelech, M., *J. Chem. Soc. Faraday Trans.*, **91**, 3389 (1995).

Stamatakis, K. and Tien, C., *AIChE J.*, **39**, 1292 (1993).

Tien, C., "Analysis of Cross-flow Filtration Based on Particle Adhesion Model", *9th World Filtration Congress*, New Orleans, Louisiana, USA (2004).

Vasseur, P. and Cox, R.G., *J. Fluid Mech.*, **78**, 385 (1976).

Zhang, W., He, G., Cao, P. and Chen, G., *Sep. Pur. Tech.*, **30**, 27 (2003).

Zydney, A.L. and Colton, C.K., *Chem. Eng. Commun.*, **47**, 1 (1986).

Index